D0847865

A Guide to
HUMAN
GENE THERAPY

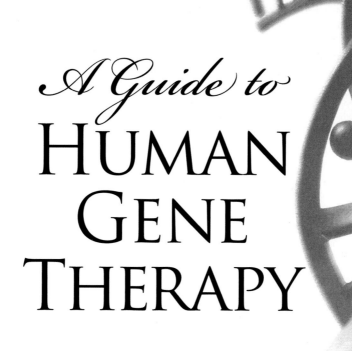

A Guide to
HUMAN
GENE
THERAPY

Editors

Roland W Herzog
Sergei Zolotukhin
University of Florida, USA

World Scientific

NEW JERSEY · LONDON · SINGAPORE · BEIJING · SHANGHAI · HONG KONG · TAIPEI · CHENNAI

Published by

World Scientific Publishing Co. Pte. Ltd.

5 Toh Tuck Link, Singapore 596224

USA office: 27 Warren Street, Suite 401-402, Hackensack, NJ 07601

UK office: 57 Shelton Street, Covent Garden, London WC2H 9HE

British Library Cataloguing-in-Publication Data
A catalogue record for this book is available from the British Library.

ISBN-13 978-981-4280-90-7
ISBN-10 981-4280-90-9

Typeset by Stallion Press
Email: enquiries@stallionpress.com

Printed by Fulsland Offset Printing (S) Pte Ltd, Singapore

Preface

Kenneth I. Berns*

Ever since the discovery of the DNA double helix by Watson and Crick tremendous advances in our knowledge of molecular genetics and cell biology have made it one of the most exciting areas of science for more than 50 years. We have now progressed beyond the realm of basic research in genetics to the development of biotechnology, which has allowed us to produce human proteins in bacteria and to use these, as in the case of insulin, for therapy of human disease. In many cases the therapy derived from biotechnology is designed as replacement therapy to make up for defects at the DNA level, which result in either the wrong protein being made or no protein at all. A more sophisticated approach than replacement therapy would be to correct a deficit at the level of the gene by either regulating gene expression or replacing or substituting for a defective gene. Once the discovery and characterization of restriction enzymes made it possible to identify, isolate, and clone specific genes the way was open to begin to think about gene therapy. This book describes the state of the art of gene therapy, which is designed to either directly introduce a good copy of a defective gene or to have the cell express a new protein or regulatory RNA which will block the deleterious effects of a defective gene.

The requirements for executing gene therapy successfully are relatively few. First the molecular nature of the defect to be corrected needs to be understood. This has been achieved for a large number of diseases, although multifactorial diseases still pose a challenge. Second, the corrective DNA sequence (i.e., the gene) needs to be determined and sufficient amounts for

---

*Correspondence: Director, University of Florida Genetics Institute, Gainesville, FL 32610.
E-mail: kberns@ufl.edu

use produced. Many genes have been cloned and are available. The main obstacle to successful gene therapy is the requirement for a vector to carry the corrective gene into the cell. A successful vector must be able to deliver the gene to the target cell, and get it through the cytoplasm and into the nucleus where the cell must be able to express the gene contained in the vector. A variety of approaches described in the introductory chapters of this book, have been tried, including naked DNA, DNA in lipid vesicles, and viruses, which have been adapted to carry "transgenes" in place of normal viral genes. To date the most effective, in terms of delivery and expression, have been vectors derived from viruses. Viruses are natural vectors which deliver their genomes into cells where the DNA or RNA is successfully expressed. Many viruses are also capable of establishing a persistent presence in infected cells. However, viruses have potential downsides as vectors. Most of the better-characterized viruses cause serious human disease and any vector derived from such viruses must be modified to minimize the likelihood of toxicity. Secondly the body recognizes viruses as foreign and mounts an immune response. Viruses have evolved to evade the host response, but it is clear from both animal models and clinical trials that the nature of the host response has to be taken into account.

Shortly after the concept of gene therapy had gained currency there was a push to attempt it in clinical trials. With one or two notable exceptions, there was little associated toxicity, but even less in the way of apparent efficacy. It was difficult to attain therapeutic levels of transgene expression and in most instances to maintain the levels of expression achieved initially after infection. What became evident was that a much better understanding of the process of viral infection at the levels of both the cell and the intact host was needed before gene therapy could become an accepted way to treat various diseases (similar considerations apply equally well to other types of vectors). Particular issues which needed elucidation included mechanisms by which the vector interacts with the host immune system, the process of cellular uptake and trafficking of the vector, regulation of transgene expression, and modification of the vector genome to prevent insertional mutagenesis (all DNAs will recombine into the genome, some slowly and others much more efficiently). Significant advances have been made in our knowledge of the basic processes and this has greatly increased the likelihood of successful gene therapy.

The notion of gene therapy was initially received with enthusiasm. Early clinical trials were closely followed by the media. However, the early optimism soon gave way to pessimism because of the lack of evident clinical success and because of adverse effects, notably the death of one trial participant. The response mirrors earlier reactions to research into various diseases. Early trials with polio vaccines gave more people polio than protective antibody. It took many years before two successful vaccines were developed, nearly simultaneously. Although we have had success with AIDS therapy, our efforts to develop a successful vaccine have still not borne fruit. As a people Americans are impatient.

We have now achieved clinical success in two diseases using gene therapy. In France and the UK, infants with a form of severe combined immunodeficiency disease (SCID-X1) that was untreatable by available therapies have been "cured" by treating cells of their bone marrow with a vector derived from Maloney Murine Leukemia virus. In France 11 of 12 infants treated were cured; however, 4 of those cured developed leukemia, directly attributable to the vector. Three of these had the leukemia successfully treated as well. This trial is clearly a success; despite the morbidity associated with the vector. Without treatment, all would likely have died. A somewhat happier result has been achieved in the past year. An adeno-associated virus vector has been used to treat patients with a rare form of retinitis pigmentosa. All of the patients had been legally blind for years and now almost all have achieved striking improvements in their eyesight. Most encouraging, were similar phase I clinical trials done by three groups nearly simultaneously, and all achieved similar positive results. Thus, although the trials were designed primarily to show the safety of the vector, efficacy seems to be pretty clear. Similarly, treatment of adenosine deaminase deficiency (ADA-SCID) continued to be successful using a protocol similar to that for SCID-X1 but without incidence of leukemia.

Thus, once again the future looks bright for gene therapy. The chapters in this book stand testament to this view and as you read on, you will probably have the same sense of optimism for what this exciting field has in store for us.

Contents

5. Herpes Simplex Virus Vectors   69

William F. Goins, David M. Krisky, James B. Wechuck,
Darren Wolfe, Justus B. Cohen and Joseph C. Glorioso

6. Adeno-Associated Viral (AAV) Vectors   87

Nicholas Muzyczka

9. Homologous Recombination and Targeted Gene
Modification for Gene Therapy   139

Matthew Porteus

16.   Gene Therapy for Duchenne Muscular
Dystrophy   261

Takashi Okada and Shin'ichi Takeda

Contributors

Oumeya Adjali
INSERM UMR 649
CHU Hôtel-Dieu
30 Boulevard Jean Monnet,
44035, Nantes, France
02 28 08 04 15/17
02 28 08 04 16
oumeya.adjali@univ-nantes.fr

Alessandro Aiuti
HS Raffaele Institute/HSR-TIGET,
San Raffaele Telethon Institute
    for Gene Therapy
DIBIT, HSR-TIGET,
Via Olgettina, 58,
20132 Milano (MI), Italy
+39 (0)2 2643 4671
+39 (0)2 2643 4688
a.aiuti@hsr.it

Kenneth I. Berns
University of Florida-Director,
Genetics Institute
Department of Molecular
    Genetics and Microbiology
Cancer/Genetics Research Complex
1376 Mowry Road, Room CG-110,
Gainesville FL 32610-3610
352-273-8100 (Tel)
352-273-8284 (Fax)
kberns@ufl.edu

Shannon Boye
University of Florida
Department of Ophthalmic
    Molecular Genetics
Academic Research Building
1600 SW Archer Road,
Room R1-236, Gainesville
    FL 32610-0284
352-846-2913
352-392-3062
shaire@ufl.edu

Barry Byrne
University of Florida
Department of Pediatrics
    Cardiology
Powell Gene Therapy Center
Academic Research Building
1600 SW Archer Road,
Room RG-165, Gainesville
    FL 32610-0296
352-273-6563
352-273-6250
bbyrne@ufl.edu

Michele Calos
Stanford University
Department of Genetics
Stanford University School
    of Medicine
Alway Building, Room M334

300 Pasteur Drive, Stanford,
CA 94305-5120
650-723-5558
650-725-1534
calos@stanford.edu

Dr. William F. Goins
University of Pittsburgh School
of Medicine
Department of Microbiology &
Molecular Genetics
W1157 BSTWR,
200 Lothrop Street
Pittsburgh, PA 15261
412-383-9751 (Tel)
412-383-9760 (Fax)
goins@pitt.edu

Roland Herzog
University of Florida
Division of Cellular
and Molecular Therapy
Cancer/Genetics Research Complex
1376 Mowry Road, Room CG-203,
Gainesville FL 32610-3610
352-273-273-8113
352-273-8342
rherzog@ufl.edu

Christof Von Kalle
National Center for Tumor
Diseases, Director
Department of Translation
Oncology
German Cancer Research
Center (DKFZ)
D-69120 Im Neuenheimer Feld 350
TPA 69120 Heidelberg,
Germany
49-6221-56-6990 ext 6991 (Tel)
49-6211-56-6967 (Fax)

Alfred S. Lewin
University of Florida
Department of Molecular Genetics
and Microbiology
Academic Research Building
1600 SW Archer Road,
Room R1-295, Gainesville
FL 32610-0266
352-392-0676
352-392-3133
lewin@ufl.edu

Cathryn S. Mah
University of Florida
Cancer/Genetics Research Complex
Division of Cellular and
Molecular Therapy
1376 Mowry Road, room 493,
Gainesville FL 32610-3610
352-273-8258
352-273-8342
cmah@peds.ufl.edu

Ashley T. Martino
University of Florida
Cancer/Genetics Research Complex
Division of Cellular and
Molecular Therapy
1376 Mowry Road, room 225,
Gainesville FL 32610-3610
352-273-8079
352-273-8342
amartino@ufl.edu

Nicholas Muzyczka
University of Florida
Department of Molecular Genetics
and Microbiology, Powell
Gene Therapy
Center Cancer/Genetics Research
Complex

1376 Mowry Road, Room CG-208,
Gainesville FL 32610-3610
352-273-8114
352 273-8284
muzyczka@mgm.ufl.edu

Stuart Nicklin
University of Glasgow
BHF GCRC,
University of Glasgow,
126 University Place,
Glasgow G12 8TA, UK
+44-(0)141-3302521 (Tel)
+44-(0)141-3306997 (Fax)
stuart.a.nicklin@clinmed.gla.ac.uk

Takashi Okada
National Institute of Neuroscience
Department of Molecular Therapy
National Center of Neurology
    and Psychiatry
4-1-1 Ogawa-Higashi, Kodaira,
Tokyo 187-8502, Japan
+81-42-341-2712 ext.2934
+81-42-346-2182
t-okada@ncnp.go.jp

Matthew Porteus
UT Southwestern Medical Center
Department of Pediatrics MC 9063
5323 Harry Hines Boulevard
Dallas, Texas 75390
214-648-7222 (Tel)
214-648-3122 (Fax)
Matthew.porteus@utsouthwestern.edu

Paul D. Robbins
University of Pittsburgh
University of Pittsburgh School
    of Medicine
Department of Microbiology and
    Molecular Genetics

W1246 BST, 200 Lothrop Street,
Pittsburgh, PA 15261
412-648-9268
412-383-8837
probb@pitt.edu

Arun Srivastava
University of Florida
Cancer/Genetics Research Complex
Division of Cellular and
    Molecular Therapy
1376 Mowry Road, Room 492,
Gainesville FL 32610-3610
352-273-8259
352-273-8342
aruns@peds.ufl.edu

Sean M. Sullivan
Vical, Inc.
Pharmaceutical Sciences
10390 Pacific Center Court,
San Diego, CA 9212
858-646-1108 (Tel)
858-646-1150 (Fax)
ssullivan@vical.com

Thierry VandenDriessche
University of Leuven and Flanders
    Institute of Biotechnology (VIB)
Vesalius Research Center
University Hospital Campus
    Gasthuisberg, Herestraat 49
3000 Leuven, Belgium
+32-16-330558 (Tel)
+32-16-3345990 (Fax)
thierry.vandendriessche@med.
    kuleuven.be

Kirsten A.K. Weigel-Van Aken
University of Florida
Division of Cellular and
    Molecular Therapy

Cancer/Genetics Research Complex
1376 Mowry Road, Room CG-204,
Gainesville FL 32610-3610
352-273-273-8151
352-273-8342
weigel@ufl.edu

Deborah Young
University of Auckland
Department of Pharmacology &
    Clinical Pharmacology
Faculty of Medical and
    Health Sciences
85 Park Road, Grafton, Auckland,
New Zealand

+649 373 7599 extension 84491
+ 649 373 7492
ds.young@auckland.ac.nz

Sergei Zolotukhin
University of Florida
Division of Cellular and
    Molecular Therapy
Cancer/Genetics Research Complex
1376 Mowry Road, Room CG-202,
Gainesville FL 32610-3610
352-273-273-8150
352-273-8342
szlt@ufl.edu

Chapter 1
Non-Viral Gene Therapy

Sean M. Sullivan*

This chapter is meant to serve as an introduction to non-viral gene transfer by highlighting therapeutic applications that have transitioned from preclinical research into the clinic. Non-viral gene therapy is the administration of plasmid DNA encoding a transgene gene locally or systemically yielding expression of a therapeutic protein, thereby correcting a disease state. Local administration of plasmid DNA results in gene transfer to cells at the site of injection. Gene transfer efficiency can be increased by applying electric current (electroporation) or sound waves (sonoporation). Alternatively, the plasmid DNA can be formulated with cationic lipids or polymers to increase gene transfer. All of these methods result in increased uptake by cells and therefore in increased gene expression. Clinical applications of this technology include: treatment of peripheral vascular disease following local administration at the sites of muscle ischemia; development of genetic vaccines resulting in immune activation against the specific expressed antigen; development of therapeutic cancer vaccines that induce surveillance and killing of tumor cells by the immune system; correction of genetic disease by expressing a functional wild type protein in cells that lack a functional protein.

1. Introduction

It is difficult to pinpoint a specific discovery that initiated the field of plasmid DNA based gene therapy. There have been several milestones that led to its development. Table 1 lists a series of events that have impacted the development of the field.

---

*Correspondence: Vical, Inc. 10390 Pacific Center Court, San Diego CA 92130.
E-mail: ssullivan@vical.com

Table 1.1. Scientific milestones that impacted the field of non-viral gene therapy.

Scientific Milestone	Year	Refs.
First Liposome Based DNA Delivery Patent filed	1983	1
First publications describing the use of cationic lipids to transfect cells	1987–89	2–4
Demonstration that "Naked DNA" can Transfect muscle cells *in vivo*	1990	5
First human clinical trial conducted for development of melanoma cancer vaccine using cationic lipid formulated plasmid DNA	1996	6, 7
First indications of clinical efficacy demonstrated for treatment of Chronic Limb Ischemia following IM administration of VEGF Naked pDNA	1996	8, 9
Electroporation yields order of magnitude increase in gene expression following local administration	1998	10
Aqua Health (Novartis) anti-viral vaccine for salmon receives approval in Canada.	2005	
Successful demonstration of efficacy for treatment of chronic limb ischemia following IM administration of pDNA expressing hepatocyte growth factor.	2007	
Merial receives conditional USDA approval of canine melanoma therapeutic genetic vaccine.	2008	

The discovery of cationic lipids was the segue into therapeutic applications of plasmid based gene delivery. It provided a methodology for getting DNA into cells resulting in expression. This was not a new gene transfer concept in that calcium phosphate and poly cationic polypeptides, such as poly-lysine and poly-L-ornithine had been used to introduce plasmid DNA and RNA into cells.[11] However, this methodology had many uncontrolled variables resulting in a high degree of variability and not being applicable for clinical development.

The transition of cationic liposomes from an *in vitro* transfection reagent to a clinical application was first realized with the testing of the first cancer vaccine where a non-self major histocompatibility antigen, HLA-B7 was encoded along with β-2 microglobulin in an expression plasmid, complexed with cationic liposomes composed of DC-Chol/DOPE and injected into tumors. Gene transfer of the foreign major histocompatibility antigen complex triggers a T cell mediated immune response that not only results in the killing of antigen expressing tumor cells but also results in the killing of non-antigen expressing tumor cells. This local priming of the immune system against tumor cells activates immune surveillance of the body to seek and destroy neoplasms distil to the initial tumor immunization site.

As in any new therapy, the initial clinical trials provided lessons that would impact the design of future clinical trials and focus improvements in the technology that increased performance and safety. Two major technology improvement categories included increased expression of the therapeutic protein and increased duration of expression. The subsequent sections will describe the basic features of the plasmid DNA, the formulations and the gene transfer techniques that have been employed to overcome technology deficiencies. Though these deficiencies have not completely been overcome, the lessons learned have been applied to yield commercialization of animal health products and produce successful late stage human clinical trials.

2. Plasmid DNA

Plasmid DNA is a closed circular double stranded helix DNA molecule. When isolated from bacteria, pDNA is in a supercoiled, dimer or concatamer form. The isolation conditions can cause single strand or double strand nicks producing relaxed or linear forms. Isolation conditions are optimized to yield the highest percentage of supercoiled pDNA and minimize the production of the other forms because there are studies that show increased supercoil content yields higher levels of transgene expression.[12] Furthermore, the FDA has deemed the percent supercoil content a product shelf life determinant. The fundamental features of pDNA are shown in Fig. 1.1. These are: the expression cassette, the origin of replication and the drug resistance gene. The origin of replication (ORI) is a DNA sequence of 13 mer and 9 mer repeats that initiates plasmid DNA replication in bacteria. The drug resistant gene, Kanamycin resistance gene or Ampicillin resistance gene, allows for the selection of plasmid transformed bacteria.

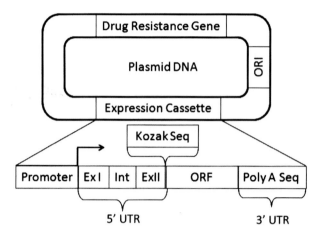

Fig. 1.1 Diagram of Plasmid DNA: Diagram shows feature of plasmid DNA. Abbeviations: ORI-Origin of Replication; ExI and ExII-Exon I and Exon II; Int-Intron; Poly A Seq-Poly Adenylation Sequence; Arrow-Transcription Initiation Sequence.

The expression cassette can be divided into the following components: the promoter; the 5'untranslated coding sequence (5'UTR) containing at least 1 intron and the kozak sequence; the open reading frame (ORF) encoding the gene to be expressed; the 3' untranslated coding sequence (3'UTR) containing the poly-adenylation sequence (PolyA). The promoter contains a DNA sequence that recruits RNA polymerase II for initiation of transcription. The promoter can contain additional sequences, known as enhancer sequences that bind proteins termed transcription factors that further facilitate the recruitment of RNA polymerase II.

The inclusion of at least one intron in the 5'UTR ensures the entry of the transcript into the pre-mRNA/mature mRNA processing pathway and export of the mRNA into the cytoplasm.[13] Located in the 5'UTR is the Kozak sequence which is a signal for the ribosome to start translation. The Kozak consensus sequence is (gcc)gccRccAUGG, where R is an adenine three bases upstream of the AUG start codon which is followed by another 'G'.[14] The AUG codon encodes methionine which is the first amino acid of the transgene. The open reading frame can be codon optimized to increase gene expression. Due to the redundancy of the genetic code, rarely used codons can be replaced with more commonly used codons, especially those codons more commonly used by the target cell; the purpose being to increase protein synthesis. This strategy can also be used to reduce the amount of CpG sequences that can activate the immune system. Activation of the immune system has been shown to reduce the amount and duration of gene expression.[15] However, in developing genetic vaccines, codon optimization can be used to increase CpG content thereby increasing immune activation.[16] The 3'UTR contains the polyadenylation sequence that is a binding site for a multi-protein complex that cleaves the end of the mRNA transcript and polyadenylate polymerase adds approximately 250 adenine nucleotide monophosphates. The poly adenylation takes place in the nucleus and promotes nuclear export of the mRNA and translation, and inhibits degradation.

Endogenous microRNA cleavage sequences approximately 20 base-pairs in length are located in the 3'UTR of endogenous mRNAs. Transfection of cells with endogenous microRNA activity against a latent target in the 3'UTR of a therapeutic gene could inhibit protein synthesis by removing the poly A tail resulting in immediate degradation of the mRNA. Databases, such as the Wellcome Trust Sanger Institute siRNA database (http://microrna.sanger.ac.uk/), are continually being updated for microRNA target sequences as they are identified. Scanning the sequence of the 3'UTR using these microRNA target sequence databases will avoid inactivation of the transcript by endogenously expressed microRNAs.

2.1 *Plasmid DNA Manufacture*

The therapeutic gene is ligated into the plasmid backbone and standard microbiological protocols are used to identify a bacterial clone that contains the plasmid DNA. The bacterial clones are also selected for the highest specific activity with regard to plasmid DNA/bacterium. Master cell banks are created using this clone. The bacteria are fermented at lab scale in a shaker flask or can be fermented using a fermenter. The bacteria are pelleted by centrifugation and resuspended in resuspension buffer. The bacteria are lysed opened by alkaline lysis, neutralized and then centrifuged. The supernatant is extracted with phenol/chloroform followed by ethanol precipitation of the pDNA. The precipitant is resuspended in buffer and double banded using CsCl equilibrium centrifugation with a vertical rotor. The ethidium bromide is extracted with buffer saturated butanol followed by dialysis of the DNA against buffer. There are commercially available kits to purify pDNA from bacteria. However, the quality of the pDNA can vary. The method outlined above yields highly purified plasmid DNA with regard to elimination of bacterial protein, RNA and endotoxin. This is especially important when formulating the pDNA with polymers and cationic lipids. Low-speed and high-speed centrifugation are not suitable for gram scale pharmaceutical manufacture. Substitution of filters for low speed centrifugation, and replacement of high speed CsCl density gradient centrifugation with anion exchange chromatography combined with hydrophobic interaction chromatography (HIC) makes this process amenable to pharmaceutical scale pDNA manufacture.[17,18] Depending upon the bacterial strain and the plasmid DNA backbone, a single 500 L fermentation run can yield 10 to 20 grams of plasmid DNA.

3. Plasmid DNA Gene Transfer Methods

3.1 *Plasmid DNA or "Naked DNA" as a Gene Delivery System*

As stated in the introduction, plasmid DNA can be injected by itself and yield gene expression. This was first discovered by intramuscular injection[5] expressing a reporter gene resulting in the marking of muscle cells at the site of injection. The expression levels are low compared to other forms of non-viral gene therapy. However, the DNA is not toxic and low levels of expression can be compensated for by increasing the dose and dosing frequency.

Two late stage clinical applications take advantage of this form of plasmid based gene delivery. Both treat peripheral vascular disease by expressing a therapeutic gene encoding for angiogenic growth factors, basic fibroblast growth factor (FGF-1)[19] or hepatocyte growth factor (HGF),[20,21] to induce new blood vessel growth in ischemic limbs. Both non-viral gene therapies have similar pDNA doses but different dosing schedules and different endpoints. For the FGF Phase 2 clinical trial, 125 patients, where revascularization surgery was not an option and had non-healing ulcers, were randomized and double blind placebo controlled for 2.5 ml injection of FGF-1 pDNA, [pDNA] = 0.2 mg/ml, on days 1, 15, 30 and 45. The primary end point was healing of at least one ulcer and secondary end points were ankle brachial index, amputation and death. The gene therapy was well tolerated. There was no significant difference in ulcer healing between treatment and placebo group. However, there was a significant reduction in risk of amputations and there was a trend in the reduced risk of death.

For expression of HGF, multiple clinical trials were conducted showing that the therapy was well tolerated and no severe complications or adverse events were observed for any of the patients.[20,21] A multicenter Phase 3 trial was conducted in Japan comprised of patients with arteriosclerosis obliterans with critical limb ischemia that could not undergo revascularization and did not respond to conventional drug therapies. Patients were randomized 2:1 therapy (55 patients) to placebo (26 patients) groups and received two intramuscular injections at the site of ischemia at 4 week intervals. Patients were followed for 8 weeks after the last administration. The primary endpoints were decreased rest pain or improvement in ischemic ulcer. The treatment group showed a 70% response in rest pain reduction and ulcer improvement whereas the placebo group showed a 30% response rate. At this time, these two therapeutic applications are in the latest stages of clinical development for plasmid DNA delivery.

Another active area of plasmid DNA therapy is the development of genetic vaccines. Purification of protein antigen can be problematic with regard to the yield and purity, often giving rise to antibody responses to the impurity rather than the intended protein antigen. This can be especially challenging for water insoluble proteins, such as integral membrane proteins. Also, combinations of different protein antigens can be prohibitive due to formulation incompatibility. Expression of the protein antigen following intramuscular, intradermal or subcutaneous administration of pDNA dramatically simplifies the immunization process and insures immune response to the native protein antigen. Genetic vaccines using pDNA alone are being developed for pandemic flu, HIV and Hepatitis C. In some cases, more than one gene is being expressed, such as HIV where six to 7 different genes are being expressed at the same time.[22-24]

One technique that has been used as a research tool is to systemically administer plasmid DNA in a large volume of vehicle, termed "hydrodynamic" gene delivery. This was initially discovered using a mouse animal model in which the pDNA is administered via the tail vein in a 1 ml injection volume rapidly (10 seconds). The result is a high gene transfer efficiency to the liver. Hydrodynamic gene delivery while not practical for clinical development of systemic non-viral gene therapies has been used to screen for biological activity of secreted proteins into the blood, such as blood clotting factors. Therapeutic applications for Naked DNA are limited by the low level of gene expression. The level of gene expression can be improved by increasing the amount of DNA that gets into the cells, enters the nucleus and is expressed. The following section will describe technologies designed to increase gene transfer efficiencies.

3.1.1 *Electroporation of Naked DNA*

Applying energy following local administration of pDNA results in large increases in gene expression. The forms of energy can be electrical — used in "electroporation" or ultrasound — used in "sonoporation". Electroporation consists of applying voltage to the site of administration in a series of electrical pulses lasting microseconds for each pulse. The hypothesis is that the electrical pulses induce a transient depolarization of the smooth muscle plasma membrane allowing the pDNA to enter the cell. The number of pulses, duration of pulses and the electrical strength of the pulse are of a particular magnitude to minimize permanent damage to the cell membrane and maximize gene transfer. Voltage was initially applied through calipers and the muscle was sandwiched between the plates. However the administration technology has been modified to use needles arranged in a hexagonal array with the electrical field alternating between opposing needles creating an electrical field around the injection site. Electroporation can increase gene expression of pDNA from one to several orders of magnitude compared to pDNA alone.[25] The fold increase is dependent on the transgene to be expressed, the administration route and the optimization of the electrical field pulse. A phase 1 clinical trial has been conducted involving the electroporation of an IL-12 expression plasmid into surface accessible tumors of melanoma patients.[26] There were 24 patients in the study. This was a dose ranging study with the most serious adverse event being pain at the site of injection. IL-12 and interferon gamma were observed at the tumor injection site. Also induction of infiltrating lymphocytes into the tumor was observed.

3.1.2 *Sonoporation of Naked DNA*

Substitution of high frequency sound waves in sonoporation, for electricity can achieve similar effects as electroporation. The hypothetical mechanism for facilitating gene transfer is similar to electroporation, in that the sound waves produce a very short lived transient breach in the integrity of the plasma membrane facilitating entry of pDNA into the cell. The key components are development of a probe that can effectively deliver the sound waves to the site of administration and application of pDNA effectively to maximize transfection efficiency.[27–29] Gene transfer efficiency can be increased by applying ultrasound contrast reagent along with the ultrasound.[30,31]

3.2 *Plasmid DNA Formulations*

3.2.1 *Cationic Lipids*

Cationic lipids are synthetic amphiphiles comprised of a hydrophobic domain (R), a linker and a hydrophilic domain (R′). For the cholesterol based cationic lipids, the hydrophobic domain is cholesterol and the cationic head group is denoted by R. Chemical structures of cationic lipids used in non-viral gene therapy clinical trials used are shown in Fig. 1.2. The lipid anchor is linked to the head group by either an ether linkage or a carbamate linkage, as shown for DC-CHOL and GL-67.

Cationic lipids bind to pDNA by electrostatic interactions between the cationic lipid moiety and the phosphate pDNA backbone or through hydrogen bonding between the amines and hydroxyl groups of the cationic lipids and the pDNA. The cationic lipids shown in Fig. 1.2 require a helper lipid, DOPE or cholesterol for the diacyl cationic lipids, and DOPE for the cholesterol based cationic lipids, to transfect cells. Cationic liposomes are prepared by first mixing the lipids in an organic solvent; the solvent is removed by evaporation producing a lipid film and the lipid film is hydrated with a buffer to form liposomes. The liposome size can be reduced by extrusion through a filter with fixed pore size, or energy can be applied to the suspension by strong bath or probe sonication to reduce the diameter of the liposomes.

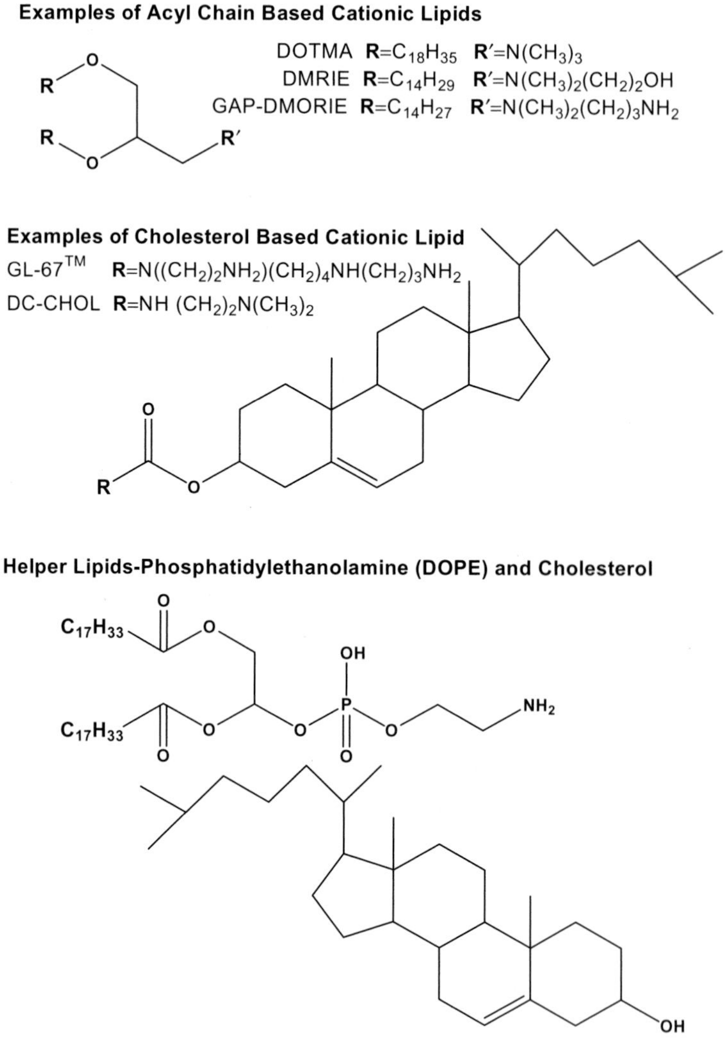

Examples of Acyl Chain Based Cationic Lipids

DOTMA R=$C_{18}H_{35}$ R'=$N(CH_3)_3$
DMRIE R=$C_{14}H_{29}$ R'=$N(CH_3)_2(CH_2)_2OH$
GAP-DMORIE R=$C_{14}H_{27}$ R'=$N(CH_3)_2(CH_2)_3NH_2$

Examples of Cholesterol Based Cationic Lipid

GL-67^{TM} R=$N((CH_2)_2NH_2)(CH_2)_4NH(CH_2)_3NH_2$
DC-CHOL R=NH $(CH_2)_2N(CH_3)_2$

Helper Lipids-Phosphatidylethanolamine (DOPE) and Cholesterol

$C_{17}H_{33}$

$C_{17}H_{33}$

Fig. 1.2 Chemical structures of cationic lipids and helper lipids used in non-viral gene therapy clinical trials.

3.2.1.1 *In vitro* transfection

For *in vitro* transfections, the pDNA is usually in 10 mM Tris, 1 mM EDTA and the commercially available liposomes are in water. Complexes are formed at 10–50 ug pDNA per ml and diluted to 1 to 5 ug/ml with serum free media and added to cells also in serum free media. After 4 hrs, the transfection complexes are aspirated off and replaced with serum containing media. The amount of cationic lipid to pDNA phosphate is titrated over a range of 2/1 to 8/1. The cationic lipids can be toxic to certain cells so there a balanced optimal ratio is required to achieve maximal gene expression with minimal toxicity. The biological activity of the transfection complexes has a short half life (hours) and maximal gene expression is achieved within 0.5 hr to 1 hr after formation.

3.2.1.2 Systemic *in vivo* gene transfer

Formulation of cationic lipids with pDNA for *in vivo* gene transfer is more complicated. First, there is very little correlation between optimal parameters, such as cationic lipid to helper lipid, ratio of cationic lipid to pDNA, liposome size or vehicle composition, developed for *in vitro* gene expression and *in vivo* gene expression. Secondly, the resulting transfection complexes are dependent on the type of cationic lipid used and how the complexes are formed. Binding of endogenous polyamines, such as spermine and spermidine, to plasmid DNA transforms supercoiled pDNA from a random coil into a toroid. Several approaches have been taken in developing cationic lipids for systemic gene transfer.

The first is to use cationic moieties known to bind to pDNA as the cationic head group. The resulting lipids are formulated with a helper lipid, such as DOPE or cholesterol, to form liposomes. The mole ratio of cationic lipid to helper lipid can vary from 100/0 to 10/90; most commonly it is set at 50:50. The mole ratio of cationic lipid to DNA phosphate can vary from 2/1 to 6/1 depending on the affinity of the cationic lipid in the context of a liposome or micelle to pDNA yielding a positively charged transfection complex. Non-reducing carbohydrates, such as sucrose, can be substituted for salt, such as NaCl, to maintain isogenicity.

An alternative approach is the synthesis of cationic lipid libraries, complex the cationic amphiphiles to the plasmid DNA and then test for gene expression.[32] Assays can use expression of reporter genes such as green fluorescent protein (GFP) or β-galactosidase (Lac-Z), or a secreted protein, such as human placental secreted alkaline phosphatase. It is impractical to screen the library other than by *in vitro* tissue culture due to the number of animals needed and the sample processing. Formulation of active cationic lipid candidates can then be optimized for *in vivo* gene transfer using the same reporter genes or a more popular *in vivo* screen is to detect expression by expression of a bioluminescent active reporter gene, such as firefly luciferase. This technique is excellent for liver, lung, spleen and heart imaging but poor for brain and muscle due to the limited biodistribution of the luciferase substrate following systemic administration.

The ideal particle size should range between 50 nm to 200 nm. Particles with diameters greater than 200 nm result in the reduction of the endocytotic index, except for monocyte derived cells such as macrophage, Kupffer cells or dendritic cells, which are specialized phagocytic cells. Once inside the endocytotic vesicle, the transfection complexes are disassembled and the pDNA must escape the internal vesicle and migrate to the nucleus for transcription of the transgene.

Modifications to the lipid composition have been made to increase systemic gene transfer efficiency. The circulation half life of the transfection complexes can be increased by covering the surface of the transfection complex with high molecular weight polyethylene glycol (2kdalt–10kdalt) inhibiting opsonization of the transfection complexes, thereby reducing non-specific uptake by the reticuloendothelium system (RES). Internalization of the transfection complex can be restricted to a specific cell type by derivatizing ligands to the surface of the transfection complex that bind to a receptor expressed on a specific cell type.[33–35] Intracellular release of the pDNA from either the endosome or lysosome can be achieved by changing the cationic lipid composition to one that is susceptible to fusion at acidic pH.[36] Lastly, nuclear uptake of the pDNA can be increased by adding nuclear localization sequences in the form of DNA sequences that bind proteins manufactured in the cytoplasm that contain nuclear localization sequences allowing the pDNA to "hitch a ride" into the nucleus.[37] There are currently no clinical trials using systemically administered cationic lipid/plasmid DNA transfection complexes.

3.2.1.3 Local administration of cationic lipid/pDNA transfection complexes

Local administration of cationic lipid/pDNA transfection complexes refers to intratumoral administration and intramuscular administration. From a safety perspective, the bulk of gene delivery and gene expression is limited to the site of administration. Therapeutic applications of intratumoral administration have focused on expression of molecules, such as cytokines[38] and self antigens[39] for the purpose of priming the immune system to kill not only the transfected cells but in so doing, prime the immune system to kill non-transfected tumors, thus creating a systemic therapy from a local administration.

With regard to the expression of self antigens, the first non-viral clinical trial was conducted by Dr. G. Nabel in collaboration with Dr. L. Huang. Intratumoral administration of cationic lipid formulated HLA B-7 pDNA was shown to be safe and tumor regression was observed. Subsequent clinical trials modified the cationic lipid formulation from the use of DC-Chol/DOPE to DMRIE/DOPE, increased the pDNA dose and tested different dosing schedules. A phase 2 dose ranging study identified a 2 mg per intratumoral injection as an effective antitumor dose and administration once a week for 6 weeks in a single tumor to be the dosing cycle. The clinical results from the phase 2 dose escalation study were used to define the patient entry criteria for a Phase 3 clinical study. The Phase 3 clinical study compares the cationic lipid/pDNA non-viral gene therapy (Allovectin-7) to Dacarbazine or Temozolomide. This therapy is currently in Phase 3 clinical trial.[40] Results from this trial should be available in 2012.

Expression of foreign antigens by administration of pDNA results in an immune response to the expressed protein creating the potential for development of genetic vaccines. However, mg quantities of pDNA injected multiple times have been required to sustain an immune response. Certain cationic lipids when formulated with pDNA not only increased transfection efficiency but were also immunostimulatory. A cationic liposome formulation composed of GAP-DMROIE and diphytanolylphosphatidylethanolamine (DPyPE) was shown to be immunostimulatory for several expressed antigens in rodent and non-human primates.[41,42] Multiple demonstrations of proof of concept advanced this formulation into a phase 1 human clinical trial for a pandemic influenza vaccine. The phase 1 trial evaluated tolerability and immunogenicity. These parameters were tested for this clinical trial included a comparison of a needle free device, Biojector 2000, vs. needle; a single plasmid expressing the H5 hemmaglutinin viral envelope protein vs. the H5 plasmid plus two additional plasmids expressing viral proteins whose amino acid sequences are highly conserved amongst multiple pandemic influenza virus clades. One hundred and three patients were enrolled and 86 were evaluable for immunogenicity. A 65% response rate was observed for neutralizing antibody equivalent to or exceeding protecting titers of 1/40. This was a durable response being observed out to 182 days (unpublished results).

3.3 *Polymer*

Polymer based plasmid DNA gene therapy can be divided into two categories: cationic polymer and neutral polymers. Examples of cationic polymers are polylysine, polyethleneimine, or panamdendrimers. Examples of neutral polymers are Polylactide glycolic anhydride (PLGA), poloxamer and polyvinylpyrolidone (PVP).

3.3.1 *Cationic Polymers*

Cationic polymers behave similarly to cationic lipids in that they contain a cationic moiety which either electrostatically bonds with the phosphate backbone of the pDNA or hydrogen bonds to the pDNA. The first cationic polymers were naturally derived from poly amino acids such as poly-L-ornithine and poly-L-lysine. Utility was derived from the virology field that first used poly-L-ornithine to facilitate viral infection of cells.[11] It was later discovered that transfection of cells with viral RNA complexed to poly-L-ornithine produced infectious virus. Lessons learned from the virology field were applied to pDNA delivery with the substitution of poly-L-ornithine for poly-L-lysine. The polymer provided a versatile backbone for chemical modification of ligands to target the complexes to cells by additional peptide sequences or chemical modification of small molecular weight ligands, such as folate or carbohydrates.[43] One application that has evolved from the bench to the clinic is the use of poly-L-lysine/pDNA complexes for the treatment of cystic fibrosis.[44] Several key developments in transitioning from the research lab to the clinic were the development of an aerosolized transfection complex, reduction in the CpG content of the plasmid to avoid secondary cytokine activation,[15] use of a polyethylene glycol-substituted 30-mer lysine peptides.[45]

Improvements in polycationic amino acids polymers have included other positively charged amino acids such as arginines and histidines.[46] The latter not only serves as an alternate DNA binding moiety but also has a pK that can impact the transfection complex when the pH environment acidifies, as is found in the lysosome and endosome, facilitating decomplexation and potentially lysing open the endocytic vacuole.[47]

Polyethyleneimine (PEI) is a completely synthetic polymer that has similar intracellular release properties as the poly-L-histidine, first introduced into the field of nonviral gene therapy by JP Behr.[48] The imine moiety of the polymer provides electrostatic bonding and hydrogen bonding to pDNA and also serves as a protonateable buffer that prevents acidification of the endosome/lysosome resulting in intracellular plasmid DNA release. This polymer has been modified with polyethylene glycol and targeting ligands. The polymer/pDNA complex is currently in preclinical development.

3.3.2 *Neutral Polymer*

Neutral polymers are defined as polymers carrying no net charge. Examples are poloxamers, polyvinylpyrolidone (PVP), polyanhydride (PLGA) polymers. The first two form simple mixtures of polymer and pDNA whereas the PLGA polymer forms nanoparticles in which the pDNA is packaged. Both PVP[49,50] and poloxamer have been used in clinical trials for local administration of pDNA. The applications have been in cancer, cardiovascular disease and vaccine development against infectious disease. For cancer applications, plasmid DNA encoding cytokines, such as IL-2, IL-12, or interferon-α in combination with IL-12,[49,50,51] is formulated with PVP and injected intratumorally resulting in expression and secretion of the cytokines from the transfected cells. The chemoattracting cytokine gradient causes immune cells to migrate to the tumor and kill the tumor cells. The therapeutic hypothesis is similar to that of the cationic lipid based immunotherapy, in that T cell mediates tumor cell killing programs and activates tumor surveillance of the immune system, thus preventing new tumor formation.

Poloxamers are block copolymers of polyethylene oxide (a) and polypropylene oxide (b) that are linked by ether linkages shown in Fig. 1.3 with molecular weights ranging from 1,000 daltons to 12,000 daltons. In

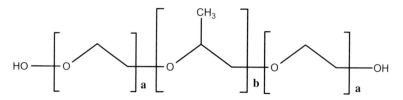

Fig. 1.3 Poloxamer chemical structure. α-Hydro-ω-hydroxypoly (oxyethylene)$_a$ poly(oxypropylene)$_b$ poly(oxyethylene)$_a$ block copolymer in which a values can range from 12 to 101 and b values from 20 to 56.

aqueous media, the polymers assemble into micelles. These micelles can be induced to aggregate to form hydrogels by increasing the temperature. The ratio of ethyleneoxide to propyleneoxide determines the gelling temperature range (cloud point). The poloxamer is formulated with plasmid DNA in saline and injected into muscle resulting in gene expression from muscle cells. Clinical applications take advantage of the local expression of therapeutic genes, specifically cardiovascular and genetic vaccines.

The human developmentally regulated endothelium locus (Del-1) gene was identified by T. Quertermous to be a potential angiogenic factor.[52] In pursuit of restoring blood flow to claudified muscle, formulation of a Del-1 expressing pDNA formulated with poloxamer 188 was shown to yield muscular gene expression that was sufficient to induce angiogenesis and be well tolerated in mice and rabbits. A Phase II multicenter clinical trial conducted with the Del-1 pDNA/poloxamer 188 formulation in 105 patients with peripheral arterial disease was shown to be safe and well tolerated.[53] The execution of the trial showed that the formulation could be scaled up to cGMP pharmaceutical manufacture, have an acceptable shelf life and meet FDA approval. Although there was no statistical significance in outcome compared to poloxamer 188 alone, these results showed that this formulation was suitable for clinical applications and identified areas for improvement to achieve therapeutic efficacy.

A modified poloxamer formulation was used for the development of a therapeutic vaccine for prevention of cytomegalovirus induced pneumonia in hematopoietic stem cell (HSC) transplant patients.[54] Poloxamer was combined with benzylalkonium chloride (BAK) to create mixed micelles that were positively charged. A series of freeze and thaw cycles were used to reduce particle size with 500 nm diameters. A mixture of 2 pDNAs encoding the surface glycoprotein B (gB) and the internal matrix protein (pp65) was added to the poloxamer/BAK and stored frozen.[55,56] A phase I clinical trial was conducted involving 22 CMV seronegative and 22 seropositive healthy subjects. The vaccine was well tolerated with no serious adverse events. Immunogenicity determined by *ex vivo* interferon (IFN)-gamma enzyme-linked immunospot assay yielded 45.5% of CMV seronegative subjects and 25% of CMV seropositive subjects in subjects receiving the full vaccine series. The safety and immunogenicity results supported further evaluation in a phase 2 clinical trial.

Conclusions

Local administration of plasmid DNA based gene therapies seem to be the most promising and have the least potential for any side effects. pDNA alone has advanced into phase 3 clinical trials for cardiovascular applications. Success of this clinical application is based upon the potent angiogenic stimulation produced by a very low level of expressed protein that remains localized at the site of administration. Improvements in gene transfer efficiency from the result of pDNA formulation with cationic liposomes and polymers are also combined with selection of expressed genes that produce a potent secondary effect, such as the case of immunostimulation against tumor antigens and pathogenic antigens. Research focused on increasing the transfection activity of plasmid DNA based gene delivery systems is very active. The hope is to combine the development of cell specific targeted gene delivery along with tissue specific promoters that further restrict gene expression to a specific cell type with the application for systemic administration.

References

1. Szoka FC, Papahadjopoulos, DP (1983). Method of inserting DNA into living cells. In *United States Patent Office*.
2. Behr JP, Demeneix B, Loeffler JP, Perez-Mutul J (1989). Efficient gene transfer into mammalian primary endocrine cells with lipopolyamine-coated DNA. *Proc Natl Acad Sci USA* **86**: 6982–6986.
3. Felgner PL, *et al.* (1987). Lipofection: a highly efficient, lipid-mediated DNA-transfection procedure. *Proc Natl Acad Sci USA* **84**: 7413–7417.

4. Felgner PL, Ringold GM (1989). Cationic liposome-mediated transfection. *Nature* **337**: 387–388.

5. Wolff JA, *et al.* (1990). Direct gene transfer into mouse muscle *in vivo*. *Science* **247**: 1465–1468.

6. Nabel GJ, *et al.* (1996). Immune response in human melanoma after transfer of an allogeneic class I major histocompatibility complex gene with DNA-liposome complexes. *Proc Natl Acad Sci USA* **93**: 15388–15393.

7. Nabel GJ, *et al.* (1993). Direct gene transfer with DNA-liposome complexes in melanoma: expression, biologic activity, and lack of toxicity in humans. *Proc Natl Acad Sci USA* **90**: 11307–11311.

8. Isner JM, *et al.* (1996). Arterial gene transfer for therapeutic angiogenesis in patients with peripheral artery disease. *Hum Gene Ther* **7**: 959–988.

9. Isner JM, *et al.* (1996). Clinical evidence of angiogenesis after arterial gene transfer of phVEGF165 in patient with ischaemic limb. *Lancet* **348**: 370–374.

10. Mir LM, Bureau MF, Rangara R, Schwartz B, Scherman D (1998). Long-term, high level *in vivo* gene expression after electric pulse-mediated gene transfer into skeletal muscle. *C R Acad Sci III* **321**: 893–899.

11. Motoyoshi F, Bancroft JB, Watts JW, Burgess J (1973). The infection of tobacco protoplasts with cowpea chlorotic mottle virus and its RNA. *J Gen Virol* **20**: 177–193.

12. Pillai VB, Hellerstein M, Yu T, Amara RR, Robinson HL (2008). Comparative studies on *in vitro* expression and *in vivo* immunogenicity of supercoiled and open circular forms of plasmid DNA vaccines. *Vaccine* **26**: 1136–1141.

13. Valencia P, Dias AP, Reed R (2008). Splicing promotes rapid and efficient mRNA export in mammalian cells. *Proc Natl Acad Sci USA* **105**: 3386–3391.

14. Lida Y, Kanagu D (2000). Quantification analysis of translation initiation signal in vertebrate mRNAs: Effect of nucleotides at positions +4 upon efficiency of translation initiation. *Nucleic Acids Symposium* **44**: 77–78.

15. Yew NS, *et al.* (2002). CpG-depleted plasmid DNA vectors with enhanced safety and long-term gene expression *in vivo*. *Mol Ther* **5**: 731–738.

16. Abdulhaqq SA, Weiner DB (2008). DNA vaccines: developing new strategies to enhance immune responses. *Immunol Res* **42**: 219–232.

17. Ferreira GNM, Monteiro GA, Prazeres DMF, Cabral JMS (2000). Downstream processing of plasmid DNA for gene therapy and DNA vaccine applications. *TIBTech* **18**: 380–388.

18. Stadler S, Lemmens R, Nyhammar Y (2004). Plasmid DNA Purification. *J Gene Med* **6**: S54–S66.

19. Nikol S, *et al.* (2008). Therapeutic angiogenesis with intramuscular NV1FGF improves amputation-free survival in patients with critical limb ischemia. *Mol Ther* **16**: 972–978.

20. Powell RJ, *et al.* (2008). Results of a double-blind, placebo-controlled study to assess the safety of intramuscular injection of hepatocyte growth factor plasmid to improve limb perfusion in patients with critical limb ischemia. *Circulation* **118**: 58–65.

21. Nakagami H, Kaneda Y, Ogihara T, Morishita R (2005). Hepatocyte growth factor as potential cardiovascular therapy. *Expert Rev Cardiovasc Ther* **3**: 513–519.

22. Seaman MS, *et al.* (2005). Multiclade human immunodeficiency virus type 1 envelope immunogens elicit broad cellular and humoral immunity in rhesus monkeys. *J Virol* **79**: 2956–2963.

23. Gudmundsdotter L, *et al.* (2006). Therapeutic immunization for HIV. *Springer Semin Immunopathol* **28**: 221–230.

24. Sandstrom E, *et al.* (2008). Broad immunogenicity of a multigene, multiclade HIV-1 DNA vaccine boosted with heterologous HIV-1 recombinant modified vaccinia virus Ankara. *J Infect Dis* **198**: 1482–1490.

25. Rizzuto G, *et al.* (1999). Efficient and regulated erythropoietin production by naked DNA injection and muscle electroporation. *Proc Natl Acad Sci USA* **96**: 6417–6422.

26. Daud AI, *et al.* (2008). Phase I Trial of Interleukin-12 Plasmid Electroporation in Patients With Metastatic Melanoma. *J Clin Oncol.*

27. Manome Y, Nakamura M, Ohno T, Furuhata H (2000). Ultrasound facilitates transduction of naked plasmid DNA into colon carcinoma cells *in vitro* and *in vivo. Hum Gene Ther* **11**: 1521–1528.

28. Huber PE, Pfisterer P (2000). *In vitro* and *in vivo* transfection of plasmid DNA in the Dunning prostate tumor R3327-AT1 is enhanced by focused ultrasound. *Gene Ther* **7**: 1516–1525.

29. Greenleaf WJ, Bolander ME, Sarkar G, Goldring MB, Greenleaf JF (1998). Artificial cavitation nuclei significantly enhance acoustically induced cell transfection. *Ultrasound Med Biol* **24**: 587–595.

30. Takeuchi D, *et al.* (2008). Alleviation of Abeta-induced cognitive impairment by ultrasound-mediated gene transfer of HGF in a mouse model. *Gene Ther* **15**: 561–571.

31. Taniyama Y, *et al.* (2002). Development of safe and efficient novel nonviral gene transfer using ultrasound: enhancement of transfection efficiency of naked plasmid DNA in skeletal muscle. *Gene Ther* **9**: 372–380.

32. Yingyongnarongkul BE, Howarth M, Elliott T, Bradley M (2004). Solid-phase synthesis of 89 polyamine-based cationic lipids for DNA delivery to mammalian cells. *Chemistry* **10**: 463–473.

33. Anwer K, Bailey A, Sullivan SM (2000). Targeted gene delivery: a two-pronged approach. *Crit Rev Ther Drug Carrier Syst* **17**: 377–424.

34. Driessen WH, Fujii N, Tamamura H, Sullivan SM (2008). Development of peptide-targeted lipoplexes to CXCR4-expressing rat glioma cells and rat proliferating endothelial cells. *Mol Ther* **16**: 516–524.

35. Hood JD, *et al.* (2002). Tumor regression by targeted gene delivery to the neovasculature. *Science* **296**: 2404–2407.

36. Li W, Huang Z, MacKay JA, Grube S, Szoka FC, Jr. (2005). Low-pH-sensitive poly(ethylene glycol) (PEG)-stabilized plasmid nanolipoparticles: effects of PEG chain length, lipid composition and assembly conditions on gene delivery. *J Gene Med* **7**: 67–79.

37. Cartier R, Reszka R (2002). Utilization of synthetic peptides containing nuclear localization signals for nonviral gene transfer systems. *Gene Ther* **9**: 157–167.

38. Li D, Shugert E, Guo M, Bishop JS, O'Malley BW, Jr. (2001). Combination nonviral interleukin 2 and interleukin 12 gene therapy for head and neck squamous cell carcinoma. *Arch Otolaryngol Head Neck Surg* **127**: 1319–1324.

39. Plautz GE, Yang ZY, Wu BY, Gao X, Huang L, Nabel GJ (1993). Immunotherapy of malignancy by *in vivo* gene transfer into tumors. *Proc Natl Acad Sci USA* **90**: 4645–4649.

40. Bedikian AY, Del Vecchio M (2008). Allovectin-7 therapy in metastatic melanoma. *Expert Opin Biol Ther* **8**: 839–844.

41. Reyes L, *et al.* (2001). Vaxfectin enhances antigen specific antibody titers and maintains Th1 type immune responses to plasmid DNA immunization. *Vaccine* **19**: 3778–3786.

42. Hartikka J, *et al.* (2001). Vaxfectin enhances the humoral immune response to plasmid DNA-encoded antigens. *Vaccine* **19**: 1911–1923.

43. Wu CH, Wilson JM, Wu GY (1989). Targeting genes: delivery and persistent expression of a foreign gene driven by mammalian regulatory elements *in vivo*. *J Biol Chem* **264**: 16985–16987.

44. Davis PB, Cooper MJ (2007). Vectors for airway gene delivery. *Aaps J* **9**: E11–E17.

45. Konstan MW, *et al.* (2004). Compacted DNA nanoparticles administered to the nasal mucosa of cystic fibrosis subjects are safe and demonstrate partial to complete cystic fibrosis transmembrane regulator reconstitution. *Hum Gene Ther* **15**: 1255–1269.

46. Putnam D, Zelikin AN, Izumrudov VA, Langer R (2003). Polyhistidine-PEG:DNA nanocomposites for gene delivery. *Biomaterials* **24**: 4425–4433.

47. Pack DW, Putnam D, Langer R (2000). Design of imidazole-containing endosomolytic biopolymers for gene delivery. *Biotechnol Bioeng* **67**: 217–223.

48. Zou SM, Erbacher P, Remy JS, Behr JP (2000). Systemic linear polyethylenimine (L-PEI)-mediated gene delivery in the mouse. *J Gene Med* **2**: 128–134.

49. Mendiratta SK, *et al.* (1999). Intratumoral delivery of IL-12 gene by polyvinyl polymeric vector system to murine renal and colon carcinoma results in potent antitumor immunity. *Gene Ther* **6**: 833–839.

50. Mendiratta SK, *et al.* (2000). Combination of interleukin 12 and interferon alpha gene therapy induces a synergistic antitumor response against colon and renal cell carcinoma. *Hum Gene Ther* **11**: 1851–1862.

51. Kaushik A (2001). Leuvectin Vical Inc. *Curr Opin Investig Drugs* **2**: 976–981.

52. Ho HK, *et al.* (2004). Developmental endothelial locus-1 (Del-1), a novel angiogenic protein: its role in ischemia. *Circulation* **109**: 1314–1319.

53. Grossman PM, *et al.* (2007). Results from a phase II multicenter, double-blind placebo-controlled study of Del-1 (VLTS-589) for intermittent claudication in subjects with peripheral arterial disease. *Am Heart J* **153**: 874–880.

54. Selinsky C, *et al.* (2005). A DNA-based vaccine for the prevention of human cytomegalovirus-associated diseases. *Hum Vaccin* **1**: 16–23.

55. Vilalta A, *et al.* (2005). I. Poloxamer-formulated plasmid DNA-based human cytomegalovirus vaccine: evaluation of plasmid DNA biodistribution/persistence and integration. *Hum Gene Ther* **16**: 1143–1150.

56. Hartikka J, *et al.* (2008). Physical characterization and *in vivo* evaluation of poloxamer-based DNA vaccine formulations. *J Gene Med* **10**: 770–782.

Chapter 2
Adenoviral Vectors

Stuart A. Nicklin* and Andrew H. Baker

Adenoviruses are one of the most widely investigated vectors for gene therapy. Their attributes include ease of genetic manipulation to produce replication-deficient vectors, ability to readily generate high titer stocks, efficiency of gene delivery into many cell types and ability to encode large genetic inserts. Adenoviruses have been utilized for a variety of therapeutic applications particularly for high level transient overexpression in cancer gene therapy, vaccine delivery and certain cardiovascular diseases. With the first licensing of clinical adenoviral gene therapy for cancer treatment the use of adenoviruses is likely to expand in the future. This chapter focuses on the history of adenovirus development, interactions with the host both for gene delivery and immunogenicity as well as potential therapeutic applications.

1. Introduction

The human adenovirus (Ad) family is divided into 6 sub-species (A-F). There are a total of 51 different serotypes identified to date, classified via differential abilities to hemagglutinate erythrocytes and characterized by differences in capsid structure and receptor use. Ads from species C, particularly serotype 5 (Ad5) are the most widely investigated viral gene transfer vectors and consequently, one of the most commonly utilized viral vectors in clinical trials. However, it is only in recent years that important findings regarding receptor usage and interactions with the host *in vivo*, particularly following intravenous (i.v.) administration, have been deciphered. There are

---

*Correspondence: BHF GCRC, Faculty of Medicine, University of Glasgow, 126 University Place, Glasgow UK.
E-mail: stuart.a.nicklin@clinmed.gla.ac.uk

still challenging hurdles to overcome, particularly Ad's inherent immuno-genicity, yet a licensed Ad gene therapy product for head and neck cancer in China, highlights its potential for advanced disease treatment.

2. Adenoviral Capsid Structure

Ads are icosahedral viruses approximately 70-90 nm in diameter with a double stranded linear 36 kb DNA genome. Virions are non-enveloped and the capsid contains three principal protein components: hexon, penton and fiber (Fig. 2.1). The 240 trimeric hexon capsomeres are the major structural proteins, with a stable structure, mediated by a series of complex extending loops and three towers extending from the top of the molecule. Individual serotype hexons differ mainly in their hypervariable regions (HVRs) at the surface of the protein, which dictates immunogenicity of the virus. Divergent HVRs are a major reason for the lack of cross-specificity of neutralizing antibodies between individual species.

At the 12 vertices of the Ad capsid are the fiber trimers, inserted into the pentameric penton base (Fig. 2.1). The fiber consists of a short tail region inserted into the penton base, a shaft domain and a globular knob domain.

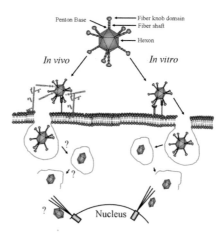

Fig. 2.1 The Ad capsid and cell entry. The icosahedral capsid is composed of 3 principal protein components, hexon, fiber and penton base. *In vitro* Ad tethers to cells via fiber binding CAR. *In vivo* Ad sequesters FX and the Ad: FX complex binds hepatocytes via HSPGs. Activation of integrins via penton base induces viral endocytosis before migration to the nucleus. It remains to be confirmed whether the *in vitro* trafficking mechanism is relevant *in vivo*.

The length of the fiber shaft is a main determinant of variation between individual serotypes, as the shortest fibers belong to the B species Ad35, while fibers from species C Ads such as Ad5 are among the longest. There are also a number of minor capsid proteins, which mainly function as "glue" for the capsid.

3. Adenoviral Cell Entry

The method for Ad infection (at least *in vitro*) is well defined, particularly for Ad5 (species C). Ads infect host cells through receptor-mediated endocytosis mediated by fiber and penton base (Fig. 2.1). Initial interaction for species C Ads is between the fiber knob domain and the Coxsackie/Ad receptor (CAR).[1,2] All Ad species, except B can interact with CAR via the fiber knob domain, although some species D Ads also use sialic acid.[3] Species B Ads utilize CD46.[4] Virus internalization is through activation of $\alpha_v\beta_3$ and $\alpha_v\beta_5$ integrins by the RGD motif in the penton base at the N-terminus of the fiber protein.[5] Once internalized to the cytoplasm via clathrin-coated endosomes the penton base mediates capsid release from the endosome via acidification and the virus migrates to the nucleus.[6]

Following i.v. Ad administration transduction involves other factors. The major site of Ad transduction following i.v. delivery is liver hepatocytes but mutation of all Ad capsid receptor binding epitopes has limited effects on blocking liver uptake.[7] This is because upon contact with blood Ad5 interacts with host proteins including coagulation factor IX (FIX), FX and complement binding protein (C4BP) to deliver virus to hepatocytes via heparan sulphate proteoglycans (HSPGs) or low density lipoprotein receptor-related protein (LRP), respectively (Figs. 2.1 and 2.3).[8,9] However, sequestration of FX and delivery of the virus: FX complex to liver via HSPGs is the primary route for i.v. delivered Ad liver transduction.[10,11] This finding defines an infectivity function for hexon, since direct interaction between hexon and FX is observed.[10,11] Additionally, FX interacts with the Ad hexon via the HVRs since exchange of the Ad5 HVRs for those of a non-FX binding Ad is sufficient to block liver gene transfer.[10] The fact that such diverse cell entry pathways are used by the virus highlights the complexity of Ad interaction with the host, providing further challenges for the development of efficient gene therapies.

4. Production of Adenoviral Vectors

The fact that Ads are one of the most widely investigated gene transfer vectors is in no small part due to the development of the human cell line, 293, transformed with the entire left hand end of the Ad5 genome.[12] This allowed the construction and study of recombinant Ads in a helper virus-free environment.

The gene products are organized into early (E1-E4) and late (L1-L5) regions, based on expression before or after initiation of DNA replication. The E1 region is the essential initiating factor for viral DNA replication as it encodes E1a, which induces expression of other early genes. E1 also ensures the cell enters S phase to enable viral replication by inhibiting key cell cycle components and blocking p53-mediated apoptosis. The E2 region encodes products required for DNA replication such as DNA polymerase, single stranded DNA binding proteins and pre-terminal protein (which are essential for viral DNA synthesis, elongation and genome packaging into the maturing viral particle). The E3 region encodes products which inhibit the cell's immune responses via blocking major histocompatibility complex (MHC) transport. E4 encodes products required for trafficking viral mRNA to the Golgi complex and for inhibiting transport of host cell mRNAs to enhance viral production. The late region transcripts encode structural capsid proteins.

First generation, replication-deficient Ad vectors contain deletions in E1, enabling insertion of expression cassettes of up to approximately 5 kb. The limit in DNA packaging is dictated by the physical size of the virion which is only able to package up to 105% (approximately 38 kb) of the native virus genome size. Generation of these vectors has been historically carried out via two plasmid transfection; the first plasmid contains sequences for the E1 region in between which is inserted the expression cassette, the second plasmid encodes the Ad genome with a large insertion in the E1 region which blocks self packaging of virus through exceeding the maximal 38 kb packaging capacity. Homologous recombination between two plasmids, using the E1 sequence acting *in trans* from 293 cells excises E1 and inserts the expression cassette, producing replication deficient Ad vectors which can be propagated in 293 cells. This can result in generation of replication competent Ad (RCA) due to recombination between the Ad genome vector and the integrated homologous E1 region in 293 cells, however modified E1-expressing cell lines and Ad generating plasmids with low homology to each other has significantly reduced RCA. Additionally, other first generation Ad production systems utilize cloning and construction of full length recombinant Ad genomes in bacteria prior to single plasmid transfection into 293 cells to package virus. To increase the size of expression cassette that first generation vectors can encode to 8 kb, E3 can also be deleted. The main limitation of first generation Ads was the discovery of leaky expression of viral genes even in the absence of E1 due to transactivation of viral genes from host transcription factors, leading to transient transgene expression. Transduced cells release cytokines and are recognized and destroyed by cytotoxic CD8$^+$ and memory CD4$^+$ T lymphocytes (CTL) via recognition of viral structural proteins presented on the cell surface by MHC class I (Fig. 2.3). Second generation vectors were developed via deletion of the E2 and/ or E4 regions potentially enabling cloning of 14 kb of foreign DNA into the vector. The initial study mutated E2a leading to increased transgene expression and decreased inflammatory responses in mouse liver and lung.[13] E4 has also been deleted to provide similar advantages,[14] however, such vectors were rapidly superseded by third generation Ads.

Third generation [also called gutless or helper dependent Ad vectors (hdAd)] are those devoid of all viral open reading frames, creating vectors containing only essential *cis* elements (inverted terminal repeats and contiguous packaging sequences). Deletion of all viral genes increases insert capacity to 36–38 kb and limits host immune responses by ablating leaky production of Ad genes, leading to persistent transgene expression. Early production of helper-dependent Ad vectors was hampered by the requirement for co-infection of cell lines with helper Ad expressing all the necessary Ad genes *in trans*, leading to contamination of hdAd stocks with wild type helper virus. New systems use recombinases to remove packaging signals from helper virus to reduce contamination.[15,16] Further refinements such as reverse insertion of the packaging signal in the helper virus to prevent recombination between helper and vector have almost eliminated helper contamination.[17] The inability to produce stable cell lines expressing all the viral helper functions, due to toxicity of viral proteins on host cells means that scale up and clinical grade production of hdAds is still a limiting factor. Nonetheless, hdAds have enormous potential for gene therapy, they show greatly reduced pro-inflammatory responses and have been shown to drive long term transgene expression in many different animal models, in certain instances for the lifetime of the animal.

5. Production of Targeted Adenoviral Vectors

Modification of the tropism of the virus would be beneficial to enable i.v. delivery to be targeted selectively to diseased tissues and away from the liver to improve safety and efficiency. To date targeting has almost exclusively focused on modifying fiber, however with the discovery that hexon is responsible for Ad5's liver tropism future Ad5 targeting research is likely to focus in this new direction. The first targeting approaches used antibodies to neutralise the knob protein (Fig. 2.2) and retargeted gene delivery via ligands such as growth factors,[18] or peptides.[19] However, successful *in vivo*

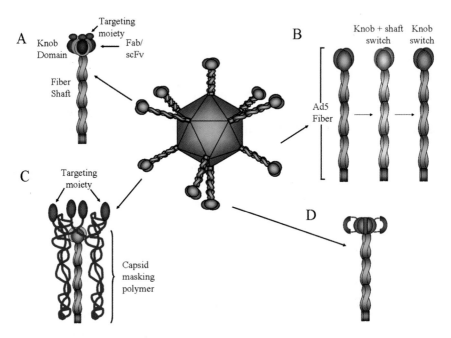

Fig. 2.2 Ad targeting. (A) Bispecific antibodies neutralize the Ad knob domain and retarget to novel/candidate receptors. (B) Serotype switching by exchanging either the Ad5 fiber knob or entire fiber for an alternative serotype fiber for new tropism. (C) Coating the Ad capsid with inert polymers covalently linked to targeting ligands. (D) Insertion of peptides into the surface exposed HI loop on the Ad fiber knob.

Ad5 retargeting using antibodies has only been reported once and additionally required a cell-specific promoter to silence hepatocyte expression, highlighting the difficulty of detargeting Ad5 from liver.[20] Alternatives to knob-neutralising antibodies include coating the Ad capsid with inert polymers, e.g. polyethylene glycol (Fig. 2.2) which coats the entire Ad, masking the immunogenic capsid enabling retargeting via, for example, growth factors.[21]

Genetic modification of the Ad fiber and/or penton base protein were the next logical steps in Ad targeting, in order to avoid potential scale up problems and additional regulatory hurdles. Although many different methods have been explored, utilization of the fiber HI loop is now used almost ubiquitously (Fig. 2.2). The HI loop is exposed outside the knob enabling incorporation of additional protein sequence without affecting fiber trimerization or knob domain folding and inserting the RGD peptide into the HI loop produces Ads that transduce cells via a CAR-independent route.[22] The ability to target Ad via ligands in the HI loop was improved considerably by identification of the amino acid residues in the fiber knob that mediated interaction with CAR, such as S408E or G, P409A, K417G or L, K420A, Y477A or T and Y491A.[23] Combining these mutations with insertion of targeting ligands into the HI loop can provide efficient, selective gene delivery *in vitro* for example.[24,25] However, translating efficient and selective *in vitro* targeting into efficient *in vivo* targeting that avoids liver transduction has been less straightforward.

Many studies have pursued detargeting of Ad5 from the liver via either single or multiple mutations in the knob domain, or by combining mutations in fiber knob, shaft and penton base with largely negative effects.[26] Clearly, blocking interaction of hexon with FX will be key to i.v. targeted Ad transduction.

The alternative to genetic modification of the Ad5 fiber to remove native receptor binding and replace with new receptor binding moieties has been the identification of novel Ad serotypes with alternative tropism to Ad5. This has enabled the use of other Ad serotype fibers via pseudotyping in which the Ad5 fiber is exchanged for another serotype fiber to provide Ad5-based vectors with alternative tropism (Fig. 2.2). The main focus of alternative serotypes has been those of species B, particular Ad35. Shayakhmetov *et al.* pseudotyped Ad5 with the Ad35 fiber to efficiently transduce hematopoietic cells refractory to Ad5 highlighting the potential of this approach.[27] Furthermore, pseudotyping Ad5 with the rare species D fiber 19p modified with renal targeting peptides in the HI loop produced significantly decreased liver transduction and retargeting to the rat kidney.[28] The prevalence of neutralizing immunity to Ad5 in patients has also driven the development of vectors based on entire alternative serotypes such as Ad35.[29] Ad35 is a relatively uncommon human Ad serotype and anti-Ad5 immunity does not cross-react with Ad35. The Ad35 receptor CD46 is a ubiquitous complement membrane protein, expressed in almost all human cells and up-regulated in certain cancers, but expressed only in the testes of rodents.[4] It can therefore be utilized for gene delivery to targets such as tumors that may be refractory to transduction with Ad5.

6. Gene Therapy Applications

As one of the commonest gene transfer vectors under investigation for human gene therapy it has been widely investigated in many disease areas. The application of gene therapy for specific diseases is covered in great depth elsewhere in this book and this section is designed to merely give an overview of Ad's potential applications.

Cancer gene therapy aims to kill tumor cells via gene delivery, e.g. by overexpressing p53 (which is absent in many tumors) to prevent tumor cell growth. In China a first generation Ad expressing p53 (Gendicine) is the first licensed gene therapy product.[30] Alternative Ad5-based cancer gene therapy include the use of suicide gene transfer in which overexpression of an enzyme converts a toxic pro-drug into an active anti-cancer agent, e.g. the herpes simplex virus thymidine kinase gene and ganciclovir. Other cytotoxic cancer gene therapy strategies have taken advantage of Ad5's mode of replication and the knowledge that many cancers have mutations in p53 or retinoblastoma pathways. Modified Ads termed conditionally replicative Ads (CRAds) or oncolytic Ads have mutations or deletions in E1. E1-deleted/ mutant Ads are unable to replicate in cells in which normal p53 function is present (as E1b 55k protein is required to inactivate it and enable cells to enter S phase) but replication takes place in tumor cells with absent p53 expression. The first example of this was the CRAd ONYX-015 with 2 mutations in the E1b gene enabling it to replicate in tumor cells.[31] The other most well investigated CRAd is Ad5-Δ24 which has a region of E1a deleted that normally mediates binding to the Rb protein to enable cells to pass the G1-S check point, thus limiting the virus to replication in tumor cells with a mutated Rb pathway.[32]

Since Ad efficiently transduces cells of the cardiovascular system, their utility for cardiovascular disease gene therapy has also been investigated. First generation Ads are disadvantaged due to transient transgene expression when, for many cardiovascular applications long term transgene expression would be beneficial. Nonetheless, first generation Ads have a niche in the development of gene therapy for cardiovascular disease where acute remodeling of blood vessels occurs via migration and proliferation of smooth muscle cells (SMC), such as in vein graft failure following coronary bypass surgery. Acute inhibition of SMC migration and proliferation in vein graft failure via Ad-mediated delivery of cell cycle inhibitors such as p53,[33] as well as inhibitors of extracellular matrix remodeling (essential for SMC migration) such as tissue inhibitors of metalloproteinases are promising approaches.[34] Additionally, Ad5 has been utilized in clinical trials for local delivery to the myocardium to overexpress pro-angiogenic factors such as fibroblast growth factor (FGF) or vascular endothelial growth factor (VEGF) to relieve symptoms of angina with some positive outcomes.[35,36]

The high immunogenicity of first generation Ad5 vectors highlights their potential for vaccination strategies. Ad5 vectors overexpressing antigens from different pathogens have been investigated in experimental models, but it is in the development of a vaccine for human immunodeficiency virus (HIV) that Ad strategies have advanced to a clinical trial.[37] Ad delivery of the HIV antigens *gag*, *pol* and *nef* in the phase 2 STEP trial assessed efficacy in lowering either HIV-1 infection rates or plasma viremia.[37] Disappointingly the trial was halted prematurely due to an unforeseen safety issue (see later). Nonetheless Ad-mediated vaccination is a promising approach for immunotherapy for a number of serious infectious diseases and cancer.

There has been major research into development of Ad as a gene therapy vector for the treatment of inherited disorders. Despite the obvious limitations of transient transgene expression from first generation Ad vectors which would limit curative treatment of an inherited disorder, the use of hdAd has provided new impetus for research in this area. As the liver is the primary site of Ad gene delivery it has been widely investigated for the treatment of liver-specific disorders, and in the case of secreted soluble gene products, as a factory for releasing therapeutic proteins to either treat distal sites, or act systemically. For example, a single i.v. delivery of hdAd expressing coagulation factor IX provided sustained improvement in coagulation time in a canine model of hemophilia B for over a year,[38] while a single hdAd delivery of low density lipoprotein receptor into ApoE knockout mice protected the mice from atherosclerosis for their entire lifetime.[39]

7. Immune Responses to Ad Vectors

The immune response to Ads is a complex process, activated rapidly upon viral administration and occuring via the innate and adaptive immune system. The Ad capsid activates the innate system independently of any viral gene expression and involves activation of mitogen activated protein kinase (MAPK), Jun N-terminal kinase (JNK) and extracellular regulated kinase (ERK) pathways, resulting in nuclear translocation of nuclear factor kappa B (NFκB) and transcription of its target genes, including a number of cytokines.[40,41] Following i.v. Ad delivery, there is rapid thrombocytopenia, cytokine production, inflammation and a rise in liver enzymes. Although transduced hepatocytes contribute, Kupffer cells, resident macrophages in the liver sinusoids, mediate the major initial host response to Ad (Fig. 2.3). Uptake of 90% of Ad by Kupffer cells takes place within minutes of delivery to eliminate it from the circulation, resulting in rapid induction of cytokines, including interleukin (IL)-1, macrophage inflammatory proteins, IL-6 and tumor necrosis factor alpha.[41] The complement system also opsonizes Ad to prime it for Kupffer cell uptake and in C3- knockout mice (a key complement protein) reduced levels of IL-6, granulocyte macrophage colony stimulating factor and IL-5 are observed after Ad delivery.[42] Ad binding to erythrocytes has been well known for many years. Classification of Ad serotypes is based on their differential agglutination of red blood cells from different species. For example, species C Ads agglutinate human and rat erythrocytes, but not those from mice, whereas species D Ads segregate into 3 subgroups; DI agglutinate rat and human erythrocytes, DII agglutinate

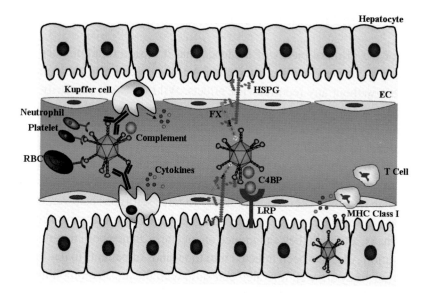

Fig. 2.3 Interactions of Ad with the host *in vivo*. Once in the bloodstream Ad5 is bound and/or phagocytosed by Kupffer cells, red blood cells (RBC), Platelets and neutrophils. Kupffer cell uptake is enhanced by antibody opsonization or complement proteins and leads to cytokine production. Ad complexes with FX or C4BP to transduce hepatocytes via HSPGs or LRP, respectively. Transduced hepatocytes present viral proteins via MHC and further release of cytokines leads to T cell mediated lysis.

rat erythrocytes only and DIII fail to agglutinate rat erythrocytes, or do but weakly. Ad interaction with erythrocytes also eliminates Ad from the bloodstream (Fig. 2.3).[43] In addition, Ad is bound by neutrophils and platelets. Together these findings suggest that host species have evolved sophisticated and orchestrated methods to overcome the invading Ad. The adaptive response to Ad takes place once Ad delivery and uptake in cells has occurred and there is a strong humoral response, which inhibits subsequent deliveries of Ad vectors of the same serotype.

Perhaps not unexpectedly as the most abundant capsid protein, the hexon is the major antigenic site of the virus,[44] and the HVRs function as serotype-specific neutralizing antigens.[45] The penton base and fiber also contribute to the generation of efficient viral neutralization as anti-fiber antibodies can mediate Ad aggregation,[46] and sera from Ad5-seropositive patients blocks FX-mediated transduction of Ad5, but not Ad5 pseudotyped with the unrelated serotype Ad45 fiber.[47]

8. Safety and Regulatory Issues

One of the primary pieces of evidence that it is safe to administer Ad to humans is the use of live, attenuated Ad as a vaccine in the American military without any reported side effects. Wild-type Ads primarily infect the upper respiratory tract, gastro-intestinal, or ocular epithelium. In healthy people they cause acute, self limiting infections, although they are a significant threat to morbidity and mortality in immunocompromised individuals. Neutralizing antibodies to types 1, 2 and 5 occur in 40–60% of individuals by the age of 15 years and by the age of 30 this rises to 70–85%. Although naturally occurring immunity to rarer serotypes is lower, there is a robust humoral response to individual serotypes. Ads are non-oncogenic in humans, although they can induce oncogenesis in newborn rodents. First generation Ads have been injected into thousands of patients to date and on the whole they are well tolerated and safe, with the most common findings being transiently elevated liver transaminases and mild fevers.

By far the most serious safety issue for Ad is that of Jesse Gelsinger, an 18 year old participant in a phase I trial for ornithine transcarbamylase deficiency (a X-linked deficiency in urea synthesis) who was administered 3.8×10^{13} viral particles into the hepatic artery and died following systemic inflammatory syndrome, disseminated intravascular coagulation and multiple organ failure.[48] The study highlighted several issues with regard to recruitment selection for clinical trials and Ad safety, triggering re-evaluation of safety and dosing regimes. Importantly, none of the pre-clinical data in rodent models or small and large non-human primate studies revealed similar side effects, highlighting the limitations of extrapolating pre-clinical data into clinical trials. This has been further emphasized in the STEP trial for Ad-mediated HIV vaccination as in individuals seropositive for Ad5 prior to vaccine administration a two-fold rise in HIV infection was observed.[37]

There are additional regulatory issues with the use of Ad gene therapy protocols, although many of these relate generally to the use of biologics. Clinical grade vector production facilities are becoming better established throughout many countries, however there are still issues related to standardization of production and titration procedures to ensure equivalent dosing and quality control criteria.

9. Conclusions

A vast amount of research over the past 50 years has lead to a detailed understanding of the biology of Ad vectors and their use as gene therapy vectors. However, it is only in recent years that important information regarding the immunological responses to Ad, both innate and adaptive, have begun to be unraveled, as well as the factors which dictate tissue uptake and transduction *in vivo*. There are still important questions to be addressed in order to be able to effectively target Ad to tissues other than the liver via the i.v. route. Combining this knowledge with avoidance of Ad interactions with the immune system, will be key to the next generation of Ad gene therapies. It is clear that Ad-mediated gene therapy has vast potential for the future treatment of human diseases.

References

1. Bergelson JM, Cunningham JA, Droguett G, Kurt-Jones EA, Krithivas A, Hong JS, *et al.* (1997). Isolation of a common receptor for coxsackie B viruses and adenoviruses 2 and 5. *Science* **275**: 1320–1323.
2. Tomko RP, Xu R, Philipson L. (1997). HCAR and MCAR: The human and mouse cellular receptors for subgroup C adenoviruses and group B coxsackieviruses. *Proc Natl Acad Sci, USA* **94**: 3352–3356.
3. Arnberg N, Edlund K, Kidd AH, Wadell G. (2000). Adenovirus type 37 uses sialic acid as a cellular receptor. *J Virol* **74**: 42–48.
4. Gaggar A, Shayakhmetov DM, Lieber A. (2003). CD46 is a cellular receptor for group B adenoviruses. *Nat Med* **9**: 1408–1412.
5. Wickham TJ, Mathias P, Cheresh DA, Nemerow GR. (1993). Integrins $\alpha 5\beta 3$ and $\alpha 5\beta 5$ promote adenovirus internalization but not virus attachment. *Cell* **73**: 309–319.
6. Seth, P. (1994). Mechanism of adenovirus-mediated endosome lysis: role of the intact adenovirus capsid structure. *Biochem Biophys Res Comm* **205**: 1318–1324.
7. Huard J, Lochmuller H, Acsadi G, Jani A, Massie B, Karpati G. (1995). The route of administration is a major determinant of the transduction efficiency of rat tissues by adenoviral recombinants. *Gene Ther* **2**: 107–115.
8. Shayakhmetov DM, Gaggar A, Ni S, Li ZY, Lieber A. (2005). Adenovirus binding to blood factors results in liver cell infection and hepatotoxicity. *J Virol* **79**: 7478–7491.
9. Parker AL, Waddington SN, Nicol CG, Shayakhmetov DM, Buckley SM, Denby L, *et al.* (2006). Multiple vitamin K-dependent coagulation zymogens promote adenovirus-mediated gene delivery to hepatocytes. *Blood* **108**: 2554–2561.
10. Waddington SN, McVey JH, Bhella D, Parker AL, Barker K, Atoda H, *et al.* (2008). Adenovirus serotype 5 hexon mediates liver gene transfer. *Cell* **132**: 397–409.

11. Kalyuzhniy O, Di Paolo NC, Silvestry M, Hofherr SE, Barry MA, Stewart PL, *et al.* (2008). Adenovirus serotype 5 hexon is critical for virus infection of hepatocytes *in vivo*. *Proc Natl Acad Sci USA* **105**: 5483–5488.

12. Graham FL, Smiley J, Russel WC, Nairu R. (1977). Characteristics of a human cell line transformed by DNA from human adenovirus type 5. *J Gen Virol* **36**: 59–72.

13. Engelhardt JF, Ye X, Doranz B, Wilson JM. (1994). Ablation of E2a in recombinant adenoviruses improves transgene persistence and decreases inflammatory response in mouse liver. *Proc Natl Acad Sci USA* **91**: 6196–6200.

14. Dedieu JF, Vigne E, Torrent C, Jullien C, Mahfouz I, Caillaud JM, *et al.* (1997). Long-term gene delivery into the livers of immunocompetent mice with E1/E4-defective adenoviruses. *J Virol* **71**: 4626–4637.

15. Parks Robin J, Chen L, Anton M, Sankar U, Rudnicki Michael A, Graham Frank L. (1996). A helper-dependent adenovirus vector system: Removal of helper virus by Cre-mediated excision of the viral packaging signal. *Proc Natl Acad Sci USA* **93**: 13565–13570.

16. Ng P, Beauchamp C, Evelegh C, Parks R, Graham FL. (2001). Development of a FLP/frt system for generating helper-dependent adenoviral vectors. *Mol Ther* **3**: 809–815.

17. Palmer D, Ng P (2003). Improved system for helper-dependent adenoviral vector production. *Mol Ther* **8**: 846–852.

18. Watkins SJ, Mesyanzhinov V, Kurochkina LP, Hawkins RE. (1997). The adenobody approach to viral targeting: specific and enhanced adenoviral delivery. *Gene Ther* **4**: 1004–1012.

19. Nicklin SA, White SJ, Watkins SJ, Hawkins RE, and Baker AH. (2000). Selective targeting of gene transfer to vascular endothelial cells by use of peptides isolated by phage display. *Circulation* **102**: 231–237.

20. Reynolds PN, Nicklin SA, Kaliberova L, Boatman BG, Grizzle WE, Balyasnikova I, *et al.* (2001). Combined transductional and transcriptional targeting improves the specificity of transgene expression *in vivo*. *Nat Biotechnol* **19**: 838–842.

21. Fisher KD, Stallwood Y, Green NK, Ulbrich K, Mautner V, Seymour LW. (2001). Polymer-coated adenovirus permits efficient retargeting and evades neutralising antibodies. *Gene Ther* **8**: 341–348.

22. Dmitriev I, Krasnykh V, Miller CR, Wang M, Kashentseva E, Mikheeva G, *et al.* (1998). An adenovirus vector with genetically modified fibers demonstrates expanded tropism via utilization of a coxsackievirus and adenovirus receptor-independent cell entry mechanism. *J Virol* **72**: 9706–9713.

23. Roelvink PW, Lee GM, Einfeld DA, Kovesid I, Wickham TJ. (1999). Identification of a conserved receptor-binding site on the fiber proteins of CAR-recognizing *adenoviridae* *Science* **286**: 1568–1571.

24. Nicklin SA, Von Seggern DJ, Work LM, Pek DCK, Dominiczak AF, Nemerow GR, *et al.* (2001). Ablating adenovirus type 5 fiber-CAR binding and HI loop insertion of the SIGYPLP peptide generate an endothelial cell-selective adenovirus. *Mol Ther* **4**: 534–542.

25. Seki T, Dmitriev I, Suzuki K, Kashentseva E, Takayama K, Rots M, *et al.* (2002). Fiber shaft extension in combination with HI loop ligands augments infectivity for CAR-negative tumor targets but does not enhance hepatotropism *in vivo*. *Gene Ther* **9**: 1101–1108.

26. Nicklin SA, Wu E, Nemerow GR, Baker AH. (2005). The influence of adenovirus fiber structure and function on vector development for gene therapy. *Mol Ther* **12**: 384–393.
27. Shayakhmetov DM, Papayannopoulou T, Stamatoyannopoulos G, Lieber A. (2000). Efficient gene transfer into human CD34+ cells by a retargeted adenovirus vector. *J Virol* **74**: 2567–2583.
28. Denby L, Work LM, Seggern DJ, Wu E, McVey JH, Nicklin SA, *et al.* (2007). Development of renal-targeted vectors through combined *in vivo* phage display and capsid engineering of adenoviral fibers from serotype 19p. *Mol Ther* **15**: 1647–1654.
29. Sakurai F, Kawabata K, Koizumi N, Inoue N, Okabe M, Yamaguchi T, *et al.* (2006). Adenovirus serotype 35 vector-mediated transduction into human CD46-transgenic mice. *Gene Ther* **13**: 1118–1126.
30. Peng Z. (2005). Current status of gendicine in China: recombinant human Ad-p53 for treatment of cancers. *Hum Gen Ther* **16**: 1016–1027.
31. Bischoff JR, Kirn DH, Williams A, Heise C, Horn S, Muna M, *et al.* (1996). An adenovirus mutant that replicates selectively in p53-deficient human tumor cells. *Science* **274**: 373–376.
32. Fueyo J, Gomez-Manzano C, Alemany R, Lee PS, McDonnell TJ, Mitlianga P, *et al.* (2000). A mutant oncolytic adenovirus targeting the Rb pathway produces anti-glioma effect *in vivo*. *Oncogene* **19**: 2–12.
33. Wan S, George SJ, Nicklin SA, Yim AP, Baker AH. (2004). Overexpression of p53 increases lumen size and blocks neointima formation in porcine interposition vein grafts. *Mol Ther* **9**: 689–698.
34. George SJ, Lloyd CT, Angelini GD, Newby AC, Baker AH. (1999). Inhibition of late vein graft neointima formation in human and porcine models by adenoviral-mediated overexpression of tissue inhibitor of metalloproteinase-3. *Circulation* **101**: 296–304.
35. Grines CL, Watkins MW, Helmer G, Penny W, Brinker J, Marmur JD, *et al.* (2002). Angiogenic gene therapy (AGENT) trial in patients with stable angina pectoris. *Circulation* **105**: 1291–1297.
36. Hedman M, Hartikainen J, Syvanne M, Stjernvall J, Hedman A, Kivela A, *et al.* (2003). Safety and feasibility of catheter-based local intracoronary vascular endothelial growth factor gene transfer in the prevention of postangioplasty and in-stent restenosis and in the treatment of chronic myocardial ischemia: phase II results of the Kuopio Angiogenesis Trial (KAT). *Circulation* **107**: 2677–2683.
37. Buchbinder SP, Mehrotra DV, Duerr A, Fitzgerald DW, Mogg R, Li D, *et al.* (2008). Efficacy assessment of a cell-mediated immunity HIV-1 vaccine (the Step Study): a double-blind, randomized, placebo-controlled, test-of-concept trial. *The Lancet* **372**: 1881–1893.
38. Brunetti-Pierri N, Nichols TC, McCorquodale S, Merricks E, Palmer DJ, Beaudet AL, *et al.* (2005). Sustained phenotypic correction of canine hemophilia B after systemic administration of helper-dependent adenoviral vector. *Hum Gene Ther* **16**: 811–820.
39. Oka K, Pastore L, Kim IH, Merched A, Nomura S, Lee HJ, *et al.* (2001). Long-term stable correction of low-density lipoprotein receptor-deficient mice with a helper-dependent adenoviral vector expressing the very low-density lipoprotein receptor. *Circulation* **103**: 1274–1281.
40. Muruve DA. (2004). The innate immune response to adenovirus vectors. *Hum Gene Ther* **15**: 1157–1166.

41. Shayakhmetov DM, Li ZY, Ni S, Lieber A. (2005). Interference with the IL-1-signaling pathway improves the toxicity profile of systemically applied adenovirus vectors. *J Immunol* **174**: 7310–7319.

42. Jiang H, Wang Z, Serra D, Frank MM, Amalfitano A. (2004). Recombinant adenovirus vectors activate the alternative complement pathway, leading to the binding of human complement protein C3 independent of anti-Ad antibodies. *Mol Ther* **10**: 1140–1142.

43. Lyons M, Onion D, Green NK, Aslan K, Rajaratnam R, Bazan-Peregrino M, *et al.* (2006). Adenovirus type 5 interactions with human blood cells may compromise systemic delivery. *Mol Ther* **14**: 118–128.

44. Sumida SM, Truitt DM, Lemckert AA, Vogels R, Custers JH, Addo MM, *et al.* (2005). Neutralizing antibodies to adenovirus serotype 5 vaccine vectors are directed primarily against the adenovirus hexon protein. *J Immunol* **174**: 7179–7185.

45. Gall JGD, Crystal RG, Falck-Pedersen E. (1998). Construction and characterization of hexon-chimeric adenoviruses: specification of adenovirus serotype. *J Virol* **72**: 10260–10264.

46. Wohlhart C. (1998). Neutralization of adenoviruses: kinetics, stoichiometry and mechanisms. *J Virol* **72**: 6875–6879.

47. Parker AL, Waddington SN, Buckley SM, Custers J, Havenga MJ, van Rooijen N, *et al.* (2009). Effect of neutralizing sera on factor X-mediated adenovirus serotype 5 gene transfer. *J Virol* **83**: 479–483.

48. Raper SE, Chirmule N, Lee FS, Wivel NA, Bagg A, Gao GP, *et al.* (2003). Fatal systemic inflammatory response syndrome in an ornithine transcarbamylase deficient patient following adenoviral gene transfer. *Mol Genet Metab* **80**: 148–158.

Chapter 3

Retroviral Vectors and Integration Analysis

Cynthia C. Bartholomae, Romy Kirsten, Hanno Glimm,
Manfred Schmidt and Christof von Kalle*

Retroviral vectors are predestined for use in gene therapy because they are
not pathogenic for humans and allow efficient long-term expression of the
therapeutic gene. Intensive research has been done on murine and human
retroviruses and derived vectors during the last few decades. In recent
years the curative potential of retroviral gene transfer into hematopoietic
stem cells has been demonstrated impressively in a number of clinical
studies. At the same time it became evident that as the therapeutic efficacy of retroviral gene transfers increased, so did the biologically relevant vector-induced side effects, ranging from *in vitro* immortalization to
clonal dominance and oncogenesis *in vivo*. Detection and monitoring of
these side effects by a genome wide inspection of the integration sites has
proven to be an effective tool to elucidate a vector's oncogenic potential.
Current research efforts aim to develop safety vectors with a reduced tendency of unwanted activation of cellular genes that ideally integrate in a
more targeted manner into the host genome. At the same time continuous
monitoring of transduced cells has to become an integral part of every
therapy using retroviral vectors.

1. Introduction

Retroviral vectors are attractive and promising gene delivery vehicles in
clinical gene therapy. They stably integrate into the host genome and allow

*Correspondence: National Center for Tumor Diseases and German Cancer Research Center (DKFZ)
Heidelberg, Otto-Meyerhof-Zentrum, Im Neuenheimer Feld 350, 69120 Heidelberg Germany.
Email: christof.kalle@nct-heidelberg.de

the long-term expression of therapeutic transgenes. As of December 2009, 1579 clinical gene therapy studies, most exclusively clinical phase I/II, have been conducted world-wide. Retroviral vectors were used in 357 trials, representing the second most commonly used gene delivery vehicles after adenoviral vectors. (http://www.wiley.co.uk/genmed/clinical/).

For a long time it was assumed that the integration of retroviral vectors into the host genome is close to random and that, consequently, the chances of integration in a critical or dangerous site are minimal. However, in recent years evidence has been accumulating which contradicts this assumption, thus making it necessary to reevaluate the risk of insertional mutagenesis when using retroviral vectors for gene delivery. Accordingly, research efforts investigating the insertion patterns of retroviral vectors were intensified so that the relationship between the benefits and risks of treating patients with this method could be more accurately evaluated.

In the first part of this chapter an overview of the structure and major characteristics of presently used retroviral gene therapy vectors is given. In the second part the most recent research on insertion mutagenesis is presented and discussed in the light of the demands for safety in gene therapy.

2. Design, Production and Mechanism of Transduction

The family of retroviridae consists of seven species, the simple α-, β-, γ-, and ε-retroviruses, and the complex δ-retro-, lenti- and spumaviruses. The viral genome is formed by two positive strand-oriented single-strand RNAs (except spumaviruses) and exhibits a 5'-cap structure and a 3'-polyadenylation. After reverse transcription of the viral RNA genome into double stranded cDNA in the cytoplasm of the host cell, the viral cDNA exhibits the same sequence at both ends, the long terminal repeat (LTR). The LTRs are essential for viral integration and expression of viral genes. The replication cycle of the retroviridae is shown in Fig. 3.1.

The LTRs consist of three structural units: U3 (U, "unique"), R (R, "redundant") and U5. The U3 region contains the promoter and enhancer elements (Fig. 3.2). Three coding regions are common to all infectious retroviruses: *gag, pol* and *env*. The *gag* gene (group-specific antigens) codes for the structural proteins, the *pol* gene (enzymatic activities) codes for the enzymes reverse transcriptase, integrase and protease and the *env* gene

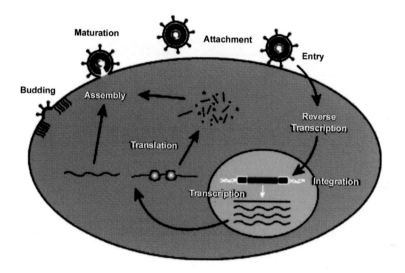

Fig. 3.1 Retroviral life cycle. The retroviral life cycle starts with adsorption on the cell membrane and the penetration of the nucleocapsid into the cell. After reverse transcription in the cytoplasm, the proviral DNA integrates into the host genome, followed by DNA transcription and translation of the virus proteins, packaging of the RNA genome in the developing particle and budding. (Source: www.clontech.com)

(glycoproteins) for the viral envelope proteins (Figs. 3.2a and 3.2b).[1] The simple retroviruses such as the murine leukemia virus (MLV) and the spleen focus forming virus (SFFV) only carry information for these domains. The complex lentiviruses (e.g. the human immunodeficiency virus, HIV) and the spumaviruses (e.g. the prototypical foamy virus) possess additional genes which code for regulatory and accessory proteins.

In clinical gene therapy, only replication deficient vectors are used as gene transfer vehicles. These vectors are produced by removing almost all of the viral genomes, replacing it with a therapeutic gene or a marker gene and regulatory elements (Fig. 3.2c). Thus, after infection of the target cells the possibility of further replication and spreading of the viral vectors can be eliminated. The transfering vector codes for the transgene, the packaging signal ψ, the primer binding site, and the retroviral 5'- and 3'-LTRs. To avoid unwanted recombination, the viral genes *gag, pol* and *env* necessary for the self-assembly of the vector are provided separately, either by using a stable packaging cell line or by transient transfection of expression plasmids.

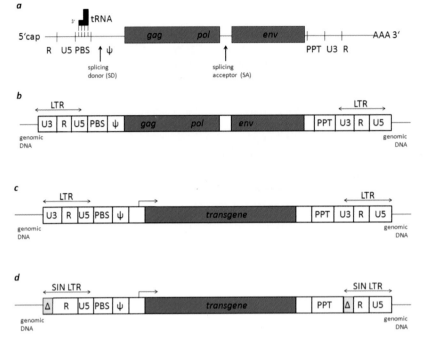

Fig. 3.2 Organization of the retroviral genome and derived vectors. Sequence elements present in the RNA genome of all infectious virus particles (a) and in the proviral genome after integration into the host genome (b). Sequence elements of a retroviral vector with complete LTR (c) and composition of a safety-optimized vector with deletions in the U3 region of LTRs (d). The arrow indicates the internal promoter that controls transgene expression. R: redundant region; U3 and U5: unique regions at 3'- or 5'- end, respectively; PPT: polypurine tract; LTR, "*Long Terminal Repeat*"; SIN: self inactivating, PBS, primer binding site; ψ, packaging signal.

Vectors derived from simple retroviruses have a transgene capacity of up to 9 kb and can only enter the nuclei of dividing cells.[2] In order to efficiently transduce resting cells *ex vivo*, e.g. hematopoietic stem cells, the cells are stimulated with cytokines in cell culture medium to induce cell division.[3] Lentivirus and foamy virus vectors are able to transduce resting, i.e. not actively dividing cells. The lentiviral vectors have a transgene capacity of 10 kb, and spumaviral vectors can carry transgenes of up to 12 kb in size. Self-inactivating (SIN)-LTR vectors have been developed to further minimize the risk of recombination and insertional mutagenesis (see below and Fig. 3.2d).

3. *In vivo* Application

In recent years the curative potential of retroviral gene transfer into hematopoietic stem cells was demonstrated impressively in a number of clinical studies.[4-8] In addition to obtain a therapeutic benefit, each transduced cell and its clonal progeny is uniquely labeled by the vector integration locus. Identification of such vector integration sites not only allows researchers to investigate the origin of a tumor recurrence and the biology of human hematopoietic regeneration, but also to follow the fate of transduced cells and to monitor their clonal behaviour over time.[6,7,9-11]

More than 15 years passed between the first description of a retroviral gene transfer in murine hematopoietic stem cells and the first successes in the clinical therapy of patients with X-chromosomal inherited severe combined immunodeficiency disorder (X-SCID).[5,6,12] The limiting factor, in previous unsuccessful clinical trials, was the low efficiency of gene transfer. This hurdle was effectively circumvented by improved transduction protocols and vectors.[13,14] Further successes followed in the gene therapeutic treatment of adenosine deaminase (ADA) SCID and chronic granulomatous disease (CGD).[4,7]

Since lentiviral vectors are capable of transducing proliferating as well as non-proliferating cells, great hopes are being set for their use in clinical applications. Initial successes have already been reported for the application of lentiviral vectors for treating neurodegenerative diseases in animal models, including the treatment of Parkinson's disease in monkeys,[14] and of metachromatic leukodystrophy, a storage disorder of the central nervous system, in a mouse model.[15]

The first clinical lentiviral gene therapy trial was initiated in 2006 for the treatment of the cerebral form of X-chromosomal linked adrenoleukodystrophy (X-ALD), a demyelinating disease of the central nervous system.[8] Mobilized hematopoietic precursor cells were harvested from two seven-year-old patients, transduced *ex vivo* with lentiviral SIN-vectors encoding the *ABCD1* transgene and reinfused into the patients. It was possible to arrest the progression of the disease in both patients through the constitutive expression of the functional *ABCD1* therapeutic transgene.

4. Side Effects in Retroviral Gene Therapy

Gene therapy using retroviral vectors is a specialized form of therapy that is mainly applied to patients for whom no therapeutic alternatives are available. Therefore, the benefit of a gene therapy always has to be opposed to the potential risks case by case.

The integration of retroviruses in the cellular genome is *per se* a mutagenic event, which can lead to disruption of the gene coding regions and/or to interference with regulatory elements. However, in contrast to the use of wild-type retroviruses and replication-competent retroviral vectors, the risk of insertional mutagenesis leading to cell transformation for a long time was assessed to be rather low when using replication-deficient oncoviral vectors that were specifically developed for clinical purposes.[17] The increased efficiency in performing *ex vivo* transduction of human CD34+ cells and, as a consequence, the greater number (up to $10^8 - 10^9$) of gene-corrected cells which can be re-infused are the factors responsible for the successful clinical treatment achieved with retroviral vectors since the beginning of 2000. At the same time it became evident that the risk of biologically relevant side effects could also increase along with the improved effectiveness of this type of treatment. Insertion-mediated mutagenesis can lead to a wide spectrum of phenomena, ranging from *in vitro* immortalization to clonal dominance and oncogenesis *in vivo*.[18–23]

Figure 3.3 illustrates one possible mechanism of how integration of MLV- and HIV-1-derived retroviral vectors (which integrate preferentially in close proximity to transcriptional start sites or gene coding regions respectively) may lead to a dysregulation of cellular genes.

4.1 *Distribution of Retroviral Integration Sites in the Cellular Genome*

The identification and sequencing of viral integration sites involves the presentation of a DNA sequence composed of the known proviral sequence and of the unknown neighboring genomic sequence. In order to characterize the unknown flanking DNA, two fundamental PCR methods were developed, inverse PCR and ligation-mediated PCR.[24,25] Although further development in these methodologies led to improvements in sensitivity, the invention of

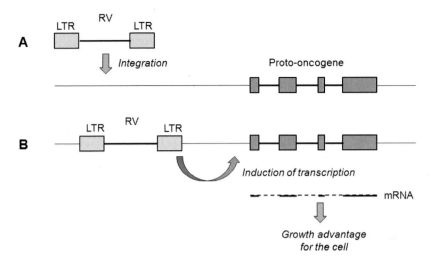

Fig. 3.3 Mechanism of proto-oncogene activation by insertional mutagenis. (A) Integration of viral genomic DNA into somatic cellular DNA in the proximity of a nontranscribed oncogene. (B) Integrated proviral DNA induces the transcription of the oncogene. RV, retroviral vector. LTR, "*Long Terminal Repeat*".

the linear amplification-mediated PCR (LAM-PCR) allowed for the first time the characterization of integration sites at the single-cell level and to monitor the polyclonal gene-modified hematopoiesis directly in limited amounts of peripheral blood leukocytes and bone marrow cells.[26,27] LAM-PCR has been shown to be the superior method to determine the clonality of the hematopoietic repopulation after transplantation in humans.

The principle of LAM-PCR is illustrated in Fig. 3.4. The first step (a) is the preamplification of the vector–genome junctions through a linear PCR with biotinylated primers hybridizing at one end of the integrated vector. The following steps (b-f) are carried out on a semisolid strepta-vidin phase in order to capture DNA strands with an incorporated biotiny-lated vector primer. After double strand synthesis (c), a restriction digest (d), and the ligation of a linker cassette (e) on the genomic end of the fragment, two exponential PCRs with nested arranged vector- and linker cassette primers are carried out in order to amplify the fragments consist-ing of linker cassette-, genomic-, and vector-sequence (g). The generated fragments are sequenced and the integration loci are determined. A further

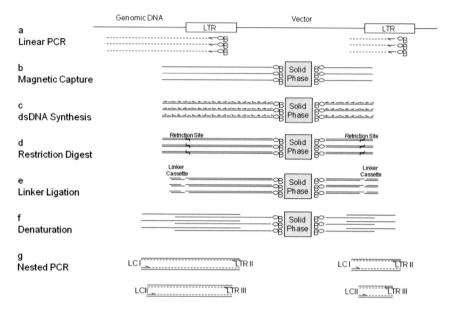

Fig. 3.4 Schematic outline of LAM-PCR to amplify 5-LTR retroviral vector-genomic fusion sequences. Steps (a-g) see text. B, biotinylated primer; LTR, long terminal repeat; LTR I/II: LTR primer for the first/second exponential PCR; LCI/II, linker cassette primer for the first /second exponential PCR; Schmidt *et al.* 2007)

development of the LAM-PCR is the non-restrictive LAM-PCR (nrLAM-PCR), which allows the genome-wide identification of retroviral integration sites in a single reaction, circumventing the detection bias caused by the use of restriction enzymes.[28] The knowledge of the genomic flanking sequence derived from a clone of interest allows the creation of primers specific for an individual integration clone. Thus, PCR analysis using these primers in combination with vector-specific primers makes it possible to monitor one single specific clone *in vivo* in different samples.[27]

Due to the availability of the complete human and murine genome sequence data the precise localization of viral integration sites in the host genome now became possible, and by the improvement of new sequencing technologies also feasible in a high-throughput manner.[29]

Severe side effects emerged in 2003, when two of the patients enrolled in the first successful gene therapy trial for the treatment of X-SCID developed a T cell acute lymphatic leukemia between 30 and 34 months after treatment. As a consequence, comprehensive investigations were conducted to observe the integration behavior of different vectors and wild-type viruses *in vitro* and *in vivo*.[30–32] These analyses provided the first concrete evidence that retroviral integration into the host genome is much less random than previously assumed, and that different vector types consistently tend to integrate into specific genomic regions.

The fact that tumorigenic chromosomal aberrations (i.e. deletions, translocations and amplifications) as well as the integration of oncogenic viruses (for instance human papilloma virus 16 (HPV16), HPV18, Hepatitis-B-Virus and Epstein-Barr-Virus) often occur in so-called "common fragile sites" (CFSs) has already been known since 2003.[33] In 2006, Bester *et al.* were able to show a significant higher integration frequency of MLV in such CFSs.[34] The integration of MLV-derived vectors preferentially occurs close to transcription start sites and in proximity to CpG islands.[32] In contrast, HIV-1-derived vectors in two thirds of cases integrate into gene coding regions over the complete length of the gene, with a preference for active genes.[30,31] Moreover, for both MLV- and HIV-derived vectors a cell type specific integration pattern was observed, due to differences in their gene expression status.[7,10,35–37] More recent investigations on the integration pattern of HIV-1 revealed a relationship between the expression of the transcription coactivator lens epithelium-derived growth factor (LEDGF) and the integration of the vector in transcribed regions in Jurkat cells.[38,39] Lewinski *et al.* showed that viral components such as integrase and group-specific antigenes influence the integration behavior of HIV, MLV and their chimeras.[40] Another group of retroviral vectors derived from Avian Sarcoma Leukosis Virus (ASLV) show neither a preference for transcription start sites nor for active genes.[30] These results make it clear that the selection of the gene transfer vector with respect to the target cell or target tissue can have a decisive influence on the clinical outcome of a gene therapy trial.

4.2 *Side Effects in Clinical and Preclinical Gene Therapy Studies*

In 2002, the first malignant transformation that developed as a result of gene transfer using replication-deficient retroviral vectors was observed.[21] Serial transplantations of gammaretroviral gene-marked CD34+ cells derived from murine bone marrow led to tumor development in secondary and tertiary recipients. All secondary transplanted animals developed alterations of their hematopoiesis after 22 weeks, and 6 out of ten animals developed acute myeloid leukemia. It was possible to show that all of these animals had the same malignant clone with a vector integrant in the *Evi1* gene locus. The functional promoter and enhancer elements in both proviral LTRs caused an overexpression of *Evi1*. The MLV vector used here encoded the transgene that had been used for clinical applications, a low-affinity nerve growth factor receptor (LNGFR). It could not be proven that the growth-mediating transgene LNGFR had a synergistic effect, but neither could such an effect be ruled out.

Gene marking studies in a non-human primate model revealed indications of a possible clonal dominance in the retrovirally transduced cells.[18] Calmels and his colleagues were able to show that clones with distinct vector integration sites (*Mds1/Evi1, Prdm16*) had a growth advantage compared to other cell clones. However, no abnormal proliferation of affected cell clones was observed, despite numerous retroviral integrations within these proto-oncogenes. Since the major part of these cell clones contributes to long-term hematopoiesis, a higher engraftment or survival probability as a result of insertional mutagenesis was discussed. Analysis of another murine gene marking study proved that proviral integration within or nearby distinct cellular genes promotes the growth of single transduced cells and contributes to their clonal expansion *in vivo*.[20]

In the context of a murine study on the retroviral expression of an ABC-transporter (*MDR1*), it was observed that the vector dose has an impact on leukemogenesis. Leukemia not only occurred due to the growth advantage of single clones but also correlated with high vector doses.[41] In 2006, the first acute myeloid leukemia in a non-human primate model was described after retroviral gene transfer in hematopoietic precursor cells. The treated animal died five years after gene therapy due to a myeloid sarcoma.[22] Two vector integrants (*Bcl2-A1, Cdw91*) were detected in the blood as well as in the tumor tissue of the animal, both of which had already achieved dominance one year after transplantation.

The first published side effect after gene transfer using a lentiviral vector was a T-cell lymphoma that occurred in the context of the constitutive expression of the therapeutic gene *Il2RG* in a murine model. The influence of the growth-related transgene on the tumor development in this study was a matter of much controversial debate.[23,42] A gene transfer-induced tumorigenesis was also observed with an adeno-associated virus (AAV) derived vector in a murine model.[43]

In three clinical gene therapy studies, integration site analyses in patient hematopoietic cells showed that the retroviral vector could affect the growth and differentiation of distinct transduced clones. Two to six years after successful gene therapy for X-SCID, 5 of 19 patients developed T-cell leukemia. In 4 of these patients, vector integrations in the proto-oncogene *LMO2* occurred causing its overexpression and expansion of these clones.[7,36,37,44,45] Also, a combinatorial effect of insertional mutagenesis and constitutive expression of the transgene *IL2RG* as a cause of lymphoproliferative disease was discussed.[23] Gene therapy for correction of X-CGD in two adult patients resulted in the restoration of oxidative antimicrobial activity in phagocytes after gene transfer. Substantial gene transfer in neutrophil cells had produced a high number of functional phagocytes. Integration analyses revealed a clonal dominance of insertion sites in the genes *MDS1-EVI1*, *PRDM16* or *SETBP1*, resulting in an expansion of gene-corrected myelopoesis.[7] A myelodysplastic syndrome developed in both patients in the course of time.[46]

5. New Strategies for Vector Biosafety in Gene Therapy

The treatment of monogenic hereditary diseases with gammaretroviral vectors led to severe side effects only in certain diseases (X-SCID, X-CGD) and only in few patients.[44–46] Therefore, the applied vector itself cannot account solely for the incidence of these side effects; in fact a number of factors seem to interact. For instance no side effects have been observed so far in the gene therapy of ADA-SCID using gammaretroviral vectors.[47] In order to minimize the risk of side effects in future gene therapy trials it will be important to choose the most suitable gene transfer vectors dependent on the disease individually and to test the usage of tissue-specific promoters or non-integrating vector systems if appropriate.

During the last years several *in vitro* and *in vivo* systems were developed, aiming to assess the genotoxic potential of different vectors. Already in 2003 it was reported that replication-competent gammaretroviruses are suitable for specific detection of cancer genes. Transplantation of transduced hematopoietic cells in mice with knocked-out *Cdkn2a* tumor suppressor gene leads to accelerated tumor growth when a cellular proto-oncogene is virally activated.[48] Thus, the most extensive database of murine genes with oncogenic potential, the mouse "*Retrovirally Tagged Cancer Gene Database*" (RTCGD) was developed.[49]

In other work, Du and colleagues showed that identification of proto-oncogenes is feasible by retroviral gene transfer in cell culture.[26] Transduction of murine bone marrow cells with replication deficient retroviruses expressing marker genes resulted in immortalized cell lines, many of which contained integrations in the *Mds1/Evi1* and *Prdm16* genes. Vector integration into the *Evi1* locus and resulting overexpression of *Evi1* appeared to be sufficient for a cell to be immortalized. Modlich and colleagues adapted and improved this method so that the results could be analyzed in a quantitative manner.[47]

In order to assess the biological safety of vectors used for the treatment of X-SCID a tumor prone mouse model was developed in which the tumor suppression gene *Arf* as well as the *Il2rg* gene had been knocked-out.[51]

Since it was proven that the activity of promoter/enhancer elements in the LTR of conventional retroviral vectors may cause insertional activation of cellular protooncogenes, the first attempts to reduce the risk of insertional mutagenesis involved the development of SIN vectors. In these vectors, the enhancer and promoter elements in the U3-region of the LTR are deleted in order to eradicate their activity after integration.[52] Expression of the transgene is driven by an internal promoter, which, in addition, can be tissue or cell-type specific for further improvement of safety (see Fig. 3.2d).

The mouse model originally developed by Lund *et al.* was suitable to perform comparative tests on the genotoxic potential of different gene transfer vectors. Mice that had received hematopoietic precursor cell transplants transduced with gammaretroviral vectors developed tumors much more rapidly than mice that had received cells harbouring lentiviral SIN-vectors with an internal human phosphoglycerate kinase (hPGK) promoter.[53] Later on it was demonstrated that retroviral vectors with SIN-LTRs show a significant reduction in genotoxicity, but when combined with strong (viral) internal promoters still have the ability to induce leukemia in a mouse model.[51]

However, the usage of physiological (cellular) promoters further reduced the risk of insertional mutagenesis.[55]

Other approaches for safety optimization include the modification of envelope proteins of vector particles in order to constrain their transduction to the target cell population.[56] Insertion of insulator elements in the LTRs aim to reduce both the influence of genomic environment on the integrated vector as well as the influence of the integrated vector on the surrounding host genome.[57]

In postmitotic tissues, no dilution of unintegrated episomal vector forms occurs, thus allowing the usage of non-integrating vector systems. Mutations in the core domain of the viral integrase prevent integration, thereby reducing the risk of insertional mutagenesis intensively. Recently, the long-term functional correction of retinal degeneration in a well-established rodent model for ocular gene therapy was reported. This was accomplished without apparent side effects using an integrase-deficient HIV-1 vector.[58]

The targeted insertion of non-integrating vectors into defined, putative safe gene loci via homologous recombination by the use of double strand break inducing enzymes (i.e. zinc finger nucleases, meganucleases) represent another promising approach for safety optimization. The decisive factor for the successful application of this method is to avoid the recognition of wrong target sequences by zinc finger nucleases that would trigger off-site double strand breaks.[59,60] Another alternative approach for targeted integration is the usage of transposons as gene ferries.[57]

References

1. Coffin JM, Hughes SH, Varmus HE (1997). Cold Spring Harbour, USA: Cold Spring Harbor Laboratory Press.
2. Roe T, *et al.* (1993). Integration of murine leukemia virus DNA depends on mitosis. *EMBO Retroviruses J* **12**(5): 2099–108.
3. Nolta JA, Smogorzewska EM, Kohn DB (1995). Analysis of optimal conditions for retroviral-mediated transduction of primitive human hematopoietic cells. *Blood* **86**(1): 101–110.
4. Aiuti A, *et al.* (2002). Correction of ADA-SCID by stem cell gene therapy combined with nonmyeloablative conditioning. *Science* **296**(5577): 2410–2413.
5. Gaspar HB, *et al.* (2004). Gene therapy of X-linked severe combined immunodeficiency by use of a pseudotyped gammaretroviral vector. *Lancet* **364**(9452): 2181–2187.

6. Hacein-Bey-Abina S, *et al.* (2002). Sustained correction of X-linked severe combined immunodeficiency by *ex vivo* gene therapy. *N Engl J Med* **346**(16): 1185–1193.

7. Ott MG, *et al.* (2006). Correction of X-linked chronic granulomatous disease by gene therapy, augmented by insertional activation of MDS1-EVI1, PRDM16 or SETBP1. *Nat Med* **12**(4): 401–409.

8. Cartier N, *et al.* (2009). Hematopoietic stem cell gene therapy with a lentiviral vector in X-linked adrenoleukodystrophy. *Science* **326**(5954): 818–823.

9. Brenner MK, *et al.* (1993). Gene-marking to trace origin of relapse after autologous bone-marrow transplantation. *Lancet* **341**(8837): 85–86.

10. Glimm H, *et al.* (2005). Efficient marking of human cells with rapid but transient repopulating activity in autografted recipients. *Blood* **106**(3): 893–898.

11. Guenechea G., *et al.* (2001). Distinct classes of human stem cells that differ in proliferative and self-renewal potential. *Nat Immunol* **2**(1): 75–82.

12. Cavazzana-Calvo M, *et al.* (2000). Gene therapy of human severe combined immunodeficiency (SCID)-X1 disease. *Science* **288**(5466): 669–672.

13. Kohn DB, *et al.* (1998). T lymphocytes with a normal ADA gene accumulate after transplantation of transduced autologous umbilical cord blood CD34+ cells in ADA-deficient SCID neonates. *Nat Med* **4**(7): 775–780.

14. Stead RB, *et al.* (1988). Canine model for gene therapy: inefficient gene expression in dogs reconstituted with autologous marrow infected with retroviral vectors. *Blood* **71**(3): 742–747.

15. Kordower JH, *et al.* (2000). Neurodegeneration prevented by lentiviral vector delivery of GDNF in primate models of Parkinson's disease. *Science* **290**(5492): 767–773.

16. Consiglio A, *et al.* (2001). *In vivo* gene therapy of metachromatic leukodystrophy by lentiviral vectors: correction of neuropathology and protection against learning impairments in affected mice. *Nat Med* **7**(3): 310–316.

17. Moolten FL, Cupples LA (1992). A model for predicting the risk of cancer consequent to retroviral gene therapy. *Hum Gene Ther* **3**(5): 479–486.

18. Calmels B, *et al.* (2005). Recurrent retroviral vector integration at the Mds1/Evi1 locus in nonhuman primate hematopoietic cells. *Blood* **106**(7): 2530–2533.

19. Du Y, Jenkins NA, Copeland NG (2005). Insertional mutagenesis identifies genes that promote the immortalization of primary bone marrow progenitor cells. *Blood* **106**(12): 3932–3939.

20. Kustikova O, *et al.* (2005). Clonal dominance of hematopoietic stem cells triggered by retroviral gene marking. *Science* **308**(5725): 1171–1174.

21. Li Z, *et al.* (2002). Murine leukemia induced by retroviral gene marking. *Science* **296**(5567): 497.

22. Seggewiss R, *et al.* (2006). Acute myeloid leukemia is associated with retroviral gene transfer to hematopoietic progenitor cells in a rhesus macaque. *Blood* **107**(10): 3865–3867.

23. Woods NB, *et al.* (2006). Gene therapy: therapeutic gene causing lymphoma. *Nature* **440**(7088): 1123.

24. Mueller PR, Wold B (1989). *In vivo* footprinting of a muscle specific enhancer by ligation mediated PCR. *Science* **246**(4931): 780–786.

25. Silver J, Keerikatte V (1989). Novel use of polymerase chain reaction to amplify cellular DNA adjacent to an integrated provirus. *J Virol* **63**(5): 1924–1928.

26. Schmidt M, *et al.* (2001). Detection and direct genomic sequencing of multiple rare unknown flanking DNA in highly complex samples. *Hum Gene Ther* **12**(7): 743–749.
27. Schmidt M, *et al.* (2007). High-resolution insertion-site analysis by linear amplification-mediated PCR (LAM-PCR). *Nat Methods* **4**(12): 1051–1057.
28. Gabriel R, *et al.* (2009). Comprehensive genomic access to vector integration in clinical gene therapy. *Nat Med* **15**(12): 1431–1436.
29. Wold B, Myers RM (2008). Sequence census methods for functional genomics. *Nat Methods* **5**(1): 19–21.
30. Mitchell RS, *et al.* (2004). Retroviral DNA integration: ASLV, HIV, and MLV show distinct target site preferences. *PLoS Biol* **2**(8): E234.
31. Schroder AR, *et al.* (2002). HIV-1 integration in the human genome favors active genes and local hotspots. *Cell* **110**(4): 521–529.
32. Wu X, *et al.* (2003). Transcription start regions in the human genome are favored targets for MLV integration. *Science* **300**(5626): 1749–1751.
33. Popescu NC (2003). Genetic alterations in cancer as a result of breakage at fragile sites. *Cancer Lett* **192**(1): 1–17.
34. Bester AC, *et al.* (2006). Fragile sites are preferential targets for integrations of MLV vectors in gene therapy. *Gene Ther* **13**(13): 1057–1059.
35. Cattoglio C, *et al.* (2007). Hot spots of retroviral integration in human CD34+ hematopoietic cells. *Blood* **110**(6): 1770–1778.
36. Deichmann A, *et al.* (2007). Vector integration is nonrandom and clustered and influences the fate of lymphopoiesis in SCID-X1 gene therapy. *J Clin Invest* **117**(8): 2225–2232.
37. Schwarzwaelder K, *et al.* (2007). Gammaretrovirus-mediated correction of SCID-X1 is associated with skewed vector integration site distribution *in vivo*. *J Clin Invest* **117**(8): 2241–2249.
38. Ciuffi A, *et al.* (2005). A role for LEDGF/p75 in targeting HIV DNA integration. *Nat Med* **11**(12): 1287–1289.
39. Marshall HM, *et al.* (2007). Role of PSIP1/LEDGF/p75 in lentiviral infectivity and integration targeting. *PLoS ONE* **2**(12): e1340.
40. Lewinski MK, *et al.* (2006). Retroviral DNA integration: viral and cellular determinants of target-site selection. *PLoS Pathog* **2**(6): e60.
41. Modlich U, *et al.* (2005). Leukemias following retroviral transfer of multidrug resistance 1 (MDR1) are driven by combinatorial insertional mutagenesis. *Blood* **105**(11): 4235–4246.
42. Pike-Overzet K, *et al.* (2006). Gene therapy: is IL2RG oncogenic in T-cell development? *Nature* **443**(7109): E5; discussion E6–7.
43. Kay MA (2007). AAV vectors and tumorigenicity. *Nat Biotechnol* **25**(10): 1111–1113.
44. Hacein-Bey-Abina S, *et al.* (2008). Insertional oncogenesis in 4 patients after retrovirus-mediated gene therapy of SCID-X1. *J Clin Invest*.
45. Howe SJ, *et al.* (2008). Insertional mutagenesis combined with acquired somatic mutations causes leukemogenesis following gene therapy of SCID-X1 patients. *J Clin Invest*.
46. Grez M, *et al.* (2007). Update on gene therapy for chronic granulomatous disease. Hum Gene Ther. European Society of Gene and Cell Therapy: Rotterdam. p. 959.
47. Aiuti A, *et al.* (2009). Gene therapy for immunodeficiency due to adenosine deaminase deficiency. *N Engl J Med* **360**(5): 447–458.

48. Lund AH, *et al.* (2002). Genome-wide retroviral insertional tagging of genes involved in cancer in Cdkn2a-deficient mice. *Nat Genet* **32**(1): 160–165.

49. Akagi K, *et al.* (2004). RTCGD: retroviral tagged cancer gene database. *Nucleic Acids Res* **32**(Database issue): D523–527.

50. Modlich U, *et al.* (2006). Cell-culture assays reveal the importance of retroviral vector design for insertional genotoxicity. *Blood* **108**(8): 2545–2553.

51. Shou Y, *et al.* (2006). Unique risk factors for insertional mutagenesis in a mouse model of XSCID gene therapy. *Proc Natl Acad Sci USA* **103**(31): 11730–11735.

52. Zufferey R, *et al.* (1998). Self-inactivating lentivirus vector for safe and efficient *in vivo* gene delivery. *J Virol* **72**(12): 9873–9880.

53. Montini E, *et al.* (2006). Hematopoietic stem cell gene transfer in a tumor-prone mouse model uncovers low genotoxicity of lentiviral vector integration. *Nat Biotechnol* **24**(6): 687–696.

54. Montini E, *et al.* (2009). The genotoxic potential of retroviral vectors is strongly modulated by vector design and integration site selection in a mouse model of HSC gene therapy. *J Clin Invest* **119**(4): 964–975.

55. Zychlinski D, *et al.* (2008). Physiological promoters reduce the genotoxic risk of integrating gene vectors. *Mol Ther* **16**(4): 718–725.

56. Waehler R, SJ Russell, DT Curiel (2007). Engineering targeted viral vectors for gene therapy. *Nat Rev Genet* **8**(8): 573–587.

57. West AG, Gaszner M, Felsenfeld G (2002). Insulators: many functions, many mechanisms. *Genes Dev* **16**(3): 271–288.

58. Yanez-Munoz RJ, *et al.* (2006). Effective gene therapy with nonintegrating lentiviral vectors. *Nat Med* **12**(3): 348–353.

59. Miller JC, *et al.* (2007). An improved zinc-finger nuclease architecture for highly specific genome editing. *Nat Biotechnol* **25**(7): 778–785.

60. Szczepek M, *et al.* (2007). Structure-based redesign of the dimerization interface reduces the toxicity of zinc-finger nucleases. *Nat Biotechnol* **25**(7): 786–793.

Chapter 4
Lentiviral Vectors

Janka Mátrai, Marinee K. L. Chuah and Thierry VandenDriessche*

Lentiviral vectors represent some of the most promising vectors for gene therapy. They have emerged as potent and versatile tools for *ex vivo* or *in vivo* gene transfer into dividing and non-dividing cells. Lentiviral vectors can be pseudotyped with distinct viral envelopes that influence vector tropism and transduction efficiency. In addition, it is possible to redirect vector transduction and generate cell-type specific targetable vectors by incorporating cell-type specific ligands or antibodies into the vector envelope. Lentiviral vectors have been used to deliver genes into various tissues, including brain, retina and liver, by direct *in vivo* gene delivery. Since they integrate stably into the target cell genome, they are ideally suited to deliver genes *ex vivo* into *bona fide* stem cells, particularly hematopoietic stem cells (HSCs), allowing for stable transgene expression upon hematopoietic reconstitution. Though there are some safety concerns regarding the risk of insertional mutagenesis, it is possible to minimize this risk by modifying the vector design or by employing integration-deficient lentiviral vectors which, in conjunction with zinc-finger nuclease technology, allow for site-specific gene correction or addition in pre-defined chromosomal loci. Lentiviral gene transfer has provided efficient phenotypic correction of diseases in mouse models paving the way towards clinical applications.

1. Basic Viral Biology

Lentiviruses are enveloped RNA viruses that are surrounded by a lipid bilayer in which the envelope proteins are embedded. The main components

*Correspondence: Flanders Institute for Biotechnology, Vesalius Research Center, University of Leuven, Leuven, Belgium.
E-mail: thierry.vandendriessche@vib-kuleuven.be

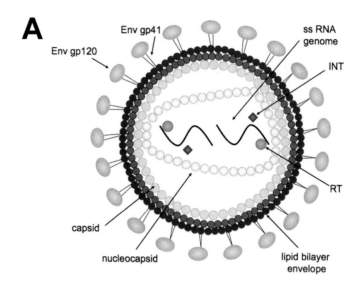

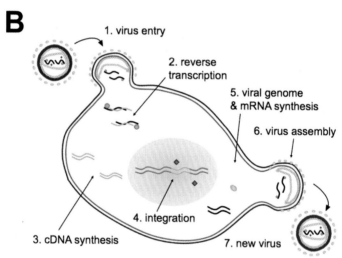

Fig. 4.1 **A.** Schematic representation of the HIV-1 virus. The figure shows the single stranded diploid RNA genome of HIV-1 embedded into the nucleocapsid encoded by the gag gene. The integrase (INT) and reverse transcriptase (RT) enzymes encoded by the *pol* gene are also indicated. The nucleocapsid is surrounded by the capsid and the lipid bilayer membrane where the glycoproteins 41 and 120 are located; **B.** Viral life cycle. The HIV-1 virus enters the cell via the cell surface receptors. The viral genome is endocytosed and reverse transcribed in the cytosol. The viral DNA subsequently crosses the nuclear membrane and finally integrates into the host genome. Using the transcription machinery of the host cell viral RNA is transcribed and transported into the cytoplasm. Here, translation and virus assembly take place and the *de novo* synthesized virus leaves the cell to initiate a new productive infection.

of the wild-type HIV-1 virus and its life cycle are shown in Figs. 4.1A and 4.1B. The envelope surrounds the viral capsid that contains the structural proteins (matrix), viral enzymes and 2 copies of a single stranded RNA genome, which codes for *gag*, *pol* and *env* genes that are common to all retroviruses (Fig. 4.1A). The *gag* gene encodes the structural matrix proteins, capsid and nucleocapsid, *pol* encodes the reverse transcriptase (RT), integrase (INT) and protease and *env* encodes the envelope glycoproteins (gp120 and gp41). Additionally, the lentiviral genome contains accessory genes encoding the regulatory proteins Tat, Rev, Vpr, Vpu Vif and Nef. Tat and Rev are essential for viral replication, whereas the other accessory genes contribute to viral pathogenicity. Tat is a potent trans-activator of HIV-1 gene expression and "jump-starts" the HIV replication cycle. Rev is a post-transcriptional trans-activator that accelerates mRNA transport from the nucleus to the cytoplasm via binding to the Rev-responsive element (RRE). Rev is essential for the expression of the *gag*, *pol* and *env* genes and for the transport of the full length RNA genome.

During infection, the Env proteins interact with their cognate cellular receptors (i.e. CD4 and the co-receptors CCR5 or CXCR4). After receptor-mediated viral entry into the cell by membrane fusion, the RNA genome is reverse transcribed in the cytoplasm by the reverse transcriptase into a double stranded proviral DNA (Fig. 4.1B). This proviral DNA is associated with different viral proteins (nucleocapsid, RT, integrase) in an intermediate protein-DNA complex designated as the pre-integration complex (PIC). The PIC also interacts with cellular proteins that facilitate the transport across the nuclear membrane into the nucleus. The intrinsic properties of the lentiviral PIC are different from that of the γ-retroviruses and endow the virus with the ability to enter the nucleus of non-dividing cells. In the nucleus, the integrase catalyzes the integration into the target cell genome via interaction with the viral long terminal repeats (LTR), particularly the *att* (attachment) sequence.

Cellular factors such as LEDGF/p75 play a key role in controlling lentiviral integration by acting as a tethering factor. Host cell factors initiate transcription from the integrated proviral LTRs resulting in production of new viral genomic RNA that encodes the different HIV-1 structural and regulatory proteins. The genomic RNA and the newly produced viral proteins are subsequently assembled as a viral particle during the budding process (Fig. 4.1B).

2. Vector Design and Production

2.1 *Vector Development*

Since HIV-1 is a human pathogen it is important to ensure that the corresponding lentiviral vectors (LV) are replication-deficient and unable to revert back to an infective wild-type HIV-1. To make functional LV particles, packaging cells need to be transfected with a plasmid containing a modified vector genome that expresses the gene of interest (transfer vector) and the helper (or packaging) plasmids that encode the essential viral proteins *in trans* (Fig. 4.2).[1,2] The gene transfer vector is devoid of all HIV-1 viral genes

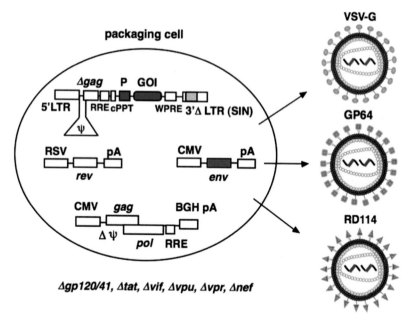

Fig. 4.2 Lentiviral vector production by trans-complementation. Packaging cells are transfected with the lentiviral vector plasmid and 3 helper (i.e. packaging) constructs encoding Gag, Pol, Rev and Env (e.g. VSV-G, GP64, RD114). Only the LV contains the packaging sequence Ψ and the gene of interest (GOI) whereas the packaging constructs are devoid of Ψ. Assembled vector particles are harvested from the supernatant and, if required, subjected to further purification and concentration. Constructs encoding different envelope genes allow for the production of distinct LV pseudotypes which exhibit different tropisms. The self-inactivating (SIN) LTR sequences that contain a partial deletion (Δ), the promoter of interest (P), Woodchuck post-transcriptional regulatory element (WPRE), central polypurine tract (cPPT), the Rous Sarcoma Virus promoter (RSV), the Cytomegalovirus promoter (CMV), Rev-responsive element (RRE) and the bovine growth hormone polyadenylation signal (BGH pA) are indicated.

encoding structural or accessory proteins. In contrast to γ-retroviral vectors, LV can typically accommodate larger transgene cassettes (up to ~ 10 kb) that are expressed using an internal promoter. The Ψ sequence is essential for packaging and ensures appropriate encapsidation of the full-length vector genome into the viral vector particles (Fig. 4.2). The LV is flanked by the 5' and 3' LTR sequences that have promoter/enhancer activity, which is essential for the correct expression of the full-length vector transcript, and play an important role in reverse transcription and integration of the vector into the target cell genome. To prevent mobilization (the packaging of the integrated LV into HIV particles) by infection with wild-type HIV a self-inactivating (SIN) vector configuration with a 400 bp deletion in the 3' LTR U3 region, including the TATA box was introduced.[3] This deletion is subsequently copied onto the 5' LTR of the proviral DNA during reverse transcription, abolishing the transcriptional activity of the LTR. Furthermore, the SIN configuration reduces the likelihood that cellular coding sequences located adjacent to the vector integration site will be aberrantly expressed, either due to the promoter activity of the 3' LTR or through an enhancer effect. Finally, the SIN design prevents potential transcriptional interference between the LTR and the internal promoter driving the transgene.[3] The introduction of chimeric CMV/HIV-5' LTR or RSV/HIV-5' LTR promoters to drive the expression of the full-length vector transcript allowed for the elimination of the HIV-1 Tat protein during vector production.

Several small genetic sequences have been introduced into the vector to enhance gene transfer efficiency. The woodchuck hepatitis virus post-transcriptional regulatory element (WPRE)[4] when placed 3' of the gene of interest leads to increased transgene expression. The RRE including a small segment of the gag gene was re-introduced into the transfer vector resulting in improved viral titers. Finally, it has been shown that the central polypurine tract (cPPT) facilitates nuclear translocation of the pre-integration complex and consequently increases lentiviral transduction.[5,6]

Of the 9 HIV-1 viral genes, only 3 are essential for the production of HIV-1 based LV namely *gag, pol,* and *rev*. The function of these genes in the LV is identical to that in the cognate wild-type HIV-1 virus. The *gag* and *pol* genes are expressed from the same helper (i.e. packaging) plasmid. Since expression of *gag* and *pol* is strictly dependent upon the Rev-RRE interaction, a separate construct encoding Rev needs to be cotransfected into the packaging cells. The generation of wild-type HIV-1 can *de facto* be excluded since none of these genes are present in the packaging cells. Additionally, LV particles require a viral envelope protein, which is essential for vector entry into the target cells. The envelope of the LV is typically encoded by a separate packaging construct that contains a heterologous *env* gene from another enveloped virus that replaces the native HIV-1 *env* gene (i.e. *pseudotyping*; see below).

2.2 *Vector Production*

The most common procedure to generate LV is to co-transfect 293T (human embryonic kidney 293) cells with the vector encoding plasmid, the *gag-pol*, *rev* and *env* packaging constructs (Fig. 4.2). The typical titer of the non-concentrated vector batches is about 10^7 transducing units/ml (TU/ml). Vector titers can further be increased by means of centrifugal filter concentration or ultracentrifugation. Concentrated vector titers typically fall in the range of $10^9 - 10^{10}$ TU/ml. Although transient transfection of 293 T cells can produce high titer LV, this method is cumbersome and difficult to scale up which poses significant manufacturing hurdles. To overcome these limitations, stable packaging cell lines were developed that conditionally express the cytotoxic Gag-Pol and VSV-G Env using an inducible promoter. This packaging cell line allows for the production of high-titer LV ($>10^7$ TU/ml) paving the way towards industrial large-scale vector production in bio-reactors for clinical trials and therapeutic applications.[7]

Typically LV titers are reported as TU/ml, which is determined with a reporter or selectable marker gene by quantifying the number of transduced cells following gene transfer. Alternatively, vector titer can be determined by assessing the number of actual vector copies in the transduced cell population by (real-time) quantitative PCR.[8] Since both of these methods for determining viral titers are based on a chosen cell line, additional methods have been developed to allow for comparison of different pseudotyped LV titers. LV particle titers can be determined by measuring the amount of the HIV-1 capsid p24 protein by enzyme-linked immunosorbent assay (ELISA), or by estimating the RNA content of the vector particles by RNA dot-blot analysis or reverse transcriptase (RT)-qPCR.

3. Gene Transfer Concepts and Potential Applications

3.1 *Target Cells and Diseases*

In contrast to γ-retroviral vectors, LV can also transduce non-dividing cells, which opened up new therapeutic avenues and applications. In particular, local administration of LV into the brain resulted in stable transduction of terminally differentiated neurons.[1,9] Stable transduction was also reported in other tissues, including skeletal muscle. Systemic administration typically resulted in relatively efficient and widespread transduction of liver and spleen. Proof of concept has been established in pre-clinical animal models that LV can result in effective gene therapy for hereditary, acquired or complex disorders that affect these various organs (e.g. neurodegenerative disease, congenital blindness, liver disease, hemophilia, metabolic diseases, etc.). LV are also ideally suited to transduce various stem/progenitor cells, particularly HSCs.[10,11] HSCs are attractive targets for gene therapy because of their capacity of self-renewal and potential life-long ability to replenish mature blood cell populations. Sustained expression of therapeutic transgenes at clinically relevant levels has been demonstrated following LV transduction of HSCs for a variety of hematologic and metabolic diseases (e.g. hemoglobinopathies, severe combined immune deficiencies, hemophilia, etc.).[12,13] By restricting expression of the therapeutic gene in a cell lineage-specific fashion using lineage-specific promoters, it has been possible to generate LV that upon transduction of HSC are only expressed in the desired lineage upon hematopoietic reconstitution.[14]

3.2 *Pseudotyping*

The most commonly used viral envelope protein used for pseudotyping LV is the vesicular stomatitis virus glycoprotein (VSV-G), but other envelopes including rabies, MLV-amphotropic, Ebola, baculovirus and measles virus envelopes can be used too (Fig. 4.2).[15] Pseudotyping HIV-1 vectors obviates safety concerns associated with the use of HIV-1 gp120, which has known pathogenic consequences. Moreover, pseudotyping has a dramatic impact on the biodistribution, vector tropism, and viral particle stability. For instance, filovirus envelope-pseudotyped LV enhance transduction of airway epithelia or endothelial cells[16] whereas baculovirus GP64 and hepatitis

C E1 and E2 pseudotyping enhances hepatic transduction.[17,18] RD114 pseudotyping favors transduction in lymphohematopoietic cells.[19] LV pseudotyped with the Edmonston measles virus (MV) glycoproteins H and F allowed efficient transduction through the MV receptors, SLAM and CD46, both present on blood T cells.[20]

3.3 Cell Type Specific Targeting

Pseudotyping also provides a means to generate cell-type specific LV. This can be accomplished by incorporating cell-type specific ligands or antibodies into the viral envelope (Fig. 4.3). Modification of the envelope with these cell type specific retargeting moieties can redirect the binding of the LV particles to the corresponding cellular receptor. However, envelope engineering can potentially adversely alter the fusion domain of Env resulting in low vector titers. It is therefore important to ensure that redirecting vector tropism does not compromise the post-binding step in vector entry. To address this concern, fusogenic envelope proteins are incorporated along side the retargeted envelope proteins into the LV particles. Typically, the fusogen is constructed by modifying viral envelope proteins, so that they lack the ability to bind to their cognate receptor but still retain the ability to trigger membrane fusion. Thus, the specificity of such a LV is then solely determined by the retargeting moiety (Fig. 4.3). A slightly different paradigm was explored to achieve more efficient lentiviral transduction into T cells and HSC. By engineering LV particles displaying T-cell activating single chain antibody peptides or IL-7, resting T cells could be efficiently transduced.[21,22] Similarly, to obtain more selective and efficient gene transfer into HSCs recombinant membrane proteins were engineered and were incorporated into LV particles to display "early acting" cytokines on their surface.

3.4 Integration-Defective Lentiviral Vectors

Since a functional integrase is required to achieve stable genomic lentiviral integration, it is possible to reduce this integration by mutational inactivation of the integrase protein. The use of integration-defective LV (IDLV) could potentially minimize concerns associated with random integration and

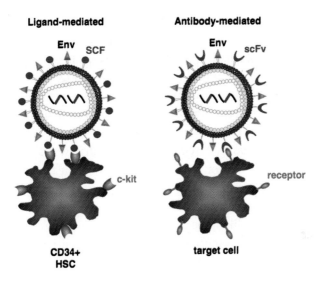

Fig. 4.3 LV targeting into specific cell types. The LV Env protein is amenable to engineering allowing the display of cell type specific ligands or single chain antibody fragment (scFv). These ligands or scFv can then bind onto cell surface receptors that are specifically expressed on the desired target cells. Some of the ligand-receptor interaction may activate the target cells and consequently enhance transduction. Some Env proteins like the amphotropic MLV Env can be engineered to display the stem cell factor (SCF) as ligand for the SCF cellular receptor (i.e. c-kit) allowing enhanced transduction of CD34+ HSC. A mutated version of the RD114 envelope is used as fusogen. An alternative retargeting paradigm is based on the display of scFv on the measles hemagluttinin envelope H protein while the native tropism of this measles Env was ablated. The measles envelope F protein acts as fusogen.

insertional oncogenesis.[23] Consequently these IDLV have a defect in integration characterized by accumulation of double stranded episomal DNA circles in the host cell nucleus. IDLV are ideally suited to achieve short-term expression of a gene of interest in dividing cells since expression declines as the cells progress through subsequent cell cycles due to the loss of non-integrated episomes. For instance, IDLV may be used for the delivery of zinc finger nucleases (ZNF) and a corrective or additive genetic sequence for site-specific integration.[24] However, in non-dividing target cells, in particular retina, brain or skeletal muscle cells, transgene expression is typically sustained from IDLV.[23,25,26] Robust functional rescue of congenital blindness could be achieved with IDLV in a rodent model of autosomal recessive retinitis pigmentosa, a severe form of Leber congenital amaurosis, resulting from a mutation in *rpe65*.[23] Though genomic integration of IDLV is clearly impaired, low frequency residual integrations cannot be excluded.

4. Immune Consequences

LV can trigger innate and adaptive immune reactions directed against the vector particles, the transduced cells and/or the transgene product encoded by the vector. Indeed, immunological responses to LV gene transfer make these vectors effective for vaccination.[27] However, untoward immune responses may curtail expression of the gene of interest, which would be undesirable when treating hereditary diseases that warrant stable transgene expression. The vector particles themselves can trigger a rapid but self-limiting pro-inflammatory response leading to a transient cytokine surge (e.g. interleukin-6)[28] and IFN$\alpha\beta$ response.[29] This pro-inflammatory immune response can likely be ascribed to efficient interaction of the LV particles with antigen-presenting cells (APCs)[6] and it has been proposed that this may involve engagement of toll-like receptor 7 (TLR7), a pattern-recognition receptor (PRR) for single-stranded RNA (ssRNA) and/or TLR9 which recognizes unmethylated CpG. Remarkably, when LV were administered to animals that lack the capacity to respond to IFNβ, there was a dramatic increase in hepatocyte transduction, and stable transgene expression was achieved. Exposure to LV will likely induce vector-specific antibodies that will neutralize the vector particles and consequently prevent gene transfer by subsequent vector readministration. It is also possible, in principle, that pre-existing antibodies to the heterologous Env protein used to pseudotype the LV may interfere with viral transduction.

Since the immune system of many patients suffering from hereditary diseases has not been tolerized to the functional transgene product that they are missing, it is possible that LV transduction may evoke antigen-specific immune reactions that could result in the elimination and/or neutralization of the transgene product by antibodies and/or the clearance of the transduced cells by cytotoxic T cells (CTLs) in an MHC class I-restricted fashion.[30,31,32] The strength of these antigen-specific humoral and cellular adaptive immune reactions following LV administration, depends on several parameters, including the transgene product, vector design, vector dose, route of vector administration, target cell type and genotype of the recipient animal or patient. In particular, LV transduction of APCs may result in ectopic expression of the transgene product. This may increase the risk of inducing humoral and/or cellular immune response that curtail long-term gene expression in particularly if LV transduction enhances the maturation of APCs. To minimize this risk it is warranted to use cell type-specific promoter/enhancers to restrict transgene expression to the target tissue while preventing inadvertent ectopic transgene expression in APCs.[31] Nevertheless, even with highly tissue-specific promoters, immune response against the transgene product could not be prevented due to "leaky" transgene expression in APCs. To ensure robust tissue-specific expression of the transgene product and prevent, ectopic, "leaky" expression in APCs, an additional layer of regulation was built into the LV by incorporating a target sequence for the hematopoietic-specific microRNA, miR-142-3p. This eliminated off-target expression in hematopoietic cells, particularly APCs, allowing for possible induction of immune tolerance and resulting in sustained expression of the gene of interest, *in casu* FIX in a hemophilia B mouse model.[32,33]

5. Safety Issues

The use of SIN vector configuration, heterologous envelopes and Tat-independent vector production has significantly improved the overall vector safety. Moreover, potential homologous overlap between vector and packaging constructs is minimized to reduce the risk of generating replication-competent lentiviruses (RCL) by homologous recombination. Integration of a γ-retroviral vector encoding the IL2Rγc gene in proximity of the *lmo2* proto-oncogene contributed to the deregulated LMO2 expression that lead to the leukemiogenesis observed in several subjects enrolled in a clinical trial for SCID-X1.[34] Consequently, one of the main potential safety concerns related to the use of LV in clinical applications is the risk of insertional oncogenesis. Both γ-retroviral and LV show an integration bias into transcriptional units indicating that integration is not strictly random.[35,36] However, γ-retroviral vectors have a predilection towards integrating in the immediate proximity of transcription start sites (TSS) and a small window around DNAse I hypersensitive sites, whereas LV are more likely to integrate further away from the TSS.[37,38] To assess the relative oncogenicity/genotoxicity of LV, hematopoietic stem cell gene transfer studies are being conducted in a tumor-prone mouse model.[39] These studies uncovered low genotoxicity of LV integration compared to when γ-retroviral vectors were used. Retroviral vectors triggered dose-dependent acceleration of tumor onset contingent on LTR activity. Insertions at oncogenes and cell-cycle genes were enriched in early-onset tumors, indicating cooperation in tumorigenesis. In contrast, tumorigenesis was unaffected by the SIN LV and did not enrich for specific integrants, despite the higher integration load and robust expression of LV in all hematopoietic lineages. Hence, the prototypical LV appeared to have low oncogenic potential.[40] Lentiviral integration profiles and/or oncogenic risks may depend on several confounding variables including the vector copy number, the target cell type, the proliferation and/or activation status of the target cells, the nature of the transgene itself, the vector design (SIN vs. non-SIN, choice of promoter/enhancers), underlying disease and possible selective advantage of rapidly growing cells, protocol-specific cofactors and finally the intrinsic genotypic variation of the model animals and the treated patients. This implies that under "permissive" conditions it may be possible to uncover insertional oncogenic events that can be ascribed to the LV.

6. Conclusions and Perspectives

The conversion of the highly pathogenic HIV-1 into an efficient and relatively safe gene delivery vector serves as testimony to the worthiness of the impressive journey that has been undertaken. LV have now become commonplace in experimental research. LV gene transfer can treat or cure disease in preclinical animal models and has now moved into the clinical arena with multiple gene therapy trials ongoing or approved. The early results from these clinical trials are encouraging and underscore the safety and potential efficacy of using LV.[41,42]

References

1. Naldini L, *et al.* (1996). *In vivo* gene delivery and stable transduction of nondividing cells by a lentiviral vector. *Science* **272**: 263–267.
2. Vigna E, Naldini L (2000). Lentiviral vectors: excellent tools for experimental gene transfer and promising candidates for gene therapy. *The Journal of Gene Medicine* **2**: 308–316.
3. Zufferey R, *et al.* (1998). Self-inactivating lentivirus vector for safe and efficient *in vivo* gene delivery. *Journal of Virology* **72**: 9873–9880.

4. Zufferey R, Donello JE, Trono D, Hope TJ (1999). Woodchuck hepatitis virus posttranscriptional regulatory element enhances expression of transgenes delivered by retroviral vectors. *Journal of Virology* **73**: 2886–2892.

5. Follenzi A, Ailles LE, Bakovic S, Geuna M, Naldini L (2000). Gene transfer by lentiviral vectors is limited by nuclear translocation and rescued by HIV-1 pol sequences. *Nature Genetics* **25**: 217–222.

6. VandenDriessche T, *et al.* (2002). Lentiviral vectors containing the human immunodeficiency virus type-1 central polypurine tract can efficiently transduce nondividing hepatocytes and antigen-presenting cells *in vivo*. *Blood* **100**: 813–822.

7. Broussau S, *et al.* (2008). Inducible packaging cells for large-scale production of lentiviral vectors in serum-free suspension culture. *Mol Ther* **16**: 500–507.

8. Geraerts M, Willems S, Baekelandt V, Debyser Z, Gijsbers R (2006). Comparison of lentiviral vector titration methods. *BMC Biotechnol* **6**: 34.

9. Naldini L, Blomer U, Gage FH, Trono D, Verma IM (1996). Efficient transfer, integration, and sustained long-term expression of the transgene in adult rat brains injected with a lentiviral vector. *Proceedings of the National Academy of Sciences of the United States of America* **93**: 11382–11388.

10. Miyoshi H, Smith KA, Mosier DE, Verma IM, Torbett, BE (1999). Transduction of human CD34+ cells that mediate long-term engraftment of NOD/SCID mice by HIV vectors. *Science* **283**: 682–686.

11. Case SS, *et al.* (1999). Stable transduction of quiescent CD34(+)CD38(-) human hematopoietic cells by HIV-1-based lentiviral vectors. *Proceedings of the National Academy of Sciences of the United States of America* **96**: 2988–2993.

12. May C, *et al.* (2000). Therapeutic haemoglobin synthesis in beta-thalassaemic mice expressing lentivirus-encoded human beta-globin. *Nature* **406**: 82–86.

13. Biffi A, *et al.* (2004). Correction of metachromatic leukodystrophy in the mouse model by transplantation of genetically modified hematopoietic stem cells. *The Journal of Clinical Investigation* **113**: 1118–1129.

14. Shi Q, *et al.* (2007). Lentivirus-mediated platelet-derived factor VIII gene therapy in murine haemophilia A. *J Thromb Haemost* **5**: 352–361.

15. Watson DJ, Kobinger GP, Passini MA, Wilson JM, Wolfe JH (2002). Targeted transduction patterns in the mouse brain by lentivirus vectors pseudotyped with VSV, Ebola, Mokola, LCMV, or MuLV envelope proteins. *Mol Ther* **5**: 528–537.

16. Kobinger GP, Weiner DJ, Yu QC, Wilson JM (2001). Filovirus-pseudotyped lentiviral vector can efficiently and stably transduce airway epithelia *in vivo*. *Nature Biotechnology* **19**: 225–230.

17. Bartosch B, Dubuisson J, Cosset FL (2003). Infectious hepatitis C virus pseudoparticles containing functional E1-E2 envelope protein complexes. *The Journal of Experimental Medicine* **197**: 633–642.

18. Kang Y, *et al.* (2005). Persistent expression of factor VIII *in vivo* following nonprimate lentiviral gene transfer. *Blood* **106**: 1552–1558.

19. Hanawa H, *et al.* (2002). Comparison of various envelope proteins for their ability to pseudotype lentiviral vectors and transduce primitive hematopoietic cells from human blood. *Mol Ther* **5**: 242–251.

20. Frecha C, *et al.* (2008). Stable transduction of quiescent T cells without induction of cycle progression by a novel lentiviral vector pseudotyped with measles virus glycoproteins. *Blood* **112**: 4843–4852.

21. Maurice M, Verhoeyen E, Salmon P, Trono D, Russell SJ, Cosset FL (2002). Efficient gene transfer into human primary blood lymphocytes by surface-engineered lentiviral vectors that display a T cell-activating polypeptide. *Blood* **99**: 2342–2350.

22. Verhoeyen E, Dardalhon V, Ducrey-Rundquist O, Trono D, Taylor N, Cosset FL (2003). IL-7 surface-engineered lentiviral vectors promote survival and efficient gene transfer in resting primary T lymphocytes. *Blood* **101**: 2167–2174.

23. Yanez-Munoz RJ, *et al.* (2006). Effective gene therapy with nonintegrating lentiviral vectors. *Nature Medicine* **12**: 348–353.

24. Lombardo A, *et al.* (2007). Gene editing in human stem cells using zinc finger nucleases and integrase-defective lentiviral vector delivery. *Nature Biotechnology* **25**: 1298–1306.

25. Apolonia L, *et al.* (2007). Stable gene transfer to muscle using non-integrating lentiviral vectors. *Mol Ther* **15**: 1947–1954.

26. Philippe S, *et al.* (2006). Lentiviral vectors with a defective integrase allow efficient and sustained transgene expression *in vitro* and *in vivo*. *Proceedings of the National Academy of Sciences of the United States of America* **103**: 17684–17689.

27. He Y, Zhang J, Mi Z, Robbins P, Falo LD, Jr. (2005). Immunization with lentiviral vector-transduced dendritic cells induces strong and long-lasting T cell responses and therapeutic immunity. *J Immunol* **174**: 3808–3817.

28. Vandendriessche T, *et al.* (2007). Efficacy and safety of adeno-associated viral vectors based on serotype 8 and 9 vs. lentiviral vectors for hemophilia B gene therapy. *J Thromb Haemost* **5**: 16–24.

29. Brown BD, *et al.* (2007). *In vivo* administration of lentiviral vectors triggers a type I interferon response that restricts hepatocyte gene transfer and promotes vector clearance. *Blood* **109**: 2797–2805.

30. Annoni A, *et al.* (2007). The immune response to lentiviral-delivered transgene is modulated *in vivo* by transgene-expressing antigen-presenting cells but not by CD4+CD25+ regulatory T cells. *Blood* **110**: 1788–1796.

31. Follenzi A, Battaglia M, Lombardo A, Annoni A, Roncarolo MG, Naldini L (2004). Targeting lentiviral vector expression to hepatocytes limits transgene-specific immune response and establishes long-term expression of human antihemophilic factor IX in mice. *Blood* **103**: 3700–3709.

32. Brown BD, *et al.* (2007). A microRNA-regulated lentiviral vector mediates stable correction of hemophilia B mice. *Blood* **110**: 4144–4152.

33. Brown BD, Venneri MA, Zingale A, Sergi Sergi L, Naldini L (2006). Endogenous microRNA regulation suppresses transgene expression in hematopoietic lineages and enables stable gene transfer. *Nature Medicine* **12**: 585–591.

34. Hacein-Bey-Abina S, *et al.* (2008). Insertional oncogenesis in 4 patients after retrovirus-mediated gene therapy of SCID-X1. *The Journal of Clinical Investigation* **118**: 3132–3142.

35. Schroder AR, Shinn P, Chen H, Berry C, Ecker JR, Bushman F (2002). HIV-1 integration in the human genome favors active genes and local hotspots. *Cell* **110**: 521–529.

36. Trono D (2003). Virology. Picking the right spot. *Science* **300**: 1670–1671.

37. Sinn PL, Sauter SL, McCray PB, Jr. (2005). Gene therapy progress and prospects: development of improved lentiviral and retroviral vectors–design, biosafety, and production. *Gene Ther* **12**: 1089–1098.
38. Baum C (2007). Insertional mutagenesis in gene therapy and stem cell biology. *Curr Opin Hematol* **14**: 337–342.
39. Montini E, *et al.* (2006). Hematopoietic stem cell gene transfer in a tumor-prone mouse model uncovers low genotoxicity of lentiviral vector integration. *Nature Biotechnology* **24**: 687–696.
40. Montini, *et al.* (2009). The genotoxic potential of retroviral vectors is strongly modulated by vector design and integration site selection in a mouse model of HSC gene therapy. *J Clin Invest* **119**: 755–758.
41. Cartier N, *et al.* (2009). Hematopoietic stem cell gene therapy with a lentiviral vector in X-linked adrenoleukodystrophy. *Science* **326**: 818–823.
42. Matrai J, Chuah MKL, VandenDriessche, T (2010). Recent advances in lentiviral vector development and applications, *Mol Ther* **18**: 477–490.

Chapter 5
Herpes Simplex Virus Vectors

William F. Goins*, David M. Krisky[†],
James B. Wechuck[†], Darren Wolfe[†], Justus B. Cohen
and Joseph C. Glorioso[‡]

Herpes simplex virus (HSV), a member of the human herpesviruses that infects humans, has been developed as a gene delivery vehicle for treating peripheral neuropathies, chronic pain and brain tumors including glioblastoma multiforme (GBM) because of the ability of the virus to readily transduce cells of the peripheral (PNS) and central nervous systems (CNS) as part of the natural biology of virus infection. In fact, HSV vectors are currently employed in human clinical trials for treating numerous forms of cancer and recently cancer-related pain. Considerable work in the design of HSV vectors has involved the engineering of vectors with reduced cytotoxicity yet are still capable of expressing sufficient amounts of the desired therapeutic gene product. In addition, methods to purify these vectors for use in pre-clinical animal studies have been developed.

1. Introduction

Herpes simplex virus type 1 (HSV-1) is a member of the herpesvirus family for which there are 8 members of the family which infect humans; HSV-1, HSV-2, varicella zoster virus (VZV), human cytomegalovirus (HCMV), Epstein-Barr virus (EBV), human herpesvirus-6 (HHV-6), human herpes virus-7 (HHV-7) and human herpes virus-8 (HHV-8) also know as Kaposi sarcoma herpes virus (KSHV). HSV-1 is responsible for 60% of the oro-facial lesion and 40% of the genital lesions.[1] Although these lesions can

Correspondence: Department of Microbiology & Molecular Genetics, University of Pittsburgh School of Medicine, Pittsburgh, PA 15261, [†]Current Address: Diamyd Inc., Pittsburgh PA 15219.
Email: *goins@pitt.edu; [‡]glorioso@pitt.edu

be painful, they are self-limiting in the immune competent host. The only instances where severe clinical HSV manifestation becomes a concern is when the virus enters the CNS ultimately leading to encephalitis, or when the virus encounters immuno-compromised patients such as transplant patients on immunosuppressive drug therapy to prevent rejection, or in AIDS patients.

A common feature of all the members of the herpesvirus family is the ability of these viruses to persist long-term in the host, even in the presence of high levels of neutralizing antibodies.[1] During initial HSV infection, replicating virus released by infected epithelial cells may come into contact with neuronal cell termini that innervate the site of the primary infection, resulting in the infection of these neuronal cells where it does not proceed through the normal lytic virus life cycle that results in the production of progeny virions and death of that cell, but instead the virus is capable of entering a latent or "quiescent" state where all the HSV lytic cycle genes are shut off and no HSV proteins are made so that the host is incapable of recognizing these neurons and clearing these infected cells.[2] The host can keep the virus in check within the neurons of the PNS where it lies dormant or latent, sometimes for the lifetime of the host. Certain stimuli such as surgery, UV light, local trauma, fatigue, immune suppression, and fever lead to the virus exiting the latent or "quiescent" state and the re-establishment of the lytic life cycle in the PNS nerve cells, also known as "reactivation".

HSV is an enveloped, double-stranded DNA-containing virus that possesses natural neurotropism as part of its life cycle within the human host. The virus particle (Fig. 5.1A) is composed of an icosahedral-shaped nucleocapsid containing the 152 kb linear dsDNA molecule encased in a lipid bilayer envelope that it acquires from the host cell when it buds from the cell as part of the egress process. The viral genome is composed of two segments, the unique long (U_L) and short (U_S) regions each of which are flanked by inverted repeats (Fig. 5.1A). Over 85 viral genes are present primarily within the unique long segment with several important regulatory genes present in the repeat sequence, and thus are present within two copies in each viral genome. The viral genes can be categorized into two groups; those that are essential for virus replication in standard cells used to grow the virus in cell culture and those that are not essential to virus

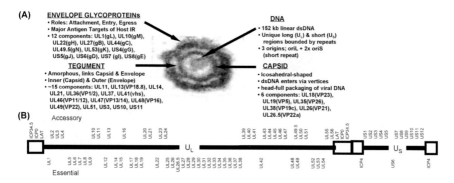

Fig. 5.1 Schematic diagram of HSV virion particle and viral genome. (A) EM photomicrograph of HSV virion particle with the four major virion components. The HSV linear dsDNA genome is contained within the icosahedral-shaped capsid that is surrounded by the tegument layer that in turn is enclosed within a lipid bilayer membrane containing the viral encoded glycoproteins. (B) The 152 kb linear dsDNA genome composed of the unique long (U_L) and short (U_S) segments, each flanked by inverted repeats. Viral genes are grouped according to whether they are essential for virus replication *in vitro* or accessory.

growth *in vitro* but play an accessory role in supporting virus growth or pathogenesis in the host (Fig. 5.1B). The HSV capsid is composed of multiple proteins all of which are encompassed by the tegument, an electron dense amorphous component of the virion that contains numerous proteins that contribute to the viral life cycle including the VP16 transactivating protein which activates the expression of all the immediate early (IE) genes,[3] and the virion host shut-off function that contributes to differential stabilities of host and viral mRNAs.[4] Surrounding the tegument layer is the envelope containing 12 virus-encoded glycoproteins that are responsible for virus attachment and entry into host cells, contribute to virus pathogenesis and serve as the major viral antigens recognized by the host immune response.

HSV entry into cells (Fig. 5.2) requires the coordinated action of four virion envelope glycoproteins, gB, gD, gH, and gL.[5] HSV attachment is mediated by the binding of gC and gB to heparan sulfate proteoglycans (HSPG) on the cell surface (Fig. 5.2A). While this binding is not required for infection,[6] it facilitates receptor engagement by gD (Fig. 5.2B). Binding of gD to a specific entry receptor is believed to activate gB and/or gH as mediators of fusion between the viral envelope and cellular membranes (Fig. 5.2C).[7] Membrane fusion can take place at the cell surface,

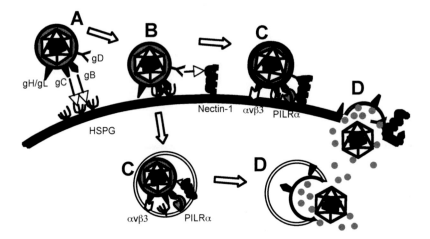

Fig. 5.2 HSV entry into the host cell. (A) The viral glycoproteins C and B (gC and gB) bind heparan sulfate proteoglycan (HSPG) in the initial attachment event. (B) Next, gD binds to either HVEM (HveA) or nectin-1 (HveC) on the cell surface or within endosomes, depending on the host cell. (C) Following a conformational change in gD accompanied by gB binding to PILRα and gH binding to αvβ3 integrin, (D) fusion occurs between the virion envelope and the cell surface or endosomal membrane with nucleocapsid release into the cytoplasm.

resulting in direct delivery of de-enveloped capsids to the cytoplasm, or at endosomal membranes following endocytic uptake of enveloped particles (Fig. 5.2D).[8]

Several receptors for HSV-1 gD have been identified. HVEM is a member of the TNF receptor family and binds to gD in its two cysteine-rich amino-terminal repeats.[9] Nectin-1 is a cell adhesion molecule belonging to the immunoglobulin superfamily where the N-terminal variable domain of nectin-1 is sufficient for gD recognition.[10] Receptor binding is believed to unlock a domain on gD referred to as the "profusion domain" resulting in activation of the fusogenic potential of gB and/or gH.[11,12]

Since the HSV receptors are ubiquitous on almost all cell types and tissues,[13] it may be necessary to target the virus to specific cell types through retargeting virus binding to specific receptors. Targeting of HSV to specific cells types requires: (i) the identification of cell-specific surface receptor(s) to which viral binding/entry can be redirected and (ii) the modification of viral glycoproteins to recognize novel receptors while eliminating the binding of these viral ligands to the natural HSV receptors, a process that ideally should be accomplished without compromising infectivity. Novel targeting ligands have been inserted into the amino terminus of gC and/or gD to achieve targeting of HSV to cell-specific receptors.[14,15] Additionally, soluble adapter molecules have been employed to target EGF receptor-bearing cells.[16]

Engineered HSV vector types can be broadly defined as (i) replication-competent vectors that are deleted for certain nonessential functions in order to achieve attenuation of their pathogenic phenotype,[17] but preserving their ability to carry out lytic infection of many dividing cell types, (ii) replication defective vectors deleted for at least one essential function,[18] limiting their ability to replicate in complementing cells engineered to provide the essential function(s) *in trans*, (iii) plasmid vectors referred to as amplicons,[19] which contain HSV packaging signals and an HSV origin of replication, creating a vector potentially devoid of all viral genes but requiring HSV helper functions supplied by helper virus or plasmids for their production, and (iv) bacterial artificial chromosome (BAC) vectors that have a bacterial origin of replication and a drug resistance gene to enable efficient genome manipulation and propagation in bacteria.[20] The replication competent vectors have been developed for the treatment of cancer relying on restricted virus replication within tumor cells. These vectors have the potential advantage of intra-tumoral virus spread and tumor cell lysis and thus have been referred to as oncolytic vectors. Thus far oncolytic HSV vectors have been used in clinical trials in patients with a variety of tumors, but have failed to show adequate efficacy due to limited replication and spread within the human tumor mass.[21] Replication defective and amplicon vectors were developed for localized applications suitable for a variety of diseases and have the potential advantages of low toxicity and large transgene payload. Recently, a replication-defective vector has been employed in a phase-I clinical trial to treat chronic pain associated with bone metastases.[22] The manufacture of the amplicon vector type has been unusually difficult with infectious titers produced in the order of 10^6–10^8 plaque-forming units (PFU) per mL compared to titers reaching 10^{12}/mL for the replication defective and competent vectors, making their use in human trials problematic. The remainder of the chapter will concentrate on the design and use of the replication defective HSV vector backbone.[18]

2. HSV Biology in the Design of Replication Defective Vectors

Although HSV has a complex genome, the genes are expressed in a well-ordered cascade in which the immediate early (IE) regulatory genes are required for activation of the early (E) and late (L) gene viral functions (Fig. 5.3), the latter requiring in addition viral DNA synthesis to become activated. Early viral genes are mostly involved in viral DNA replication while late genes largely encode viral structural proteins, the virus capsid,

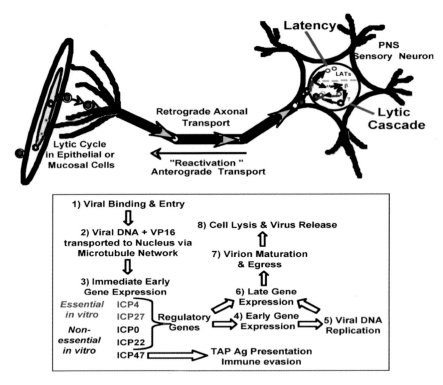

Fig. 5.3 HSV lytic and latent life cycles. Lytic HSV infection results in the onset of the three-phase cascade of HSV gene expression. The five immediate early (IE) genes are the first genes transcribed, and regulate the expression of the HSV early class genes involved in viral DNA synthesis. Following DNA replication, these IE gene products activate late gene class expression, which enables virion formation. Virion particles released from these cells encounter nerve terminals that innervate the site of primary infection. The particle travels by retrograde axonal transport to the nerve cell nucleus. Instead of entering the lytic phase, the viral genome circularizes, and all viral genes are silenced except for the latency-associated transcripts (LATs), the hallmark of HSV latency.

tegument and the envelope glycoproteins (Fig. 5.3). The interdependency of viral gene expression can be demonstrated by the removal of just one essential IE gene in order to render the virus defective for both early and late viral gene expression.

There are five IE genes in HSV-1 (Fig. 5.3), only two of which are essential for virus replication. Infected cell protein 4 (ICP4) is an essential IE gene that transactivates expression of early and late genes by directly recruiting transcription factors to these viral promoters.[23] ICP27 is an essential IE gene whose function involves post-transcriptional activation of early and late genes.[24] ICP0, a non-essential gene, mediates the disaggregation of nuclear structures termed PML bodies by degrading certain member components, and arrests the cell cycle without the induction of apoptosis.[25] Thus ICP0 plays an important role in establishing an intracellular milieu favorable for viral replication and consequently is essential for virus reactivation from latency. ICP22 is involved in redirecting RNAP II to sites of viral transcription allowing the virus to expropriate the cellular transcription machinery for viral transcription.[26] ICP47 inhibits the loading of immunogenic peptides onto MHC class I molecules, thereby transiently mitigating immune recognition of infected cells.[27] With the exception of ICP47, the individual IE genes are highly cytotoxic to most cell types and thus these genes must be removed in order to produce safe non-toxic vectors. Cells of the nervous system are a notable exception since ICP0 is rapidly degraded in neurons and there is reduced expression of cytotoxic viral functions.[28]

Wild–type HSV is capable of establishing a latent infection in sensory neurons of the TG or DRG depending on the site of initial virus interaction with the host (Fig. 5.3). During latency viral lytic gene expression is silenced and only a single region of the genome remains transcriptionally active (Fig. 5.4A) producing a family of non-coding latency-associated transcripts (LATs).[29] The major 2.0 Kb LAT is neither capped nor polyadenylated and represents stable introns that result from splicing of a large 8.3 Kb unstable primary transcript. Latent viral genomes are partially methylated and sequestered into an episomal minichromosome-like structure bound by nucleosomes in basically inactive heterochromatin.[30] Moreover, these latent viral genomes have no discernible effect on function of the host cell. Thus, HSV is particularly well suited for the delivery of genes to sensory

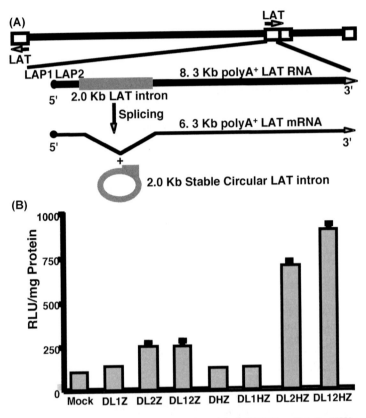

Fig. 5.4 HSV latent gene expression. (A) The location of the LAT RNAs within the HSV genome is depicted. The large 8.3 Kb LAT primary transcript has the 2.0 Kb LAT intron removed by splicing. The location of the two latency active promoters, LAP1 and LAP2, are depicted in relation to the LATs. (B) Replication defective recombinant vectors deleted for ICP4 and ICP27 [D] containing either LAP1 [L1], LAP2 [L2], LAP1+2 [L12], HCMV [H] or chimeric combinations driving expression of the lacZ reporter gene [Z] were used to infect the trigeminal ganglia of mice (N = 10) following corneal scarification and gene expression assayed using a kit is shown as relative light units (RLU) per mg of total protein.

neurons and provides an excellent vehicle to target ectoptic expression of transgenes to specific neuronal populations.

The LAT promoter/regulatory region has been shown to be composed of two latency-active promoters, LAP1,[31] a TATA box promoter with numerous cis-regulatory sites, and LAP2,[32] a TATA-less promoter that possesses CT-rich repeats that contribute to the ability of this promoter region to remain active when the genome is silenced.[32] LAP2 is moveable and active

in expressing transgenes for prolonged periods from both replication competent and defective vectors;[32] LAP1 does not possess this capability in
the absence of LAP2 (Fig. 5.4B).[33] Moreover, LAP2 has the ability to keep
normally transiently active promoters (HCMV) active in the context of replication defective vector backbones (Fig. 5.4B), suggesting that LAP2 could
be employed to drive long-term transgene or therapeutic gene expression
when other viral or cellular promoters are non-functional.

3. HSV Vector Design Technology

The remainder of the chapter will focus on the creation of replication defective HSV-1-based vectors. The goal for the replication defective vectors
was to identify genomic configurations that provide vigorous transgene
expression while eliminating attendant cytotoxicity. Pertinent parameters
of vector characterization include target cell transduction efficiency, relative abundance and longevity of transgene expression, and cell viability.
The initial first generation replication defective HSV vectors evolved from
single gene deletions, such as ICP4 (SOZ.4 in Fig. 5.5). Since some of these
vectors were missing a gene product that is essential to the lytic life cycle,
the only way to propagate these vectors was by using complementing cell
lines which express the deleted gene product *in trans*.[23] Although these
vectors have been effective for gene delivery to PNS neurons, they have
shown high levels of toxicity in most non-neuronal cell types. Therefore,
second generation replication defective vectors were generated to further
decrease the inherent toxicity to these cell types. Deletion of two IE regulatory genes such as ICP4 and ICP27 (DOZ.1 in Fig. 5.5) lead to vectors
which displayed increased levels of transgene expression with little to no
reduction in the cytotoxicity profile depending on the IE gene combination
deleted, again suggesting that the deletion of additional IE genes would
be required. The third generation vectors deleted the three IE genes ICP4,
ICP27 and ICP22 (TOZ.2 in Fig. 5.5). These vectors displayed only 50%
toxicity, a definite improvement over the second-generation counterparts,
and there was an increase in reporter gene expression. However, since these
vectors still retained significant cytotoxicity in non-neuronal cell lines, the
next goal was to eliminate the expression of ICP0 in this background, as it
appeared to display the greatest toxicity. Therefore, the fourth generation

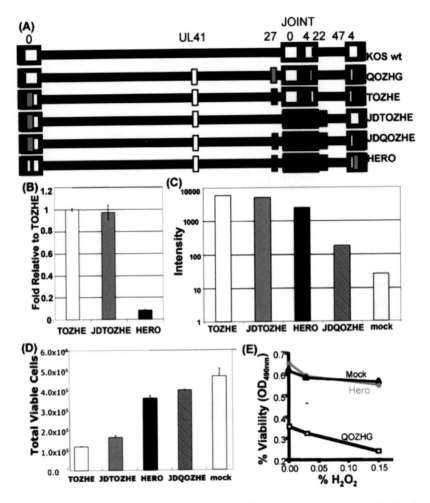

Fig. 5.5 Design of HSV replication defective vectors. Five subsequent generations of replication defective HSV vectors deleted for single or multiple IE genes are depicted; grey boxes represent deleted gene sequences. All of the vectors possess the same expression cassette composed of the ICP0 promoter driving expression of the lacZ reporter gene in the UL41 locus so that the levels of transgene expression can be directly compared in the various mutant vectors. Toxicity of the vectors (Clear Bars) and transgene expression (Black Bars) plotted as relative light units (RLU) in Vero cells at 2 days.

vectors deleted just the single copy of ICP0 present within the internal joint sequences, leaving an intact copy of ICP0 within the terminal repeat at the far left-hand end of the viral genome (JDTOZ in Fig. 5.5). This vector showed further reduced toxicity ($\sim$79% of control) while providing robust transgene expression. In an effort to further reduce vector-associated toxicity, the final copy of ICP0 was deleted from the fourth generation vector to provide a replication defective vector devoid of all the four IE regulatory gene products. This vector (JDQ0Z in Fig. 5.5) displayed no measurable toxicity differences from uninfected cells, yet the loss of the second copy of the ICP0 gene product resulted in a tremendous loss in transgene expression. This result was similar to what has been reported in other studies where all the IE regulatory genes were deleted.[34] Some ICP0 expression is required to enable transgene expression, however, ICP0 has some inherent cytotoxicity that may be tolerated depending on the target cell type. Overall, these experiments underscore the role of ICP0 in maintaining a transcriptionally active genome and suggest that insight into the mechanism of virus genome silencing will be essential to providing a vector capable of transgene expression in the total absence of vector toxicity.

4. Gene Transfer/Therapy Applications

Due to the ability of HSV to readily transduce neurons of the PNS as part of its natural life cycle, the virus becomes the intuitive choice for treating diseases of the PNS. Replication defective vectors deleted for single or multiple IE genes have been used in animal models of various (i) peripheral neuropathies caused by diabetes[35] and chemotherapeutics[36] via delivery of neurotrophic factors, (ii) pain[37] with the delivery of enkephalin or GAD, and (iii) also the delivery of neurotrophins for erectile dysfunction.[38] Vector mediated therapeutic gene expression either prolonged neuronal survival measured by neuroanatomical methods or altered the behavior of the animals. Transgene expression could be targeted to specific neuronal populations by simply selecting the site of vector inoculation.

HSV has been employed as a vector to treat cancer, primarily those cancers of the CNS such as GBM in which both replication-defective vectors expressing HSV-TK and other anti-tumor agents,[39] and the conditionally replicating oncolytic HSV vectors have shown promise[21] In fact, numerous HSV oncolytic vectors have been used in phase-I human clinical trials and have displayed excellent safety profiles and have in some patients shown efficacy.[40,41] In addition to the above-mentioned PNS and CNS disorders,

HSV replication defective vectors have been employed in approaches to treat muscle disease, arthritis, and have also been used as vaccine vectors.

5. Immunology

A major issue in any gene transfer approach concerns the immune response to the virus and/or transgene. One needs to determine the effects of the host innate and adaptive immune response to the virus vector as 50–80% of the human population is seropositive for HSV. The response by NK cells, macrophages, CD8 and CD4 cells, as well as the presence of virus neutralizing antibodies determines the extent of initial virus infection and spread as well as the ability of the host immune system to keep the virus in a latent state. This plays an important role in cases where the approach may require multiple or repetitive dosing to achieve a therapeutic response.

The host immune response to replication-defective genomic and oncolytic HSV vectors has been studied in great detail. Different replication competent oncolytic vectors have been examined.[42,43] In all instances except one,[43] there was no observable difference in transduction, vector-mediated gene expression or when measured therapeutic efficacy in pre-immune animals versus naïve animals following administration of one of the above oncolytic HSV vectors. Moreover, while increased levels of neutralizing Abs, NK cells, macrophages and T-cells in the sero-positive animals did not result in reduced transduction at the site of vector injection, virus dissemination to off-target sites was limited by this host response.[42] Similar results were seen using a replication-defective HSV vector,[44] where virus transduction was limited to the site of injection.

One indication that HSV vectors may be able to be re-dosed is that multiple injections of an oncolytic vector performed over a period of days to weeks did not lead to reduced transduction of animals, but rather resulted in increased efficacy in pre-immune mice.[42] Similar results have also been seen in clinical trials[40] where multiple injections of vector into HSV sero-positive patients did not show increased toxicity but seemed to yield improved tumor transduction and efficacy. Finally, re-dosing has also been achieved using replication-defective HSV vectors expressing different and diverse therapeutic transgenes.[37]

6. Safety and Regulatory Issues

Three main issues exist for the use of HSV vectors in human clinical trials. Firstly, will virus introduced into specific tissues or organs spread within that tissue, or spread to other tissues within the body? Second, is the issue of the immune status of a patient in relation to the potential inflammatory response to the vector or vector-mediated expression of the transgene. Third, a major concern exists that the introduction of the HSV vector will cause reactivation of a resident latent virus in sero-positive patients that could lead to recombination event between the vector genome and the genome of the reactivated wild-type virus.

To address the first issue, toxicology and bio-distribution studies have been performed using conditionally replicating oncolytic and replication defective HSV vectors in both rodent and non-human primate animal models.[44,46] No vector-mediated toxicity was observed and there was no evidence of vector spread past the injection site either using qPCR for viral genomes or HSV-specific IHC staining. As final support for the safety of HSV vectors, multiple phase-I and I/II studies using oncolytic HSV vectors demonstrated no evidence of viral encephalitis or reactivation of wild-type virus.[40,41] Furthermore, no adverse events could be unequivocally attributed to the vectors. Finally the ability to reactivate latent endogenous virus was not observed in phase-I human clinical trials with replication competent oncolytic viruses.[41] Moreover, pre-clinical studies in rodents and primates pre-infected with wild-type HSV to establish latent infections did not detect reactivation by super-infection with the oncolytic vectors.[45]

7. Summary

Considerable progress has been made in developing safe, non-toxic HSV vectors to express therapeutic gene products that can be tailored to a variety of applications, primarily to those affecting the nervous system such as neuropathy, pain and cancer. In addition, methodologies to produce and purify these vectors have enabled their production to high titers under cGMP conditions as required by the FDA. Initial phase-I trials using these vectors have shown an excellent safety profile, yet overall efficacy may require other improvements to the vector delivery system as a whole to improve the clinical outcome.

References

1. Fatahzadeh M, Schwartz RA (2007). Human herpes simplex virus infections: epidemiology, pathogenesis, symptomatology, diagnosis, and management. *J Am Acad Dermatol.* **57**: 737–763.
2. Stevens JG (1989). Human herpesviruses: a consideration of the latent state. *Microbiol. Rev.* **53**: 318–332.
3. Herrera FJ, Triezenberg SJ (2004). VP16-dependent association of chromatin-modifying coactivators and underrepresentation of histones at immediate-early gene promoters during herpes simplex virus infection. *J Virol.* **78**: 9689–9696.
4. Esclatine A, Taddeo B, Evans L, Roizman B (2004). The herpes simplex virus 1 UL41 gene-dependent destabilization of cellular RNAs is selective and may be sequence-specific. *Proc Natl Acad Sci USA.* **101**: 3603–3608.
5. Campadelli-Fiume G, Amasio M, Avitabile E, Cerretani A, Forghieri C, Gianni T, *et al.* (2007). The multipartite system that mediates entry of herpes simplex virus into the cell. *Rev Med Virol.* **17**: 313–326.
6. Laquerre S, Argnani R, Anderson DB, Zucchini S, Manservigi R, Glorioso JC (1998). Heparan sulfate proteoglycan binding by herpes simplex virus type 1 glycoproteins B and C, which differ in their contributions to virus attachment, penetration, and cell-to-cell spread. *J Virol.* **72**: 6119–6130.
7. Lazear E, Carfi A, Whitbeck JC, Cairns TM, Krummenacher C, Cohen GH, *et al.* (2008). Engineered disulfide bonds in herpes simplex virus type 1 gD separate receptor binding from fusion initiation and viral entry. *J Virol.* **82**: 700–709.
8. Nicola AV, Straus SE (2004). Cellular and viral requirements for rapid endocytic entry of herpes simplex virus. *J Virol.* **78**: 7508–7517.
9. Montgomery RI, Warner MS, Lum BJ, Spear PG (1996). Herpes simplex virus 1 entry into cells mediated by a novel member of the TNF/NGF receptor family. *Cell.* **87**: 427–436.
10. Geraghty RJ, Krummenacher C, Cohen GH, Eisenberg RJ, Spear PG (1998). Entry of alphaherpesviruses mediated by poliovirus receptor-related protein 1 and poliovirus receptor. *Science.* **280**: 1618–1620.
11. Bender FC, Samanta M, Heldwein EE, de Leon MP, Bilman E, Lou H, *et al.* (2007). Antigenic and mutational analyses of herpes simplex virus glycoprotein B reveal four functional regions. *J Virol.* **81**: 3827–3841.
12. Galdiero S, Falanga A, Vitiello M, Browne H, Pedone C, Galdiero M (2005). Fusogenic domains in herpes simplex virus type 1 glycoprotein H. *J Biol Chem.* **280**: 28632–28643.
13. Simpson SA, Manchak MD, Hager EJ, Krummenacher C, Whitbeck JC, Levin MJ, *et al.* (2005). Nectin-1/HveC Mediates herpes simplex virus type 1 entry into primary human sensory neurons and fibroblasts. *J Neurovirol.* **11**: 208–218.
14. Menotti L, Cerretani A, Hengel H, Campadelli-Fiume G (2008). Construction of a fully retargeted herpes simplex virus 1 recombinant capable of entering cells solely via human epidermal growth factor receptor 2. *J Virol.* **82**: 10153–10161.

15. Zhou G, Roizman B (2006). Construction and properties of a herpes simplex virus 1 designed to enter cells solely via the IL-13alpha2 receptor. *Proc Natl Acad Sci U S A.* **103**: 5508–5513.

16. Nakano K, Asano R, Tsumoto K, Kwon H, Goins WF, Kumagai I, *et al.* (2005). Herpes simplex virus targeting to the EGF receptor by a gD-specific soluble bridging molecule. *Mol Ther.* **11**: 617–626.

17. Markert JM, Parker JN, Buchsbaum DJ, Grizzle WE, Gillespie GY, Whitley RJ (2006). Oncolytic HSV-1 for the treatment of brain tumours. *Herpes.* **13**: 66–71.

18. Krisky DM, Wolfe D, Goins WF, Marconi PC, Ramakrishnan R, Mata M, *et al.* (1998). Deletion of multiple immediate-early genes from herpes simplex virus reduces cytotoxicity and permits long-term gene expression in neurons. *Gene Ther.* **5**: 1593–1603.

19. Oehmig A, Fraefel C, Breakefield XO, Ackermann M (2004). Herpes simplex virus type 1 amplicons and their hybrid virus partners, EBV, AAV, and retrovirus. *Curr Gene Ther.* **4**: 385–408.

20. Sciortino MT, Taddeo B, Giuffre-Cuculletto M, Medici MA, Mastino A, Roizman B (2007). Replication-competent herpes simplex virus 1 isolates selected from cells transfected with a bacterial artificial chromosome DNA lacking only the UL49 gene vary with respect to the defect in the UL41 gene encoding host shutoff RNase. *J Virol.* **81**: 10924–10932.

21. Currier MA, Gillespie RA, Sawtell NM, Mahller YY, Stroup G, Collins MH, *et al.* (2008). Efficacy and safety of the oncolytic herpes simplex virus rRp450 alone and combined with cyclophosphamide. *Mol Ther.* **16**: 879–885.

22. Wolfe D, Mata M, Fink DJ (2009). A human trial of HSV-mediated gene transfer for the treatment of chronic pain. *Gene Ther.* **16**: 455–460.

23. DeLuca NA, McCarthy AM, Schaffer PA (1985a). Isolation and characterization of deletion mutants of herpes simplex virus type 1 in the gene encoding immediate-early regulatory protein ICP4. *J. Virol.* **56**: 558–570.

24. Sandri-Goldin RM (2008). The many roles of the regulatory protein ICP27 during herpes simplex virus infection. *Front Biosci.* **13**: 5241–5256.

25. Everett RD, Parsy ML, Orr A (2009). Analysis of the functions of herpes simplex virus type 1 regulatory protein ICP0 that are critical for lytic infection and de-repression of quiescent genomes. *J Virol.* **83**: 4963–4977.

26. Bastian TW, Rice SA (2009). Identification of sequences in herpes simplex virus type 1 ICP22 that influence RNA polymerase II modification and viral late gene expression. *J Virol.* **83**: 128–139.

27. Jugovic P, Hill AM, Tomazin R, Ploegh H, Johnson DC (1998). Inhibition of major histocompatibility complex class I antigen presentation in pig and primate cells by herpes simplex virus type 1 and 2 ICP47. *J Virol.* **72**: 5076–5084.

28. Chen X, Li J, Mata M, Goss J, Wolfe D, Glorioso JC, *et al.* (2000). Herpes simplex virus type 1 ICP0 protein does not accumulate in the nucleus of primary neurons in culture. *J Virol.* **74**: 10132–10141.

29. Spivack J, Fraser N (1988). Expression of herpes simplex virus type 1 latency-associated transcripts in trigeminal ganglia of mice during acute infection and reactivation of latent infection. *J. Virol.* **62**: 1479–1485.

30. Deshmane SL, Fraser NW (1989). During latency, herpes simplex virus type 1 DNA is associated with nucleosomes in a chromatin structure. *J. Virol.* **63**: 943–947.

31. Dobson AT, Sederati F, Devi-Rao G, Flanagan WM, Farrell MJ, Stevens JG, *et al.* (1989). Identification of the latency-associated transcript promoter by expression of rabbit β-globin mRNA in mouse sensory nerve ganglia latently infected with a recombinant herpes simplex virus. *J. Virol.* **63**: 3844–3851.

32. Goins WF, Sternberg LR, Croen KD, Krause PR, Hendricks RL, Fink DJ, *et al.* (1994). A novel latency-active promoter is contained within the herpes simplex virus type 1 U_L flanking repeats. *J. Virol.* **68**: 2239–2252.

33. Perez MC, Hunt SP, Coffin RS, Palmer JA (2004). Comparative analysis of genomic HSV vectors for gene delivery to motor neurons following peripheral inoculation *in vivo*. *Gene Ther.* **11**: 1023–1032.

34. Samaniego LA, Neiderhiser L, DeLuca NA (1998). Persistence and expression of the herpes simplex virus genome in the absence of immediate-early proteins. *J Virol.* **72**: 3307–3320.

35. Goss JR, Goins WF, Lacomis D, Mata M, Glorioso JC, Fink DJ (2002). Herpes simplex-mediated gene transfer of nerve growth factor protects against peripheral neuropathy in streptozotocin-induced diabetes in the mouse. *Diabetes.* **51**: 2227–2232.

36. Chattopadhyay M, Goss J, Wolfe D, Goins WC, Huang S, Glorioso JC, *et al.* (2004). Protective effect of herpes simplex virus-mediated neurotrophin gene transfer in cisplatin neuropathy. *Brain.* **127**: 929–939.

37. Goss JR, Mata M, Goins WF, Wu HH, Glorioso JC, Fink DJ (2001). Antinociceptive effect of a genomic herpes simplex virus-based vector expressing human proenkephalin in rat dorsal root ganglion. *Gene Ther.* **8**: 551–556.

38. Kato R, Wolfe D, Coyle CH, Huang S, Wechuck JB, Goins WF, *et al.* (2007). Herpes simplex virus vector-mediated delivery of glial cell line-derived neurotrophic factor rescues erectile dysfunction following cavernous nerve injury. *Gene Ther.* **14**: 1344–1352.

39. Niranjan A, Wolfe D, Tamura M, Soares MK, Krisky DM, Lunsford LD, *et al.* (2003). Treatment of rat gliosarcoma brain tumors by HSV-based multigene therapy combined with radiosurgery. *Mol Ther.* **8**: 530–542.

40. Hu JC, Coffin RS, Davis CJ, Graham NJ, Groves N, Guest PJ, *et al.* (2006). A phase I study of OncoVEXGM-CSF, a second-generation oncolytic herpes simplex virus expressing granulocyte macrophage colony-stimulating factor. *Clin Cancer Res.* **12**: 6737–6747.

41. Markert JM, Liechty PG, Wang W, Gaston S, Braz E, Karrasch M, *et al.* (2009). Phase Ib trial of mutant herpes simplex virus G207 inoculated pre-and post-tumor resection for recurrent GBM. *Mol Ther.* **17**: 199–207.

42. Lambright ES, Kang EH, Force S, Lanuti M, Caparrelli D, Kaiser LR, *et al.* (2000). Effect of preexisting anti-herpes immunity on the efficacy of herpes simplex viral therapy in a murine intraperitoneal tumor model. *Mol Ther.* **2**: 387–393.

43. Herrlinger U, Kramm CM, Aboody-Guterman KS, Silver JS, Ikeda K, Johnston KM, *et al.* (1998). Pre-existing herpes simplex virus 1 (HSV-1) immunity decreases, but does not abolish, gene transfer to experimental brain tumors by a HSV-1 vector. *Gene Ther.* **5**: 809–819.

44. Wolfe D, Niranjan A, Trichel A, Wiley C, Ozuer A, Kanal E, *et al.* (2004). Safety and biodistribution studies of an HSV multigene vector following intracranial delivery to non-human primates. *Gene Ther.* **11**: 1675–1684.

45. Sundaresan P, Hunter WD, Martuza RL, Rabkin SD (2000). Attenuated, replication-competent herpes simplex virus type 1 mutant G207: safety evaluation in mice. *J Virol.* **74**: 3832-3841.

Chapter 6
Adeno-Associated Viral (AAV) Vectors

Nicholas Muzyczka*

Adeno-associated Virus (AAV) is a non-pathogenic virus with wide tissue and species tropism. AAV vectors have proven to be safe and efficient for gene transfer to non-dividing cells, can generate long term gene expression in a variety of animal models and can accommodate a variety of tissue specific and inducible promoter elements. Progress in understanding the biology of the virus has helped in the development of efficient and scalable production methods, and the simplicity of the viral structure suggests that it might be possible to design targeted vectors. Recent clinical trials also suggest that AAV will be useful in treating human disease.

1. Introduction

AAV is a parvovirus, which are small, non-enveloped viruses that contain a linear, single-stranded DNA (ssDNA) that is about 5 kb.[1] AAVs have been isolated from a wide variety of vertebrate species, including humans, monkeys, cows, horses, birds, and sheep. Over a hundred variants have been isolated from primate and human tissues and a number of these have been characterized for gene therapy. Most humans ($\sim$80%) are positive for AAV antibodies, but no human disease has been associated with AAV infection. The lack of disease association is one of the major safety features of AAV vectors. Over the last 10 years recombinant AAV (rAAV) vectors have become widely used as a delivery vehicle for modeling the treatment of various human diseases. More recently, rAAV has shown partial success

*Correspondence: Powell Gene Therapy Center, University of Florida College of Medicine.
E-mail: muzyczka@mgm.ufl.edu

in clinical trials[2,3] and promises to be an important therapeutic tool. This chapter will briefly summarize the biology of AAV, the vector technologies used to produce and purify rAAV, the characteristics of the vector system and recent attempts to develop the next generation of rAAV vectors that incorporate tissue targeting.

2. Biology of AAV

AAV requires a helper virus for productive infection.[1] Because AAV is often found as a contaminant in human adenovirus isolates, it is believed that human adenovirus (Ad) is the natural AAV helper. However, herpes simplex virus (HSV) can also provide complete helper function. In the absence of a helper virus, it is believed that AAV can produce a persistent infection, which can be rescued when cells are infected with a helper virus. In cell culture, AAV has been shown to integrate preferentially into a specific human chromosomal location, 19q2, and it has been suggested that this may be a mechanism for AAV persistence. More recent evidence, however, suggests that wild type AAV persists in humans as an episome.[4]

The linear AAV genome (Fig. 6.1) contains two inverted terminal repeats (ITRs) that flank two open reading frames, *rep* and *cap*[1] Three promoters, p5, p19, and p40 initiate spliced and unspliced mRNAs that code for two large Rep proteins, Rep78 and 68, two short Reps, Rep52 and 40, and three

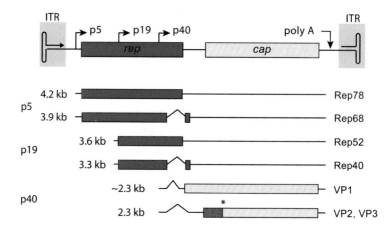

Fig. 6.1 AAV genetic map. Only the most abundant mRNA species are shown.

capsid proteins, VP1, 2, and 3. The two shorter capsid proteins, VP2 and VP3 are encoded by the same mRNA. VP2 (gray + yellow) is initiated from an upstream ACG codon, while VP3 is initiated from the first ATG codon (*). VP1 is coded by an alternatively spliced mRNA that uses an alternative upstream splice acceptor site. The relative efficiencies of initiating translation from the two initiation codons in the VP2/3 mRNA and the relative efficiency of the splice acceptor site usage accounts for the fact that the VP3:VP2:VP1 ratio is 10:1:1. The Rep proteins are synthesized from two promoters, p5 and p19 and their mRNAs use the same splice donor and acceptor sites as the capsid mRNA, but are spliced less often than the capsid mRNA. This accounts for the relative ratio of unspliced to spliced Rep proteins (about 2:1). The p19 promoter is approximately 5 times stronger than the p5 promoter and this accounts for the relative amounts of Rep52/40 to Rep78/68 (5:1). All of the Rep and capsid proteins contain overlapping amino acid sequences in common from their respective open reading frames. The ITRs are 145 bases long. They contain two short palindromes flanked by a long palindrome, and an additional non-palindromic sequence that is repeated at both ends. The ITRs are the only AAV sequences required *in cis* for the production of rAAV vectors; they are the origins for DNA replication, packaging signals, and chromosome maintenance signals.[5] The ITRs are also Rep dependent enhancers for AAV transcription from all three promoters, and in some tissues have been shown to have weak promoter activity by themselves.

The helper functions provided by Ad have been studied extensively; they are the E1a, E1b 55K and 19K proteins, E4 orf 6, E2a DNA binding protein (DBP), and VA RNA genes.[1] Most of these genes by themselves can provide partial helper function that produces semipermissive conditions for AAV replication, but completely permissive conditions require all of these genes. The effect of these genes is to induce transcription from the AAV p5 promoter (E1a and E2a), induce cellular S phase (E1a), inhibit apoptosis (E1b, E4), inhibit mitosis, promote viral mRNA transport to the cytoplasm (E1b, E4), inhibit the PKR antiviral response (VA), and inhibit non-homologous end joining of AAV DNA (E1b, E4).[6] The initial effect of the Ad helper functions appears to be to push a quiescent cell into S phase, freeze it in S phase by preventing mitosis, and prevent apoptosis, thereby making cellular replication enzymes available for AAV DNA synthesis.

When Ad is not present, Rep represses its own p5 promoter. Because Rep is required for activating the other two AAV promoters, there is no detectable wild-type AAV gene expression in the absence of Ad E1a genes. Ad E1a turns on host S phase and activates p5 via YY1 and MLTF sites upstream of p5; Rep then turns on p19 and p40. This results in a 300 fold increase in AAV transcription once Ad genes are expressed.[7] Autorepression of the p5 promoter by Rep protein and derepression by Ad E1 also explains why attempts to make rAAV producer cells in some established cell lines have failed. Integrated AAV genomes are stable in HeLa and KB cells, but not 293 cells because expression of E1 genes in 293 cells presumably activates Rep expression, thereby initiating AAV replication.

AAV relies primarily on cellular replication factors for DNA synthesis in the presence of Ad, and *in vitro* reconstruction of AAV DNA replication has identified a minimum set of protein complexes;[8] these are pol δ, RFC, PCNA, MCM. All of these proteins are highly conserved enzymes in eukaryotes and this may account for the fact that rAAV can be produced equally well in human and insect cells using Ad, herpes or baculovirus as helper viruses.[9–11]

The minimum helper functions for herpes infected cells have not been completely characterized. Both the HSV DBP, and the helicase/primase complex, are necessary. In addition, the herpes DNA polymerase and early genes (ICP0, ICP4, ICP22) appear to provide partial helper function under some conditions.[12] Like Ad, HSV pushes quiescent cells into S phase, suggesting that both helper viruses may provide genes that create the same permissive environment for AAV. To date, nothing is known about potential baculovirus helper functions.

AAV replicates by a modified rolling circle mechanism,[1] whose basic elements are illustrated in Fig. 6.2. The 3′OH of the input ssDNA forms a primer for the synthesis of the complementary strand. In a process called terminal resolution, the covalently closed hairpin of the monomer turnaround (mT) genome is then cleaved at a unique position (*trs*, terminal resolution site) so as to form a new 3′ primer, which is used to repair the end to a normal duplex configuration (mE). The ITR of this intermediate is then reconfigured to a double hairpin (reinitiation) and the 3′OH primer is used to displace a ssDNA genome and new mT form which again undergoes terminal resolution in the next cycle. A new ssDNA progeny molecule is

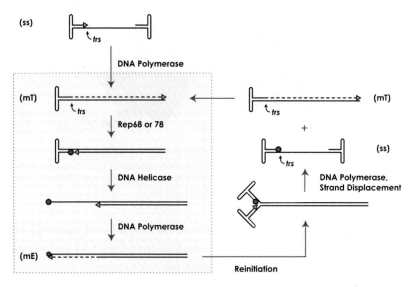

Fig. 6.2 AAV DNA replication. The yellow box isolates the steps in terminal resolution, during which the covalently joined hairpin end (mT, monomer turnaround) is resolved to a normal duplex end (mE, monomer extended). When Rep nicks at the *trs*, it forms a covalent linkage with the 5′OH end of the DNA (red circle). See text for more details.

made during each cycle and packaged. To accomplish terminal resolution, Rep78 or 68 bind specifically to a 22 bp imperfect GAGC repeat within the ITR called the Rep binding element (RBE). The helicase activity of Rep then unwinds the nearby *trs* sequence and Rep cleaves on the appropriate strand forming a covalent phosphotyrosine linkage with the 5′OH end of the nick.[13,14] An accessory element called RBE' stimulates the Rep helicase activity when Rep is bound to a hairpin as opposed to a linear ITR, thereby providing specificity for nicking the hairpin.[14] Although the RBE and *trs* sequences within the ITR are optimal for nicking, degenerate sequences that are present in the p5 promoter and in human chromosome 19 are also recognized and cleaved albeit at a reduced level. This accounts for the fact that ITR negative genomes that contain an RBE and a *trs* with the appropriate spacing between them have been observed to undergo limited DNA amplification when Rep is expressed and can be packaged into AAV capsids.[15]

The AAV capsid is one of the simplest found in nature (Fig. 6.3). Sixty polypeptides consisting of VP1, VP2 and VP3 in a 1:1:10 ratio form a $T = 1$

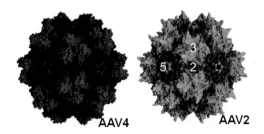

Fig. 6.3 AAV capsid. The crystal structures of AAV4 and AAV2 are shown to 3.2 Å resolution. Both capsids are centered on a 2 fold axis flanked by 5 fold pores horizontally and 3 fold axes vertically. Color intensity indicates the height of the surface. Note that although AAV4 and AAV2 are only 55% identical in amino acid sequence, they have similar core structures.

icosahedral capsid. It is generally believed that capsids are assembled first as empty particles and that Rep 52 is used to drive the 3′OH end of newly synthesized DNA into the capsid through the 5 fold pore.[16]

Although it was long believed that the ITR contained a packaging signal, studies of ITR negative AAV genome amplification suggest that ITR negative molecules can also be packaged provided they have an RBE and *trs*.[15] This suggests that the Rep78/68 protein, which is covalently attached to the newly synthesized ssDNA progeny molecule, may be the packaging signal. Consistent with this, Rep78 and 68 form complexes with capsid assembly intermediates and with newly packaged full capsids. The full capsids have an external Rep protein that is still covalently attached to ssDNA and is accessible to both proteases and DNAse. The capsids are extraordinarily stable to pH ($<$pH4), heat (65°, 1 hr), detergents and proteases. X-ray crystal structures of several different serotypes have shown that essentially the same residues are used at the 2-, 3-, and 5- fold axes to provide capsid integrity.[17]

AAV uses at least two cell surface receptors, one for binding to the cell surface, usually a glycan (heparan sulfate or sialic acid), and a second receptor that promotes clathrin mediated endocytosis,[1,18] (e.g., fibroblast growth factor receptor). To date, only the AAV2 heparan sulfate binding site has been mapped on the capsid surface; it consists of a minimum of 5 amino acids in a basic patch at the 3 fold symmetry axis.[1] Once the virus binds to the cell surface, activation of an endosomal uptake receptor triggers clathrin mediated uptake into an early endosome.[18] Receptor binding, uptake or subsequent pH changes in the early endosome appear to trigger conformational changes in the capsid surface that lead to the extrusion of the N terminal regions of VP1 and VP2.[19] These contain nuclear localization signals (NLS) and the phospholipase A2 activity (PLA) that is unique to VP1. Genetic studies have shown that both the PLA and NLS are necessary for efficient infection.[19,20] The role of the phospholipase domain is not clear; it may be involved in release of capsids from endosomes or the subsequent entry of capsids into the nucleus. Evidence exists for immediate release of virus from the early endosome, as well as trafficking to recycling endosomes, lysosomes and Golgi.[18]

Proteosome inhibitors generally increase transduction, although to different levels in different cell types, suggesting that virus released into cytoplasm may be functional for transduction.[18] This is consistent with experiments in which injection of neutralizing antibodies into cells inhibits transduction.[19] Recently, phosphorylation of tyrosines on the capsid surface has been implicated in AAV trafficking and transduction.[21] Finally, there is also no definitive agreement on whether intact virus particles enter the nucleus prior to uncoating of the nucleic acid.

3. Vector Technology

Long term, persistent expression from an AAV vector was first achieved in tissue culture cells by substituting a neomycin cassette driven by a heterologous SV40 promoter in place of the AAV capsid gene.[10] The recombinant was packaged by a plasmid cotransfection system in which the missing capsid gene was supplied *in trans*. Subsequently, it was shown that all of the AAV coding regions could be removed and supplied *in trans*, only the 145 bp terminal repeat sequences were essential *in cis* for vector production.[5] Little attention was paid to AAV vectors until two groups demonstrated that efficient transduction of mouse muscle tissue produced robust expression of the transgene that persisted for up to a year with no apparent decrease in expression.[22,23] These early reports were followed by demonstrations that rAAV could be used to efficiently transduce the eye, brain, and liver.[24] AAV transduction was unique among viral vectors in that transduction was efficient, and appeared to persist unabated for the lifetime of the animal. The longest uninterrupted expression engineered with AAV vectors thus far has been in the eye, where expression of the rpe65 gene has been demonstrated in the deficient dog model for greater than 8 years with no loss in the level of expression. This has made AAV the vector of choice for achieving long term, persistent expression in animal models.

Interest in AAV vectors continued to lag until several groups solved the problem of making helper free rAAV.[25,26] Small amounts of research grade vectors are typically made by cotransfecting two or three plasmids into human tissue culture cells (Fig. 6.4). The process takes advantage of the discovery that AAV and rAAV plasmids are infectious.[27] Following transfection into cells, the AAV ITR-containing genome is rescued from the plasmid and undergoes standard AAV DNA replication. The rep and cap gene products as well as the Ad helper function genes are supplied on one or two additional plasmids (Fig. 6.3) that are co-transfected.[25,26] This method generally produces approximately 10^3 mature viral particles per cell, much lower than the yield from a wild type AAV infection, 10^5 particles per cell. However, it requires the construction of only one rAAV plasmid containing the transgene and is easily achieved in most laboratories. The method allows the rapid screening of transgene mutants as well as tissue specific promoters that might be appropriate.

For clinical and commercial applications, several groups have developed scalable methods for large scale virus production. These methods generate much better yields of virus ($>10^5$ vector genomes per cell). They also rely on virus infection rather than DNA transfection, and therefore, can be scaled to large fermentor or bioreactor preparations using defined media. A variety of methods have been described but the two that currently appear to be the most robust use baculovirus or HSV as the helper viruses.[11,28−30] When two HSVs, one expressing rep and cap, and the other carrying the rAAV, are coinfected in human cells, the yield of rAAV is similar to that seen in wild-type

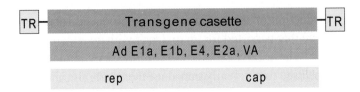

Fig. 6.4 rAAV production. The most commonly used method for producing rAAV vectors that are free of helper virus is to transfect three plasmids that contain (1) the transgene cassette flanked by the AAV ITRs, (2) a plasmid that expresses the Ad helper genes, and (3) a plasmid that expresses the AAV rep and cap genes but is lacking the ITRs. Often the second and third plasmids are combined into one.

AAV infections. To minimize the production of contaminating HSV virus, a replication defective HSV is used. The use of baculovirus vectors in insect cells is equally efficient for rAAV production. The development of these scalable production methods should make clinical therapeutic applications possible in the future.

Several methods have been used for rAAV purification, including CsCl and iodixanol density gradient purification, ion exchange chromatography, and affinity chromatography.[31,32] In addition, large scale production techniques have been developed that use only column chromatography for virus purification, thereby eliminating inefficient centrifugation methods.[33] Finally, the development of accurate and reproducible methods for determining the titer of full and empty particles and the relative infectivity of vector preparations has been an equally important aspect of advances in AAV vector technology, and this is discussed extensively in a recent review.[34]

Until recently only a limited number of AAV serotypes had been identified. However, Gao *et al.* and others have now isolated over a hundred different serotypes from human, primate and other mammalian species.[35] AAV serotypes do not appear to be species specific, so all of these serotypes appear to be capable of transducing rodents, primates and other mammalian species. Currently, serotypes 1-11 are routinely available and experiments from many laboratories have demonstrated that each serotype has a unique tissue tropism when injected into animals. The different serotype tropisms are presumably due in part to the fact that each serotype capsid uses a different combination of cell surface receptors. However, it is also clear the different serotypes may use different trafficking strategies or have different rates of uncoating.[18,36] Fortunately, it is easy to screen multiple serotypes using the same transduction cassette. Typically, a vector construct containing AAV2 ITRs can be packaged into capsids from serotypes 1-10 by substituting the appropriate capsid in trans.[32]

The most severe limitation on AAV vectors is the packaging size. Typically, only 5 kilobases can be packaged into a rAAV capsid. Two strategies have been proposed for increasing the packaging limit. The first takes advantage of the splicing machinery and the fact that AAV DNA typically forms head to tail concatemers during transduction.[37] Thus, a gene can be interrupted in an intron and the two halves (one carrying a splice donor and a portion of the intron and the other carrying a splice acceptor, and the remainder of the intron can be packaged in separate vectors. When the two vectors are simultaneously used to transduce cells, the resulting concatemers can synthesize an intact mRNA that contains an AAV ITR junction within the intron. Splicing of the intron produces a viable mRNA and removes the ITR sequence. The second strategy is similar but incorporates an intein, a protein sequence that self splices or ligates two polypeptides after they have been synthesized.[38] It is not yet clear whether either of these split gene–two vector approaches will provide a viable solution for delivery of genes that are too large to package into standard AAV capsids.

4. Vector Characteristics *In Vivo*

When first developed it was expected that AAV vector would integrate into host chromosomes and that this was the mechanism for achieving persistent expression. In early cell culture experiments, rAAV was found to be integrated randomly throughout host chromosomes.[5,39] Flotte *et al.* suggested that this might not be the case *in vivo* by showing that rAAV carrying the cystic fibrosis gene persisted in rabbit airway epithelial cells as a monomer linear episome.[40] Subsequently, several groups have demonstrated that rAAV persists as an extrachromosomal piece of DNA.[4,41,42] This in part explains why the long term transduction frequency in rapidly dividing hematopoietic cells tends to be low, in contrast to what is seen in organs containing non-dividing cells, such as brain, eye, liver and muscle. Studies in mouse muscle could find no evidence of integration at a sensitivity that would have detected 1 integrated vector in 200 that persisted in cells.[42] Similar studies in mouse liver also revealed that most of the rAAV genomes were extrachromosomal with only a low level of integration ($<10\%$).

Although it is now reasonably well established that rAAV genomes persist as episomes, the reason for the persistent expression of AAV vectors, compared to non-integrating Ad vectors, herpes vectors, plasmid DNA transduction vectors, and integrating retroviral vectors is not clear. The fact that rAAV does not integrate may spare it from a variety of host epigenetic mechanisms, such as DNA methylation and histone acetylation, that have been implicated in shutoff of other viral vectors. An alternative possibility is that the AAV ITR may have signals that specifically prevent elimination or silencing of the AAV genome.

AAV vectors display low immunogenicity.[43] AAV vectors do not activate key components of innate immunity, including toll-like receptors, type 1 IFN responses or other cytokines. AAV capsids also blunt activation of the complement cascade. In addition, because AAV vectors are completely deleted for viral genes, the frequency of generating a cytotoxic T lymphocyte (CTL) response to rAAV-expressed transgenes is reduced. This is in part due to the fact that antigen presenting cells are less prone to be infected by some AAV serotypes. There is also evidence that rAAV targeted to the liver can induce tolerance to transgenes, which is apparently mediated by regulatory T cells. As expected, infection with rAAV produces circulating neutralizing antibodies to the AAV capsid, which prevent gene expression when the same serotype is re-administered. Even low neutralizing titers can prevent transduction. In contrast, some serotypes appear to be capable of re-administration in partially immuno-protected organs, such as the eye and brain with no evidence of inflammatory response or loss of expression on second administration.

Because most animal models have not been exposed to human and primate AAV serotypes, there are neither neutralizing antibodies nor CTLs to affect the initial application of vector. Most humans, however, have experienced a prior exposure to AAV, most likely along with an Ad infection, leading to the development of both circulating antibodies and a CTL response to AAV capsid. This is believed to have led to the CTL induced loss of gene expression in a clinical trial to restore expression of Factor IX.[43] In contrast, a recent clinical trial for Alpha-1-antitrypsin produced long term expression of the transgene inspite of inducing both circulating antibodies and a CTL response to capsid protein.[3] It is therefore, still not clear whether the previous exposure that most individuals have had to AAV will make it difficult to use rAAV for human gene therapy.

Most AAV vectors display a slow onset of gene expression *in vivo* that takes 1-3 months to reach the maximum level, after which expression plateaus and remains constant. Two explanations have been proposed for this behavior. rAAV vectors can persist as intact particles for extended periods of time in infected cells,[36] implying that uncoating is inherently slow. In addition, once the nucleic acid material is released, the single stranded DNA must synthesize the complementary strand before gene expression can occur. This most likely occurs by *de novo* synthesis of the second strand.

Srivastava and colleagues demonstrated that a phosphorylated cellular protein, FKBP52, was an inhibitor of second strand synthesis.[44] This protein bound to the newly exposed ssDNA just downstream of the *trs*.

To avoid the problem of second strand synthesis McCarty *et al.* have developed a method for packaging an rAAV genome that consists of a full length duplex DNA in which one end is covalently joined, i.e., a full length inverted repeat.[45] The disadvantage of this method is that half of the packaging capacity of the rAAV capsid is lost; only cassettes that are 50% of AAV genome size can be packaged. However, once the vector genome is uncoated within the cell, it immediately "snaps back" on itself to produce a duplex template for transcription. This produces dramatic increases in the efficiency of transduction (up to 100 fold) and eliminates the delay in onset of gene expression.

The AAV ITR does not have a typical TATA like promoter element and in most applications no expression is seen unless a promoter is included in the transgene cassette. A wide variety of promoter and enhancer sequences have been used successfully in rAAV. In general, tissue specific promoters have behaved as expected and provided the same tropism that is seen in transgenic mouse experiments. Although the packaging capacity of AAV puts a serious constraint on the size of the promoter that can be used, some small tissue specific promoters have been developed that provide cell type specificity. The CMV enhancer/promoter and a CMV-chicken β actin (CBA) hybrid promoter have been the most widely used constitutive promoters. Of these the CBA promoter is the strongest promoter reported and has not been shown to shut off in any tissue to date. Several groups have also demonstrated the *in vivo* use of inducible promoters in an rAAV background.

5. Next Generation Vectors

All rAAV serotypes are promiscuous; they infect multiple organs and tissues when vector is delivered systemically. Because the virus has a simple capsid structure, a number of groups have explored the possibility of targeting AAV to specific tissues. Two general strategies have been followed. The first uses a directed approach in which novel cell surface ligands are genetically inserted into the capsid sequence. Girod *et al.*[46] were the first of several groups to show that a short peptide could be inserted at aa 587 without affecting capsid viability. This approach has successfully increased the amount of rAAV that was taken up by one or another cell type following systemic administration. Peptides inserted at this position have the additional property that they split the R585,588 heparan sulfate binding site, thereby reducing AAV2 binding to its primary receptor.

An approach designed to accommodate large ligands has been suggested by Warrington *et al.*[47] This group has shown that GFP can be fused to the N terminus of the minor capsid protein VP2 with no significant affect on titer or viral stability. In this approach a ligand up to 35 kDa can be used to decorate the outside of the capsid at 3-6 copies.

Although binding of the capsid to the surface of the correct cell is important, other aspects of viral entry also presumably affect the efficiency of transduction, including receptor endocytosis, trafficking and uncoating. To identify viruses that are optimized for all steps in the viral entry process, several groups have used a random library approach to identifying useful capsid variants.[48] These groups have screened capsid libraries that have been shuffled from multiple existing serotypes and further mutagenized by using error prone PCR. By selecting libraries with pooled human serum, it is also possible to enrich for new capsid variants that do not react with neutralizing antibody.

6. Conclusions and Outlook

rAAV has been a useful tool for engineering long term expression in animals. Its safety profile has been excellent and the recent partial success in two clinical trials suggests that it may be useful for curing a variety of human diseases.

References

1. Berns KI, Parrish CR (2007). Parvoviridae. In *Fields Virology* (Knipe DM and Howley, PM Eds.), pp. 2437–2477. Lippincott Williams and Wilkins, New York.
2. Cideciyan AV, Aleman TS, Boye SL, Schwartz SB, Kaushal S, Roman AJ, *et al.* (2008). Human gene therapy for RPE65 isomerase deficiency activates the retinoid cycle of vision but with slow rod kinetics. *Proc Nat Acad Sci USA* **105**: 15112–15117.
3. Brantly ML, Chulay J, Wang L, Mueller C, Humphries M, Spencer LT, *et al.* (2009). Sustained Expression of Alpha-1 Antitrypsin Despite Effector T Lymphocyte

Responses in a Phase 1 Clinical Trial of Intramuscular rAAV1-AAT. *J Clin Invest* [submitted].

4. Schnepp BC, Jensen RL, Clark KR, Johnson PR (2009). Infectious molecular clones of adeno-associated virus isolated directly from human tissues. *J Virol* **83**: 1456–1464.

5. McLaughlin SK, Collis P, Hermonat PL, Muzyczka N (1988). Adeno-associated virus general transduction vectors: analysis of proviral structures. *J Virol* **62**: 1963–1973.

6. Berk AJ (2007). Adenoviridae: The viruses and their replication. In *Fields Virology* (Knipe DM and Howley PM Eds.), pp. 2355–2394. Lippincott Williams and Wilkins, New York.

7. Pereira DJ, McCarty DM, Muzyczka N (1997). The adeno-associated virus (AAV) Rep protein acts as both a repressor and an activator to regulate AAV transcription during a productive infection. *J Virol* **71**: 1079–1088.

8. Nash K, Chen W, Muzyczka N (2008). Complete *in vitro* reconstitution of adeno-associated virus DNA replication requires the minichromosome maintenance complex proteins. *J Virol* **82**: 1458–1464.

9. Conway JE, Zolotukhin S, Muzyczka N, Hayward GS, Byrne BJ (1997). Recombinant adeno-associated virus type 2 replication and packaging is entirely supported by a herpes simplex virus type 1 amplicon expressing Rep and Cap. *J Virol* **71**: 8780–8789.

10. Hermonat PL, Muzyczka N (1984). Use of adeno-associated virus as a mammalian DNA cloning vector: transduction of neomycin resistance into mammalian tissue culture cells. *Proc Nat Acad Sci USA* **81**: 6466–6470.

11. Urabe M, Ding C, Kotin RM (2002). Insect cells as a factory to produce adeno-associated virus type 2 vectors. *Hum Gene Ther* **13**: 1935–1943.

12. Alazard-Dany N, Nicolas A, Ploquin A, Strasser R, Greco A, Epstein AL, *et al.* (2009). Definition of herpes simplex virus type 1 helper activities for adeno-associated virus early replication events. *PLoS pathogens* **5**: e1000340.

13. Im DS, Muzyczka N (1990). The AAV origin binding protein Rep68 is an ATP-dependent site-specific endonuclease with DNA helicase activity. *Cell* **61**: 447–457.

14. Brister JR, Muzyczka N (2000). Mechanism of Rep-mediated adeno-associated virus origin nicking. *J Virol* **74**: 7762–7771.

15. Nony P, Chadeuf G, Tessier J, Moullier P, Salvetti A (2003). Evidence for packaging of rep-cap sequences into adeno-associated virus (AAV) type 2 capsids in the absence of inverted terminal repeats: a model for generation of rep-positive AAV particles. *J Virol* **77**: 776–781.

16. Bleker S, Pawlita M, Kleinschmidt JA (2006). Impact of capsid conformation and Rep-capsid interactions on adeno-associated virus type 2 genome packaging. *J Virol* **80**: 810–820.

17. Nam HJ, Lane MD, Padron E, Gurda B, McKenna R, Kohlbrenner E, *et al.* (2007). Structure of adeno-associated virus serotype 8, a gene therapy vector. *J Virol* **81**: 12260–12271.

18. Ding W, Zhang L, Yan Z, Engelhardt JF (2005). Intracellular trafficking of adeno-associated viral vectors. *Gene Ther* **12**: 873–880.

19. Sonntag F, Bleker S, Leuchs B, Fischer R, Kleinschmidt JA (2006). Adeno-associated virus type 2 capsids with externalized VP1/VP2 trafficking domains are generated prior to passage through the cytoplasm and are maintained until uncoating occurs in the nucleus. *J Virol* **80**: 11040–11054.

20. Grieger JC, Snowdy S, Samulski RJ (2006). Separate basic region motifs within the adeno-associated virus capsid proteins are essential for infectivity and assembly. *J Virol* **80**: 5199–5210.

21. Zhong L, Li B, Mah CS, Govindasamy L, Agbandje-McKenna M, Cooper M, *et al.* (2008). Next generation of adeno-associated virus 2 vectors: point mutations in tyrosines lead to high-efficiency transduction at lower doses. *Proc Nat Acad Sci USA* **105**: 7827–7832.

22. Kessler PD, Podsakoff GM, Chen X, McQuiston SA, Colosi PC, Matelis LA, *et al.* (1996). Gene delivery to skeletal muscle results in sustained expression and systemic delivery of a therapeutic protein. *Proc Nat Acad Sci USA* **93**: 14082–14087.

23. Xiao X, Li J, Samulski RJ (1996). Efficient long-term gene transfer into muscle tissue of immunocompetent mice by adeno-associated virus vector. *J Virol* **70**: 8098–8108.

24. Buning H, Perabo L, Coutelle O, Quadt-Humme S, Hallek M (2008). Recent developments in adeno-associated virus vector technology. *J Gene Med* **10**: 717–733.

25. Grimm D, Kern A, Rittner K, Kleinschmidt JA (1998). Novel tools for production and purification of recombinant adenoassociated virus vectors. *Hum Gene Ther* **9**: 2745–2760.

26. Xiao X, Li J, Samulski RJ (1998). Production of high-titer recombinant adeno-associated virus vectors in the absence of helper adenovirus. *J Virol* **72**: 2224–2232.

27. Samulski RJ, Berns KI, Tan M, Muzyczka N (1982). Cloning of adeno-associated virus into pBR322: rescue of intact virus from the recombinant plasmid in human cells. *Proc Nat Acad Sci USA* **79**: 2077–2081.

28. Conway JE, Rhys CM, Zolotukhin I, Zolotukhin S, Muzyczka N, Hayward GS, *et al.* (1999). High-titer recombinant adeno-associated virus production utilizing a recombinant herpes simplex virus type I vector expressing AAV-2 Rep and Cap. *Gene Ther* **6**: 986–993.

29. Kang W, Wang L, Harrell H, Liu J, Thomas DL, Mayfield TL, *et al.* (2008). An efficient rHSV-based complementation system for the production of multiple rAAV vector serotypes. *Gene Ther* in press.

30. Aslanidi G, Lamb K, Zolotukhin S (2009). An inducible system for highly efficient production of recombinant adeno-associated virus (rAAV) vectors in insect Sf9 cells. *Proc Nat Acad Sci USA* in press.

31. Zolotukhin S, Byrne BJ, Mason E, Zolotukhin I, Potter M, Chesnut K, *et al.* (1999). Recombinant adeno-associated virus purification using novel methods improves infectious titer and yield. *Gene Ther* **6**: 973–985.

32. Zolotukhin S, Potter M, Zolotukhin I, Sakai Y, Loiler S, Fraites TJ Jr., *et al.* (2002). Production and purification of serotype 1, 2, and 5 recombinant adeno-associated viral vectors. *Methods (San Diego, Calif)* **28**: 158–167.

33. Kaludov N, Handelman B, Chiorini JA (2002). Scalable purification of adeno-associated virus type 2, 4, or 5 using ion-exchange chromatography. *Hum Gene Ther* **13**: 1235–1243.

34. Aucoin MG, Perrier M, Kamen AA (2008). Critical assessment of current adeno-associated viral vector production and quantification methods. *Biotechnol Adv* **26**: 73–88.

35. Gao G, Vandenberghe LH, Wilson JM (2005). New recombinant serotypes of AAV vectors. *Curr Gene Ther* **5**: 285–297.

36. Thomas CE, Storm TA, Huang Z, Kay MA (2004). Rapid uncoating of vector genomes is the key to efficient liver transduction with pseudotyped adeno-associated virus vectors. *J Virol* **78**: 3110–3122.

37. Yan Z, Lei-Butters DC, Zhang Y, Zak R, Engelhardt JF (2007). Hybrid adeno-associated virus bearing nonhomologous inverted terminal repeats enhances dual-vector reconstruction of minigenes *in vivo*. *Hum Gene Ther* **18**: 81–87.

38. Li J, Sun W, Wang B, Xiao X, Liu XQ (2008). Protein trans-splicing as a means for viral vector-mediated *in vivo* gene therapy. *Hum Gene Ther* **19**: 958–964.

39. Nakai H, Montini E, Fuess S, Storm TA, Grompe M, Kay MA (2003). AAV serotype 2 vectors preferentially integrate into active genes in mice. *Nat Genet* **34**: 297–302.

40. Afione SA, Conrad CK, Kearns WG, Chunduru S, Adams R, Reynolds TC, *et al.* (1996). *In vivo* model of adeno-associated virus vector persistence and rescue. *J Virol* **70**: 3235–3241.

41. Nakai H, Yant SR, Storm TA, Fuess S, Meuse L, Kay MA (2001). Extrachromosomal recombinant adeno-associated virus vector genomes are primarily responsible for stable liver transduction *in vivo*. *J Virol* **75**: 6969–6976.

42. Schnepp BC, Clark KR, Klemanski DL, Pacak CA, Johnson, PR (2003). Genetic fate of recombinant adeno-associated virus vector genomes in muscle. *J Virol* **77**: 3495–3504.

43. Zaiss AK, Muruve DA (2008). Immunity to adeno-associated virus vectors in animals and humans: a continued challenge. *Gene Ther* **15**: 808–816.

44. Qing K, Hansen J, Weigel-Kelley KA, Tan M, Zhou S, Srivastava A (2001). Adeno-associated virus type 2-mediated gene transfer: role of cellular fkbp52 protein in transgene expression. *J Virol* **75**: 8968–8976.

45. McCarty DM (2008). Self-complementary AAV vectors; advances and applications. *Mol Ther* **16**: 1648–1656.

46. Girod A, Ried M, Wobus C, Lahm H, Leike K, Kleinschmidt JA, *et al.* (1999). Genetic capsid modifications allow efficient re-targeting of adeno-associated virus type 2. *Nat Med* **5**: 1438.

47. Warrington KH Jr., Gorbatyuk OS, Harrison JK, Opie SR, Zolotukhin S, Muzyczka N (2004). Adeno-associated virus type 2 VP2 capsid protein is nonessential and can tolerate large peptide insertions at its N terminus. *J Virol* **78**: 6595–6609.

48. Kwon I, Schaffer DV (2008). Designer gene delivery vectors: molecular engineering and evolution of adeno-associated viral vectors for enhanced gene transfer. *Pharm Res* **25**: 489–499.

Chapter 7
Regulatory RNA in Gene Therapy

Alfred. S. Lewin*

Regulatory RNAs include *trans*-acting molecules such as miRNAs, siRNAs, ribozymes and *cis*-acting elements within mRNA molecules. *Trans*-acting RNAs may be delivered either as modified oligonucleotides or via gene therapy vectors, such as plasmids or viruses. Oligonucleotide therapy requires re-administration of the therapeutic RNA, while vector mediated delivery may provide long-term expression of the siRNA or ribozymes. Even though ribozymes have been studied for 25 years, RNA interference using siRNAs has become the major tool for suppressing gene expression for therapy. Non-specific effects of RNA mediated regulation demands careful control experiments and testing of multiple therapeutic molecules for each target gene. Adding or modifying sequences in the 3' untranslated regions of mRNAs provides a powerful tool for controlling mRNA stability, localization and translation.

1. Introduction

Regulatory RNAs may be broadly defined as non-protein coding RNA sequences that in some way control the expression of genes. RNA molecules such as *Xist* or naturally occurring siRNAs may lead to stabilization of heterochromatin and shut off entire chromosomes or chromosome regions.[1] For gene therapy, in contrast, one wishes to control the expression of a single gene and for this purpose specificity is required. For gene silencing, gene therapists have therefore turned to antisense RNA, RNA decoys or

---

*Correspondence: Department of Molecular Genetics and Microbiology, University of Florida, Gainesville, FL 32610.
E-mail: lewin@ufl.edu

aptamers, ribozymes, artificial siRNAs or siRNAs disguised as miRNAs. For gene regulation, *cis*-acting RNA elements may be added or modified.

Antisense RNA molecules are RNA molecules that are complementary to the coding sequence of a messenger RNA. They may block gene expression by interfering with translation. Antisense molecules are most effective, however, when delivered as DNA oligonucleotides that can stimulate RNAse H-mediated decay of mRNAs to which they bind. Alternatively, morpholino based antisense molecules and PNAs (peptide nucleic acids) can form stable duplexes with target mRNAs and prevent their translation into protein.[2] Although these are powerful experimental and therapeutic approaches, they do not fall under the heading of regulatory RNA, and will not be discussed in this chapter.

RNA decoys are simply segments of RNA that contain naturally-occurring binding sites for viral or cellular proteins. If delivered and expressed at a high level, they may sequester a protein essential for the pathogenic process. An example is an HIV TAR decoy, which has been embedded in a small nucleolar RNA and used to sequester the Tat protein from the HIV promoter.[3] RNA aptamers are products of reiterative RNA selection methods designed to produce high-affinity binding sequences. Aptamers usually bind proteins via the tertiary structure of the small RNA rather than the primary sequence, and affinities can be quite high, comparable to antigen antibody interactions. An aptamer to VEGF is commercially available (as Macugen™) for the treatment of choroidal neovascularization, a complication of age related macular degeneration.[4]

Ribozymes are catalytic RNA molecules. RNA catalysis was discovered in the context of RNAse P and self-splicing introns and almost simultaneously was found to be a feature of self-processing satellite RNAs of plant pathogens.[5] Other RNA enzymes have been discovered in a variety of contexts, and novel catalysts consisting of RNA or DNA have been artificially selected in the test tube. Pre-clinical gene therapy has made the most use of two short ribozymes derived from tobacco ringspot virus satellite RNA, the hammerhead and the hairpin ribozyme.[6] The RNA subunit of RNAse P can cleave targeted mRNAs, using a guide RNA to direct the ribozyme to a particular mRNA. Once ribozyme mediated digestion separates the 5′ cap structure from the polyA tail, mRNAs are rapidly degraded by cellular exonucleases. Like protein enzymes, ribozymes exhibit catalytic turnover, and each ribozyme may cleave many potential target RNAs. While their turnover is not dramatic relative to protein catalysts, on the order of 10 per minute, most mRNAs are not abundant in the steady state. Therefore, ribozymes can be useful tools for elimination of specific RNAs.

RNA interference (RNAi) is mediated by small interfering RNAs (siRNAs), which are 19-25 nucleotide RNA duplexes with sequence homology to cellular mRNAs.[7,8] One strand of an siRNA is exactly complementary to a target mRNA and is termed the *guide strand*, while the other strand has the same sense as the mRNA and is called the *passenger strand*. Naturally occurring siRNAs are processed from larger double stranded RNAs by an endonuclease called Dicer. Like other nucleases of its class (ribonuclease III), Dicer leaves 5′ phosphate groups and 3′ hydroxyls, with 2 nucleotide overhangs at the 3′ ends. Since mammalian cells are intolerant of long double stranded RNA molecules (see below), for gene therapy, siRNAs are either delivered as short dsRNA molecules with 3′ overhangs, or they are expressed from DNA vectors as small hairpin RNAs that are processed by Dicer into siRNAs. Once siRNAs are processed and delivered to the cytoplasm, they associate with the RNA Induced Silencing Complex (RISC), where they unwind and one RNA strand becomes incorporated into the complex, which then directs cleavage of the complementary mRNA.

An alternative approach for delivery of siRNAs is to express them as miRNAs.[9] Like siRNAs, miRNAs are short RNA molecules that bind to mRNAs in the context of RISC. However, miRNAs are not fully complementary to their target RNAs and, in general, bind to the mRNAs in the 3′ untranslated region (3′UTR). MicroRNAs block gene expression at the level of protein synthesis and do not lead to immediate degradation of mRNA, though RNA half-life may decrease. There are hundreds of natural miRNAs in mammalian cells and each may regulate the translation of many mRNAs. Because of this complexity, miRNAs themselves have not been advanced as tools for gene therapy, except in the case of certain tumors in which miRNA genes are defective. However, an RNA with exact complementarity to the coding region of a mRNA can be placed within the context of a miRNA precursor. This precursor will be processed in the nucleus and exported by the same mechanism as native miRNAs, providing a mechanism to express an siRNA using RNA polymerase II.

Manipulation of *cis*-acting regulatory RNA sequences has not developed as rapidly as the use of siRNAs for regulation of gene expression. Nevertheless, this approach is more subtle and discriminating than simple RNA knockdown. Sequences in the 3′ UTR of genes delivered for therapy have been modified to limit their expression in unintended tissues,[10] and to direct mRNAs to a particular location in the target cell.[11] In addition, small RNAs can be delivered in *trans* to affect the activity of *cis*-acting sequences that govern alternative splicing.

2. Delivery of Therapeutic RNAs

Delivery is a major concern for those trying to use small RNA molecules as tools for gene therapy. There are two basic approaches for delivering therapeutic RNA to animals and humans (Fig. 7.1). The first is to deliver RNA molecules directly and the second is to deliver DNA vectors that drive the transcription of the therapeutic RNA. The development of oligonucleotides as drugs has received support from the pharmaceutical industry. The efficacy of siRNAs as inhibitors allows access to many cellular targets that were until recently thought to be "undruggable". Annotated genome sequences facilitate design of siRNAs and identification of potential unintended mRNA targets. There are two key advantages to delivering RNA as a *gene-directed* therapy, rather than as a gene therapy. The first is that RNA drugs can be "dosed to effect" for a specific condition and a specific patient. The second is that therapy can be halted upon adverse reaction, and the inherent instability of RNA will lead to rapid clearance from the body. Delivering RNA has several drawbacks, however. For one, RNA drugs are not orally available, so that delivery may require infusion or injection. Intravenously injected RNA molecules tend to be filtered by the kidney, liver, spleen and lungs. If these are not target organs, then bioavailability may be limited. Finally, because of the susceptibility of RNA to digestion by cellular and extracellular enzymes and the clearance of nucleic acids from the circulation, RNA drugs must be continuously reapplied in order to treat chronic medical conditions. This may be seen as a disadvantage to a patient, though not necessarily to an investor.

Because RNA is unstable under physiologic conditions, RNA therapists rely on a series of modifications to the ribose-phosphate backbone to

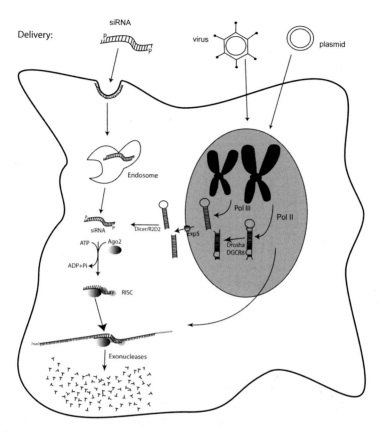

Fig. 7.1 Delivery of siRNA. Small interfering RNA may be delivered either as double stranded oligonucleotides or as DNA vectors (plasmids or virus vectors). Oligonucleotides are delivered via liposomes or nanoparticles and are released from the endosome before recognition by Argonaute proteins (Ago) in the cytoplasm. DNA vectors are expressed in the nucleus as small hairpin RNAs or as microRNAs before export to the cytoplasm and processing by Dicer. Both pathways converge on the RNA induced silencing complex (RISC) and lead to degradation of complementary mRNA.

prevent spontaneous and nuclease catalyzed degradation.[12] These modifications include blocking the 3′ termini, substituting phosphate groups with phosphorothioate groups and modifying the 2′ position of the ribose sugars. Locked nucleic acid (LNA) residues, which contain bonds between the 2′ and 4′ carbons of ribose have been particularly effective in stabilizing siRNAs without blocking their activity. The trick is that most of these modifications must be made in the passenger strand of the siRNA in order to retain guide strand directed cleavage of mRNA in the presence of RISC.

In addition, a compromise must be made between stability and activity in terms of the number of modifications that can be made. Typically, however, the half-life of siRNA in serum can increase from seconds to hours using modifications that do not reduce inhibitory activity substantially. Ribose and phosphorothioate modifications are also compatible with ribozyme activity. Again, a balance must be drawn between maximizing stability and retaining catalytic activity: modifying purine groups in the catalytic core of the hammerhead, for example, eliminates ribozyme function.

Both liposomes and synthetic nanoparticles have been employed to deliver regulatory RNAs *in vivo*.[13] Such particles can be directed to specific cell types by including carbohydrate or protein ligands within the particle. These approaches to nucleic acid administration are discussed elsewhere in this book. A major advance in siRNA delivery was made by Soutschek *et al.*, who covalently modified the 5′ ends of an siRNAi directed against the mRNA for apolipoprotein B mRNA.[14] This modification led to silencing of ApoB mRNA in the liver and jejunum, and opened the door for direct targeting of siRNAs to tissues of interest following intravenous administration by adding cell penetrating peptides or single chain antibodies directly to the RNA.[15]

Delivering therapeutic RNAs as DNA permits long-term expression of the RNA molecules and may allow more precise delivery to a particular organ or cell type. As for other therapeutic genes, control of RNA expression is determined by the specific virus and virus serotype employed, by the site of injection and by the promoter sequence used. If longevity of expression is an advantage, it may also be detrimental, if treatment leads to harmful side effects. As for other gene therapy applications, thorough pre-clinical safety testing is required before human testing begins. For sequence specific inhibitors such as siRNAs and ribozymes, however, animal testing may not reveal unintended targets whose knockdown can cause human complications.

RNA delivery has employed both viral vectors and liposome formulated plasmids. These delivery systems are described in other chapters in this volume. Synthesis of ribozymes or siRNAs can be directed by either RNA polymerase III promoters or RNA polymerase II promoters. RNA pol III normally produces small nuclear RNAs, transfer RNAs and 5S ribosomal RNA. While pol III promoters are powerful, much of the small hairpin transcripts produced by pol III, remains in the nucleus, separated from mRNA targets. In addition, pol III promoters are generally unregulated and expressed in all cell types. The need for nuclear export signal and regulated promoter systems has persuaded many investigators to embed siRNA sequences within miRNAs. MicroRNAs are typically produced by RNA pol II in the context of larger precursors containing signals for processing by the nuclear enzyme Drosha and transport to the cytoplasm by Exportin 5.[16] They are often made as part of long transcripts that may contain multiple miRNAs or they may be encoded in the introns of protein coding genes. Therefore, it is possible to insert a miRNA disguised siRNA in the intron or the untranslated region of another gene. Details of such constructs are discussed below.

3. Ribozymes

Ribozymes have been used primarily to suppress the expression of disease related genes.[6] As noted above, the hammerhead and the hairpin ribozymes are the easiest to design and have target sequences (NUX for hammerheads and BNGUC for hairpins, where N is any nucleotide, B is any nucleotide but adenosine and X is any nucleotide but guanosine) that can be found in any mRNA.[5] To design a hammerhead ribozyme, for example, one looks for NUX triplets (typically GUC, CUC or UUC) in the target mRNA and designs an antisense RNA of 12 or 13 nucleotides surrounding the cytosine, which does not form a conventional base pair with the ribozyme. The catalytic core of the ribozyme is then inserted within the antisense sequence, dividing it into two hybridizing arms which form mini-helices with the target mRNA. Since product release may be rate limiting for ribozyme turnover, these helices are kept short, typically 5–7 base pairs. Design of the hairpin is somewhat more complex, with an 8 nucleotide loop inserted between helices formed between ribozyme and target. To use RNase P for RNA knockdown, one uses the endogenous enzyme and designs a guide RNA resembling the acceptor stem of a transfer RNA when it is paired with the mRNA to be digested.[17] For all ribozymes, it is necessary to identify accessible target sequences: cleavage sites cannot be buried in stable internal helices within the mRNA. Computer algorithms may help identify such stems if used judiciously to analyze short stretches of sequence.

An advantage of ribozymes relative to small interfering RNAs is that ribozymes can be tested in cell-free reactions in order to identify which are most catalytically active. Such testing should be performed at relatively low magnesium concentration (10 mM) in order to eliminate ribozymes that will not function efficiently in cells. The most active ribozymes can then be tested in transfected cells to verify that they can cleave full length mRNA within the cellular environment. Expressing ribozymes in human cells also provides at least a crude assessment of "off-target" effects mediated by ribozymes: if a ribozyme unintentionally retards the cell cycle or leads to apoptosis, then it is time to choose a new target sequence. Finally, those ribozymes that pass the *in vitro* tests can be cloned in a vector and tested in animal models of human disease. Ribozymes have been expressed from viral vectors using either pol II promoters or pol III promoters, in which case the ribozymes may be expressed in tandem with tRNAs. Alternatively, RNA modifications can be introduced, including phosphorothioate nucleotides at the 5′ end of the ribozyme, an inverted abasic nucleotide at the 3′ end and 2′ amino and 2′O-methyl substitutions at certain residues in the core of the ribozyme.[18] Such modifications will lower cleavage rate *in vitro* by 5–10 fold, but may increase stability in cells by a 1000 fold. Modified ribozymes can be delivered directly to the circulation or to affected tissues.

Therapeutic targets for ribozymes have included dominant oncogenes, RNA viruses and retroviruses and autosomal dominant disease genes. While ribozymes have been successful in reducing tumor growth in animal models, there have been few clinical trials of ribozyme gene therapy. Hammerhead ribozymes directed against HIV-1 targets have been tested in lymphocytes or progenitor cells infused into AIDS patients.[19,20,21] A ribozyme directed at proliferating cell nuclear antigen mRNA was tested, unsuccessfully, as preventative of proliferative vitreoretinopathy.[22] Despite over 20 years of research, no therapeutic ribozyme has been brought to market as a disease therapy. Problems with ribozymes include their relatively low affinity for target RNAs and the lack of specific targeting mechanisms for target binding: ribozymes can be embedded in pol II transcripts for export from the nucleus, but unlike siRNAs they do not associate with a protein complex that promotes mRNA recognition. In addition, the active conformation of *cis*-acting hammerhead ribozymes is stabilized by tertiary loop-bulge interactions that are hard to recapitulate in *trans*.[23] Nevertheless, highly active ribozymes can reduce mRNA levels as effectively as siRNAs, and ribozymes can be partnered with small hairpin RNAs, as in the HIV example, without competing for the same nuclear export system. Relative to the RNAi user community, however, the ribozyme user community is small, and commercial support for ribozyme technology is non-existent. For this reason alone, the search for effective RNA-based inhibitors of gene expression should begin with siRNA.

4. RNAi for Gene Therapy

As with ribozymes, siRNAs have been employed primarily to suppress gene expression by inducing mRNA degradation. Therapeutic targets have included dominant oncogenes, viruses, growth factors and almost any gene that is induced or mis-regulated in disease.[24] For example therapeutic siRNAs are being clinically tested for treatment of respiratory syncytial virus using an inhaled formulation of a modified siRNA, for the down regulation of VEGF in the eye, and for pachyonchia congenita, an inherited skin disease.[25] While it is possible to produce allele-specific siRNAs for the treatment of dominantly inherited disease, single base changes may not be sufficient to provide discrimination between the mutant and the wild-type mRNA. One method to overcome this problem is to deliver an siRNA that targets both mutant and wild-type mRNA. This siRNA may be delivered in conjunction with an mRNA containing silent nucleotide changes rendering it resistant to attack by the siRNA and RISC.[26] Such an approach is also feasible with ribozymes.

Effective siRNAs generally conform to certain design principles: siRNA sequences should be short (19–21 bp); they should be of moderate G+C content (30–60%); they should avoid runs of 4 adenosine or 4 guanosine residues; they should contain an A-U pair at position 10 relative to 5′ end of the antisense strand, and if possible, they should have a lower thermal stability at the 5′ end of the antisense strand than at the 3′ end.[27] The last requirement reflects the asymmetric loading of RNA strands onto RISC. While either strand of an siRNA can be associated with the complex, RNA is loaded in the 5′ to 3′ direction. Therefore, if the 5′ end of the guide (anti-sense) strand unwinds first, it will preferentially load into RISC. Keeping the siRNA short is important. While longer siRNAs (25 nt) are more effective in directing cleavage of mRNA, they are more likely to stimulate pathways of innate immunity and to block the export of endogenous miRNAs.[28] As with ribozymes, it is important to avoid siRNA target sites in regions of stable secondary structure. Fortunately, one does not have to remember all of the design principles in selecting siRNA target sites, as several good design algorithms are available on line.[27] Some of these sites are sponsored by biotechnology companies that sell dsRNA or vectors expressing short hairpin RNA. In addition, several companies sell pre-validated siRNAs or shRNA plasmids for human, mouse and rat mRNAs. These commercial sources are the best choice for initiating an RNAi related project.

It is important to identify siRNAs of the greatest potency, so that the level of siRNA delivered can be minimized. Small interfering RNAs are assayed in tissue culture before testing in animal models for disease. This testing may be done by measuring the expression level of an endogenous gene using RT-PCR, immunoblots or ELISA. Alternatively, siRNAs may be co-transfected as RNA or as plasmids expressing small hairpin RNAs (shRNAs) together with a plasmid expressing the target mRNA. It is convenient to insert a segment of the mRNA containing the target sequence within the transcript of an easily measured marker gene such as luciferase, GFP or secreted alkaline phosphatase. This permits rapid screening of the inhibitory activity of the siRNAs. It is important to recognize that for therapeutic applications, one or two siRNAs per target are not enough. Non-specific effects of siRNAs and pharmacodynamic properties of small RNAs are frequently sequence specific. An siRNA that is potent in HeLa cells may be useless *in vivo*.

For gene therapy, siRNAs are usually delivered as small hairpin RNAs.[29] Their expression is driven from DNA vectors using pol III promoters such as U6 and H1. Alternatively, modified pol II promoters may be employed. To limit overhanging sequences, these have been engineered to provide the start site for RNA synthesis just before the shRNA coding sequence and a polyadenylation site just afterwards. As mentioned above, an alternative approach employs pre-miRNA sequences in which an siRNA is embedded, replacing the sequence containing the mature miRNA. These precursors contain processing sites for Drosha and Dicer, so that the siRNA can be expressed and exported from the nucleus by the normal miRNA pathway. Small interfering RNAs disguised as pre-miRNAs may be expressed from either pol II or pol III promoters. As noted, pol II promoters are more easily regulated and may be cell type specific. On the other hand, pol III promoters are more robust: they should yield high level expression of the miRNA or shRNA in any cell type.

Off target effects are an important consideration in employing RNAi for therapy.[30] Off target effects lead to non-specific suppression of gene expression and are caused in several ways: (1) An siRNA (or ribozyme for that matter) can bind to and stimulate the degradation of an unintended target RNA. (2) An siRNA can pair imperfectly with the 3′ UTR of an mRNA and suppress its translation into protein. (3) A small hairpin RNA can interfere with the trafficking of endogenous miRNA, and (4) siRNA can stimulate the dsRNA response. The dsRNA response protects animal cells from invading viruses, retrotransposons and the transcription of repetitive sequences by activating PKR (dsRNA-dependent protein kinase) and 2′, 5′-oligoadenylate synthetase. In response to dsRNA, PKR autophosphory-lates and then phosphorylates its substrates, including eukaryotic initiation factor 2, leading to arrest of protein synthesis. PKR also initiates a signal transduction pathway that leads to induction of interferons. Interferon production results in the up-regulation of antiviral and antiproliferative genes and mediators of apoptosis. The activation of 2′, 5′-oligoadenylate synthetase results in the formation of oligoadenylates that bind to and activate RNaseL, which cleaves RNA non-specifically, resulting in inhibition of translation. Stimulation of the interferon response by siRNAs appears to be mediated by Toll-like receptor (TLR) 7/8 or by TLR3.[31,32] Incorporation of 2′-O-methyl nucleotides into the siRNA suppresses the TLR7/8 pathway but may not inhibit activation of the TLR3 pathway. In one study of VEGF siR-NAs used to inhibit laser-induced choroidal neovascularization in a mouse model of age related macular degeneration, it was found that inhibition of neovascularization could be attributed solely to TLR3-mediated induction of interferon (Fig. 7.2).[31]

Off target effects can be minimized by several means. Sequence search-ing programs may help to identify unintended targets with annotated mRNAs. It is best to avoid using siRNAs that can form 15 or more consecu-tive base pairs with another RNA that is expressed in the organ(s) of interest.

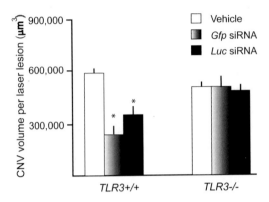

Fig. 7.2 *Gfp* siRNA or *Luc* siRNA led to a modest suppression of choroidal neovascularizartion (CNV) in *Tlr3*$^{+/+}$ but not *Tlr3*$^{-/-}$ mice. $P < 0.05$ compared to vehicle (buffer). From Kleinman *et al.* (Ref 31).

It is somewhat harder to exclude miRNA-like activity of siRNAs, because base-pairing with the 7 base seed region (residues 2–8) of the siRNA may be sufficient to inhibit translation of an unintended mRNA.[33] Inhibition of miRNA transport and stimulation of the interferon and 2′, 5′ oligoadenylate pathways can be avoided by keeping the siRNA short. Inhibition of miRNA transport requires over-expression of shRNAs of 25 bp or more.[28] While siRNAs of 19bp do not typically stimulate the dsRNA response, this response is sequence specific, and some short interfering RNAs may turn on the interferon pathway. Indeed, it has recently been suggested that many of the early reports of antiviral activity of siRNAs are attributable to the stimulation of interferon rather than the digestion of viral RNA by RISC.[32] The best way to control for non-specific effects of siRNA is to have several independent siRNAs for the target mRNA and to demonstrate that each has the same physiologic impact. These positive controls should be combined with several negative control siRNAs that together demonstrate that digestion of the intended mRNA leads to a particular physiologic response.

5. Gene Therapy Using miRNA

Individual miRNAs regulate variety of genes; some modulate the translation of hundreds of mRNAs.[34] We know little about the regulation of miRNA expression in human tissues. Therefore, it seems unlikely that delivery of native pri-miRNAs or pre-miRNAs using gene therapy vectors will be used in the near future, since they are likely to lead to unforeseen consequences. For now, pre-miRNAs are being used as vehicles for siRNAs (see above). Nevertheless, coding sequences for some miRNAs are missing or defective in specific tumors, so that it is conceivable to use gene therapy vectors to replace these missing genes.[35] Similarly, it is possible to inhibit the activity of specific miRNAs using antisense sequences or surrogate target RNAs. These can be used to decrease the impact of over-expressed miRNAs that may lead to disease. Careful assessment of non-specific effects must be made in developing these therapies.

6. Aptamers, Decoys and Bi-Functional RNAs

RNA aptamers are short oligonucleotides (usually less than 40 nt) with a stable tertiary structure that permits high-affinity binding (Kd in the range of 10^{-10} to 10^{-12}) to specific proteins or other target molecules. Consequently, aptamers have been developed as inhibitors of pathologic processes. Aptamers are generated by a sequential selection method called SELEX (systematic evolution of ligands by exponential enrichment).[36] In this process RNA libraries containing randomized regions are incubated with the ligand of interest and RNA molecules that bind the target compound are separated from those that do not bind. Binders are amplified by reverse transcription PCR and transcribed into RNA for another round of selection (binding). After multiple rounds of selection, high affinity aptamers are analyzed and counter selection with structural analogs of the target ligand may be used to produce aptamers of greater selectivity.

For therapy, aptamers are delivered as oligonucleotides. As with siRNAs and ribozymes they may be protected using ribose modifications, but these modifications may affect the affinity of the aptamer, since binding involves the RNA (or DNA) backbone as well as the nucleotide bases. As noted, one aptamer has already found clinical application for treatment of macular degeneration and it is delivered by direct and repeated ocular injections. It is conceivable that aptamers could be delivered using DNA vectors for sustained expression, but the folding of RNA sequences during transcription may not recapitulate the folding of aptamer sequences during the selection process *in vitro*, and flanking RNA sequences may affect the tertiary structure as well.

Decoy oligonucleotides contain the recognition sites for nucleic acid-binding proteins or RNAs. DNA decoys have been used to interfere with the binding of transcription factors to promoter sequences. Rossi and colleagues have used an RNA decoy delivered by a retrovirus vector, to block the interaction of TAR with the HIV-1 promoter.[3] Decoy RNAs have also been employed to block the activity of specific miRNAs, by providing multiple copies of a fully-complementary binding site. One type of miRNA inhibitor has been called "antagomir".[37] These are synthetic and chemically stabilized (LNA or 2'-O-methyl) oligonucleotides that inhibit the function of endogenous microRNAs by complementary base pairing. Antigomirs may also have modifications, such as cell penetrating peptides or cholesterol, which promote their uptake by cells. They have been shown to inhibit miRNA *in vivo* following intravenous injection. An alternative approach to miRNA inhibition is to express repeated miRNA binding sites from a strong pol II promoter. Such constructs are termed "microRNA sponges" and they appear to block miRNA activity without lowering the level of target miRNAs.[38]

Bi-functional RNAs represent a more complex approach to using recognitions sites for RNA binding proteins. In bi-function molecules, the binding sites for regulatory proteins are linked to antisense RNA in order to direct the regulatory protein to a particular region of the target mRNA. Such bi-functional oligonucleotides have been used to regulate alternative splicing of pre-mRNA by recruiting SR proteins or hnRNP proteins to alternative exons in order to either increase or decrease the utilization of an adjacent splice junction. For example, Baughan *et al.* used AAV to deliver a chimeric RNA to attract SR proteins (SF2/ASF, SC35, and hTra2h1) to the 5' terminus of exon 7 of the SMN2 gene.[39] Infection of fibroblasts from patients with spinal muscular atrophy led to an increase in exon inclusion and therefore a significant increase in SMN2 protein.

7. Modification of Cis-Acting Regulatory RNA Sequences

Mature mRNAs are bound by proteins from their time of synthesis, through nuclear transport and their translation into protein. Sequences within the mRNA govern which proteins bind and consequently determine mRNA stability and transport. In a few instances, *cis*-acting sequences have been used to control gene expression for gene therapy. However, in most cases, gene therapists are constrained by the limitation of viral packaging limits or available restriction sites, and they clip off important regulatory elements in the 5' and 3' UTRs.

The stability of a particular mRNA is often determined by sequences within the 3' UTR of that mRNA, though RNA destabilizing sequences have been identified in other locations.[40] The best known stability element in the 3' UTR is designated the ARE or AU rich element. Cellular proteins that bind the ARE may either stabilize that mRNA (e.g. HuR) or destabilize the mRNA (AUF1). Some of these *trans*-acting proteins are cell type specific, while others are ubiquitous. The localization of mRNA is one mechanism to achieve protein sorting in asymmetric eukaryotic cells. While this process has been best described in yeast and in *Drosophila* oocytes, mRNA sorting is extremely important in the nervous system, where local protein synthesis in neural dendrites may explain how patterns of neural activity lead to changes in synaptic connectivity in development and in adaptive responses in the brain.[41] In oligodendrocytes, mRNAs encoding proteins required for myelin formation are localized. Disruption of mRNA sorting and localized mRNA protein synthesis is a contributing factor of diseases such as fragile X mental retardation and spinal muscular atrophy.

To date gene therapists have made little use of either stability elements or localization elements to obtain more precise control of gene expression. There have been exceptions, however. In order to attain tumor selective gene expression for virotherapy using adenovirus, Säkioja *et al.* included an ARE element in the 3' UTR of a passenger gene (luciferase) and thereby suppressed expression in non-cancer cells but not in malignant cells.[42] Corral-Debrinski *et al.* have made an interesting use of mRNA targeting to optimize allotropic gene expression of human mitochondrial genes. In yeast, they discovered that mRNAs translated on mitochondrial-bound polyribosomes contain common elements in their 3'UTRs. They included one of these sequence elements in the 3' UTR of a transgene encoding the mitochondrial ND1 and ND4 genes rendered in the nuclear genetic code and fitted with a mitochondrial protein import sequence.[43] Including the mRNA localization element increased the fraction of the allotropic mitochondrial mRNA that was bound to the target organelle.

Of course the 3′ UTR is also the site of regulation by microRNAs, and the target sites for miRNAs can be used to improve specificity of viral mediated gene expression. To suppress the expression of coagulation factor IX (F.IX), Brown *et al.* modified a lentivirus vector to contain 4 tandem copies of a 23 nucleotide sequence with perfect complementarity to the hematopoietic-specific miRNA, miR-142-3p, in the untranslated region (3′ UTR) of the transgene expression cassette (Fig. 7.3).[44] The vector contained an hepatocyte-specific promoter, but without the miR-142 target sites, expression was detected both in the liver and the spleen of intravenously infected mice. Inclusion of the miR-142-3p target sequences prevented unintentional transgene expression from tissue-specific promoter specifically within hematopoietic cells. Furthermore, miR-142-3p–regulated lentivirus mediated stable human F.IX gene expression and correction of hemophilia B mice. Since the immune response to the product of the transgene is a concern for all gene therapists, this approach is likely to be widely imitated.

Ribozymes can also be used as *cis*-acting regulatory elements in the context of gene therapy vectors. Yen *et al.* developed a modified a hammerhead ribozyme from *Schistosoma mansoni* that is capable of extremely efficient self-cleavage in mammalian cells.[45] Placing this in the 5′ UTR of a marker gene, they used high throughput screening to identify compounds that could regulate ribozymes activity, and thus expression of the

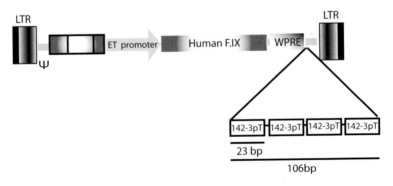

Fig. 7.3 Map of a lentiviral vector used developed by Brown *et al.* to obtain liver specific expression of Factor IX. ET is a chimeric, liver specific promoter (based on Ref 44).

gene. Unfortunately, the most efficient regulators were potentially toxic nucleoside analogues, toyocamycin and 5-fluorouridine. However, Win and Smolke have developed *cis*-acting hammerhead ribozymes that can either be activated or repressed based on an appended theophylline binding aptamer.[46] This approach has the potential to produce regulatory riboswitches that may be controlled by easily tolerated and approved drugs.

8. Conclusions

While gene therapy is usually designed to deliver genes for proteins, delivery of regulatory RNAs has the capacity to suppress endogenous gene expression in order to attenuate disease. Protein-directed therapy is well suited for recessive inherited disease, but gene silencing by siRNAs or ribozymes can be used to treat a variety of acquired, dominantly inherited and infectious diseases. In addition, modification of *cis*-acting RNA elements can make the expression of protein genes more precise and, therefore, more effective. The major dichotomy in the field concerns delivery: whether regulatory RNAs are delivered as stabilized oligoribonucleotides or whether they are administered via gene therapy vectors. Obviously, the application will determine the most suitable approach. For short-term applications, such as suppression of scarring or treating acute infections, direct RNA delivery should be better. For chronic diseases, however, gene therapy will ultimately replace the use of oligonucleotides. The major cautionary note in the field of RNA-based gene therapy is the potential for off-target effects, and the major method to overcome this obstacle is to design and test many potential RNA-inhibitors in order to develop a therapeutic RNA that has high activity for the intended mRNA and a low profile of undesirable side effects.

It is now appreciated that almost the entire genome is transcribed, but the role of the low-level transcripts produced from most of the genome is undefined. There are certainly more regulatory RNAs to discover. Recently, for example, short RNAs homologous to promoter regions were found both to inhibit and to activate gene expression at the level of transcription.[47] These promoter-directed short RNAs require Argonaute proteins and may target promoter-specific antisense transcripts. While these so-called antigene RNAs have not been adapted for gene therapy, they are likely to become useful tools for allele independent suppression of mutant genes.

As the functions of other minor transcripts are discovered, gene therapists must stay aware of developments in the study of regulatory RNA.

References

1. Grewal SI, Elgin SC (2007). Transcription and RNA interference in the formation of heterochromatin. *Nature* **447**: 399–406.
2. Juliano RL, Dixit VR, Kang H, Kim TY, Miyamoto Y, Xu D (2005). Epigenetic manipulation of gene expression: a toolkit for cell biologists. *J Cell Biol* **169**: 847–857.
3. Li MJ, Bauer G, Michienzi A, Yee JK, Lee NS, Kim J, Li S, Castanotto D, Zaia J, Rossi JJ (2003). Inhibition of HIV-1 infection by lentiviral vectors expressing Pol III-promoted anti-HIV RNAs. *Mol Ther* **8**: 196–206.
4. Ng EW, Shima DT, Calias P, Cunningham ET Jr., Guyer DR, Adamis AP (2006). Pegaptanib, a targeted anti-VEGF aptamer for ocular vascular disease. *Nat Rev Drug Discov* **5**: 123–132.
5. Doudna JA, Cech TR (2002). The chemical repertoire of natural ribozymes. *Nature* **418**: 222–228.
6. Lewin AS, Hauswirth WW (2001). Ribozyme gene therapy: applications for molecular medicine. *Trends Mol Med* **7**: 221–228.
7. McManus MT, Sharp PA (2002). Gene silencing in mammals by small interfering RNAs. *Nat Rev Genet* **3**: 737–747.
8. Meister G, Tuschl T (2004). Mechanisms of gene silencing by double-stranded RNA. *Nature* **431**: 343–349.
9. Chang K, Elledge SJ, Hannon GJ (2006). Lessons from Nature: microRNA-based shRNA libraries. *Nat Methods* **3**: 707–714.
10. Brown BD, Gentner B, Cantore A, Colleoni S, Amendola M, Zingale A, Baccarini A, Lazzari G, Galli C, Naldini L (2007). Endogenous microRNA can be broadly exploited to regulate transgene expression according to tissue, lineage and differentiation state. *Nat Biotech* **25**: 1457–1467.
11. Corral-Debrinski M (2007). mRNA specific subcellular localization represents a crucial step for fine-tuning of gene expression in mammalian cells. *Biochim Biophys Acta* **1773**: 473–475.
12. Behlke MA (2008). Chemical modification of siRNAs for *in vivo* use. *Oligonucleotides* **18**: 305–319.
13. Juliano R, Alam MR, Dixit V, Kang H (2008). Mechanisms and strategies for effective delivery of antisense and siRNA oligonucleotides. *Nucleic Acids Res* **36**: 4158–4171.
14. Soutschek J, Akinc A, Bramlage B, Charisse K, Constien R, Donoghue M, Elbashir S, Geick A, Hadwiger P, Harborth J, John M, Kesavan V, Lavine G, Pandey RK, Racie T, Rajeev KG, Rohl I, Toudjarska I, Wang G, Wuschko S, Bumcrot D, Koteliansky V, Limmer S, Manoharan M, Vornlocher HP (2004). Therapeutic silencing of an endogenous gene by systemic administration of modified siRNAs. *Nature* **432**: 173–178.
15. Kumar P, Ban HS, Kim SS, Wu H, Pearson T, Greiner DL, Laouar A, Yao J, Haridas V, Habiro K, Yang YG, Jeong JH, Lee KY, Kim YH, Kim SW, Peipp M, Fey GH,

Manjunath N, Shultz LD, Lee SK, Shankar P (2008). T cell-specific siRNA delivery suppresses HIV-1 infection in humanized mice. *Cell* **134**: 577–586.

16. Filipowicz W, Jaskiewicz L, Kolb FA, Pillai RS (2005). Post-transcriptional gene silencing by siRNAs and miRNAs. *Current Opinion in Structural Biology* **15**: 331–341.

17. Werner M, Rosa E, Nordstrom JL, Goldberg AR, George ST (1998). Short oligonucleotides as external guide sequences for site- specific cleavage of RNA molecules with human RNase P. *RNA* **4**: 847–855.

18. Beigelman L, McSwiggen JA, Draper KG, Gonzalez C, Jensen K, Karpeisky AM, Modak AS, Matulic-Adamic J, DiRenzo AB, Haeberli P (1995). Chemical modification of hammerhead ribozymes. Catalytic activity and nuclease resistance. *J Biol Chem* **270**: 25702–25708.

19. Amado RG, Mitsuyasu RT, Symonds G, Rosenblatt JD, Zack J, Sun LQ, Miller M, Ely J, Gerlach W (1999). A phase I trial of autologous CD34+ hematopoietic progenitor cells transduced with an anti-HIV ribozyme. *Hum Gene Ther* **10**: 2255–2270.

20. Macpherson JL, Boyd MP, Arndt AJ, Todd AV, Fanning GC, Ely JA, Elliott F, Knop A, Raponi M, Murray J, Gerlach W, Sun LQ, Penny R, Symonds GP, Carr A, Cooper DA (2005). Long-term survival and concomitant gene expression of ribozyme-transduced CD4+ T-lymphocytes in HIV-infected patients. *J Gene Med* **7**: 552–564.

21. Wong-Staal F, Poeschla EM, Looney DJ (1998). A controlled, Phase 1 clinical trial to evaluate the safety and effects in HIV-1 infected humans of autologous lymphocytes transduced with a ribozyme that cleaves HIV-1 RNA. *Hum Gene Ther* **9**: 2407–2425.

22. Schiff WM, Hwang JC, Ober MD, Olson JL, Dhrami-Gavazi E, Barile GR, Chang S, Mandava N (2007). Safety and efficacy assessment of chimeric ribozyme to proliferating cell nuclear antigen to prevent recurrence of proliferative vitreoretinopathy. *Arch Ophthalmol* **125**: 1161–1167.

23. Khvorova A, Lescoute A, Westhof E, Jayasena SD (2003). Sequence elements outside the hammerhead ribozyme catalytic core enable intracellular activity. *Nat Struct Biol* **10**: 708–712.

24. Lieberman J, Song E, Lee SK, Shankar P (2003). Interfering with disease: opportunities and roadblocks to harnessing RNA interference. *Trends Mol Med* **9**: 397–403.

25. Haussecker D (2008). The business of RNAi therapeutics. *Hum Gene Ther* **19**: 451–462.

26. O'Reilly M, Millington-Ward S, Palfi A, Chadderton N, Cronin T, McNally N, Humphries MM, Humphries P, Kenna PF, Farrar GJ (2008). A transgenic mouse model for gene therapy of rhodopsin-linked Retinitis Pigmentosa. *Vision Res* **48**: 386–391.

27. Pei Y, Tuschl T (2006). On the art of identifying effective and specific siRNAs. *Nat Methods* **3**: 670–676.

28. Grimm D, Streetz KL, Jopling CL, Storm TA, Pandey K, Davis CR, Marion P, Salazar F, Kay MA (2006). Fatality in mice due to oversaturation of cellular microRNA/short hairpin RNA pathways. *Nature* **441**: 537–541.

29. Cullen BR (2006). Induction of stable RNA interference in mammalian cells. *Gene Ther* **13**: 503–508.

30. Cullen BR (2006). Enhancing and confirming the specificity of RNAi experiments. *Nat Methods* **3**: 677–681.

31. Kleinman ME, Yamada K, Takeda A, Chandrasekaran V, Nozaki M, Baffi JZ, Albuquerque, RJC, Yamasaki S, Itaya M, Pan Y, Appukuttan B, Gibbs D, Yang Z, Kariko K, Ambati BK, Wilgus, TA, DiPietro LA, Sakurai E, Zhang K, Smith JR,

Taylor EW, Ambati J (2008). Sequence- and target-independent angiogenesis suppression by siRNA via TLR3. *Nature* **452**: 591–597.

32. Robbins M, Judge A, Ambegia E, Choi C, Yaworski E, Palmer L, McClintock K, Maclachlan I (2008). Misinterpreting the therapeutic effects of siRNA caused by immune stimulation. *Hum Gene Ther* **19**(10): 991–9

33. Jackson AL, Burchard J, Schelter J, Chau BN, Cleary M, Lim L, Linsley PS (2006). Widespread siRNA "'off-target" transcript silencing mediated by seed region sequence complementarity. *RNA* **12**: 1179–1187.

34. Lim LP, Lau NC, Garrett-Engele P, Grimson A, Schelter JM, Castle J, Bartel DP, Linsley PS, Johnson JM (2005). Microarray analysis shows that some microRNAs downregulate large numbers of target mRNAs. *Nature* **433**: 769–773.

35. Marquez RT, McCaffrey AP (2008). Advances in MicroRNAs: Implications for Gene Therapists. *Hum Gene Ther* **19**: 27–38.

36. Marshall KA, Ellington AD (2000). *In vitro* selection of RNA aptamers [In Process Citation]. *Methods Enzymol* **318**: 193–214.

37. Krutzfeldt J, Rajewsky N, Braich R, Rajeev KG, Tuschl T, Manoharan M, Stoffel M (2005). Silencing of microRNAs *in vivo* with antagomirs. *Nature* **438**: 685–689.

38. Ebert MS, Neilson JR, Sharp PA (2007). MicroRNA sponges: competitive inhibitors of small RNAs in mammalian cells. *Nat Methods* **4**: 721–726.

39. Baughan T, Shababi M, Coady TH, Dickson AM, Tullis GE, Lorson CL (2006). Stimulating Full-Length SMN2 Expression by Delivering Bifunctional RNAs via a Viral Vector. *Mol Ther* **14**: 54–62.

40. Bolognani F, Perrone-Bizzozero NI (2008). RNA-protein interactions and control of mRNA stability in neurons. *J Neurosci Res* **86**: 481–489.

41. Bramham CR, Wells DG (2007). Dendritic mRNA: transport, translation and function. *Nat Rev Neurosci* **8**: 776–789.

42. Sarkioja M, Hakkarainen T, Eriksson M, Ristimaki A, Desmond RA, Kanerva A, Hemminki A (2008). The cyclo-oxygenase 2 promoter is induced in nontarget cells following adenovirus infection, but an AU-rich 3′-untranslated region destabilization element can increase specificity. *J Gene Med* **10**: 744–753.

43. Bonnet C, Augustin S, Ellouze S, Benit P, Bouaita A, Rustin P, Sahel JA, Corral-Debrinski M (2008). The optimized allotopic expression of ND1 or ND4 genes restores respiratory chain complex I activity in fibroblasts harboring mutations in these genes. *Biochim Biophys Acta* **1783**: 1707–1717.

44. Brown BD, Cantore A, Annoni A, Sergi LS, Lombardo A, la V, Patrizia, D'A, Armando, Naldini L, (2007). A microRNA-regulated lentiviral vector mediates stable correction of hemophilia B mice. *Blood* **110**: 4144–4152.

45. Yen L, Magnier M, Weissleder R, Stockwell BR, Mulligan RC (2006). Identification of inhibitors of ribozyme self-cleavage in mammalian cells via high-throughput screening of chemical libraries. *RNA* **12**: 797–806.

46. Win MN, Smolke CD (2007). A modular and extensible RNA-based gene-regulatory platform for engineering cellular function. *Proc Natl Acad Sci* 0703961104-.

47. Janowski BA, Hu J, Corey DR (2006). Silencing gene expression by targeting chromosomal DNA with antigene peptide nucleic acids and duplex RNAs. *Nat Protoc* **1**: 436–443.

Chapter 8
DNA Integrating Vectors (Transposon, Integrase)

Lauren E. Woodard and Michele P. Calos*

Plasmid DNA vectors provide a way to sidestep many of the limitations of viral vectors, such as capsid immunogenicity. However, gene expression from unintegrated plasmid DNA is typically transient. DNA vectors that integrate into the chromosomes via either an integrase or transposase provide permanent gene addition. Two such systems are under intensive development. The Sleeping Beauty transposase integrates into TA dinucleotides in an essentially random fashion, while ϕC31 integrase is a sequence-specific recombinase that integrates at a more limited number of endogenous sites in mammalian genomes. Both systems have mediated successful gene therapy in numerous animal models. The features of these DNA integrating systems are described, including safety issues and efforts to increase target site specificity.

1. Basic Vector Biology

While viral vectors were the first to be developed for clinical use in gene therapy, they have often proven to be less successful than originally anticipated. Viral vectors are effective at overcoming cellular barriers such as the plasma membrane. However, immune defenses against viral vectors may result in an ineffective therapy or even death. Other major concerns include oncogene activation and accidental production of a self-replicating vector. For these reasons, plasmid DNA may provide a simpler and safer alternative

*Correspondence: Department of Genetics, Stanford University School of Medicine, 300 Pasteur Dr., Stanford, CA 94305.
E-mail: calos@stanford.edu

that is also easier to engineer and produce. In particular, integrating DNA vectors can permanently place the therapeutic gene into the genome, preventing short-term gene expression due to plasmid loss. Integration has been achieved by using transposon and integrase systems.

1.1 *Transposon Systems*

Mobile DNA elements, or transposons, were first discovered because they excise, replicate, and/or insert their DNA back into the genome, causing mutations. Retrotransposons and DNA transposons are the two major classes of mobile DNA elements. The transposons that have been applied to gene therapy applications thus far have been DNA transposons. Transposons have been grouped into families according to their similarities. Most transposons used for gene therapy, including *Sleeping Beauty* (SB), are from the largest such family, the *Tc1/mariner* family.

Because SB is by far the most developed of all of the transposons that have been tested for gene therapy, we will focus on the mechanism of this transposon. By analyzing inactive transposon sequences found in a fish genome, a consensus sequence was discovered that, when synthesized, was found to be an active transposon.[1] Because it was "awakened" from the genome, it was named *Sleeping Beauty*. The system consists of both the DNA sequence to be inserted (transposon) and the enzyme that accomplishes integration into the genome (transposase). Natural transposons code for the transposase and contain the sequences required for its insertion, called terminal inverted repeat (IR)/direct repeat (DR) elements. Only what is between these elements will be permanently placed in the genome. In engineered systems, the transgene cassette is placed between the two IR/DR elements so that it will be inserted (Fig. 8.1). The transposase gene may either be placed on the outside of the IR/DR elements so that it will be left behind after insertion, or it may be placed on a separate plasmid. Once the cell makes the transposase, it can insert the transposon into the genome. Two molecules of SB transposase bind to each of the two IR/DR elements that are located on the transposon DNA molecule. Then, the transposase cuts the transposon DNA at the IR/DR elements in a staggered manner, creating GTC overhangs on the transposon. It also cuts the host genomic DNA in a staggered manner at a TA dinucleotide, creating TA overhangs. In the final step, these overhangs are repaired by host cofactors. This process

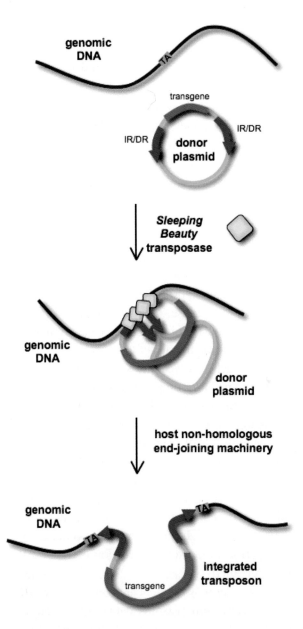

Fig. 8.1 Transposition using *Sleeping Beauty* for gene therapy. The transposase binds IR/DR elements in the transposon and mediates recombination at TA dinucleotides in the genome.

is evidenced by the duplication of the sequence GTCTA after transposon insertion.

1.2 *Integrase Systems*

An integrase is a recombinase enzyme that is capable of integrating DNA, usually that of a virus, into another piece of DNA, usually the host chromosome. Bacteriophages are omnipresent viruses that infect bacteria. In some cases, bacteriophage DNA can be integrated into the host chromosome by an integrase. If no cofactors are required, these integrases can also be utilized in foreign organisms to incorporate genes into the genomic DNA. Microbial recombinases can be broadly grouped into two categories dependent on the amino acid responsible for cleaving the DNA, either a tyrosine or serine residue. The serine integrases have been found to be the most useful for integration into the genomes of foreign organisms. Integrases recombine two distinct sequences to create hybrid sites that can no longer be recognized and recombined by the integrase, making integration irreversible. Of the integrases studied thus far ϕC31 integrase has been found to have the most favorable properties for gene therapy.

ϕC31 integrase *att*aches to specific sequences in both the *p*hage and *b*acterial genomes, termed *attP* and *attB*, respectively. The sites are somewhat palindromic, meaning that one half of the site is partially similar to the other half. The minimal site length is 34 bp for *attB* and 39 bp for *attP*.[2] It is believed that one molecule of ϕC31 integrase binds to each half of the *att* site and that the two sites come together through formation of a tetrameric complex (Fig. 8.2). The DNA is cleaved at a core sequence in the *att* sites, and the DNA-integrase complex rotates to form the recombination products. The resulting hybrid *att* sites are termed *attL* and *attR*, because they are found to the left (5′) and right (3′) of the phage genome. Wild-type ϕC31 integrase cannot perform the excision reaction between *attL* and *attR*, so integration is irreversible.[3]

ϕC31 integrase is relatively efficient at recombining at pseudo *attP* sites when given an *attB*-containing plasmid, but poor at recombining at pseudo *attB* sites.[4] When used in mammalian cells for gene addition, the ϕC31 integrase system is comprised of two plasmids. One plasmid expresses the active integrase enzyme. Another plasmid contains the *attB* sequence to allow for sequence-specific integration, the transgene, and often a selectable

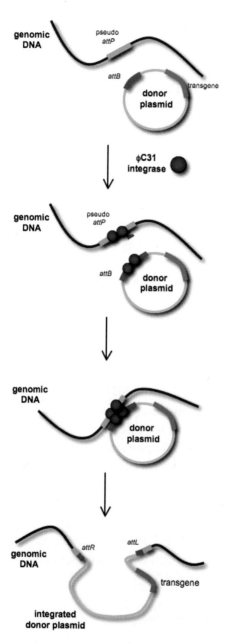

Fig. 8.2 Integration using φC31 integrase for gene therapy. The integrase binds to both the *attB* sequence supplied on the donor plasmid and native pseudo *attP* sequences in the genome, then recombines the DNA to integrate the transgene permanently.

marker that can be used to eliminate cells that have not received integration events.

2. Vector Design and Production

2.1 *Design of Transposon Systems*

There are many factors that can affect transposition. A phenomenon called overexpression inhibition has been observed with SB, which means that too much of the transposase actually inhibits the transposition reaction.[5] This complication necessitates the optimization of transposon to transposase DNA ratios for every new application or vector. Both the size of the transgene and the transposon-containing plasmid are inversely correlated with transposition efficiency in the SB system.[6] This correlation is even more pronounced if DNA of prokaryotic, rather than eukaryotic, origin is used as the transgene. If the distance between transposon ends is decreased, transposition increases.[7] Another way to decrease the physical distance between the transposon ends is to methylate the DNA, which causes DNA condensation. Methylation of the transposon gave an 11-fold increase in overall transposition rates.[8]

The first SB transposase was SB10, and the original transposon was called pT. This system mediated transposition in only 0.3% of cultured human cells, which was an increase of 20 to 40-fold over the background rate of random integration. Improvements in the IR/DR sequences created the pT2, pT3, and pMSZ transposons, which are 2-4 times more effective than pT when combined with SB10.[9–12] The enzyme SB10 has also been improved many times. A recent improvement, HSB17 transposase, together with the pMSZ transposon gave a 30-fold improvement in cell culture over the original SB10/pT system.[13] However, in mice the most effective enzyme was the HSB2 enzyme, which was 7-fold improved over SB10 when used with pT or pT3 transposons.[9]

2.2 *Design of Integrase Systems*

Some attractive features of the integrase system include that the donor plasmid has no size limit, there is no known overexpression inhibition phenomenon, and the integrase does not require cell division, at least in the liver where long-term integration events were found in cells that did not undergo replication.[14] Mutants of ϕC31 integrase have been identified that have increased activity. One such mutant, P3, appeared to have increased specificity for the wild-type attP site, making it useful for applications in which a wild-type *attP* is present, such as plasmid excision or integration into an engineered cell line.[15] For gene therapy, a more useful mutant is P2, which demonstrated two-fold increased integration efficiency at pseudo *attP* sites in cell culture and mouse liver. There was no obvious change in specificity between P2 and wild-type integrase, but more extensive sequencing of pseudo *attP* sites is necessary.[15]

For gene therapy, ϕC31 integrase has generally been expressed under control of the strong constitutive cytomegalovirus (CMV) promoter. Although adding a C-terminal nuclear localization signal was found to aid in a chromosomal excision reaction,[16] it hampered integration into pseudo *attP* sites *in vitro* and did not aid integration in mouse liver.[14] Also, codon optimization was found to be helpful for an excision reaction in the early mouse embryo.[17] For increased control over enzyme expression, an inducible integrase has been constructed.[18]

Donor plasmid construction is fairly simple. Generally the full-length ~300 bp *attB* site is placed on the plasmid between coding regions for an antibiotic resistance gene and the therapeutic gene. In tissue culture, the 34 bp site was found to be functional but somewhat inferior, with about 75% activity of the full-length *attB* site at integration into pseudo *attP* sites.[19] Depending on the application, the investigator may wish to use tissue-specific promoters or insulator elements to aid expression. While there is no size limit for integration, it is easier to construct and produce DNA plasmids that are of a moderate size. Bacterial artificial chromosomes may be used if the therapeutic gene is very large.[20]

2.3 Production of Plasmid DNA

Plasmids must have an origin of replication for *E. coli* and a selectable marker to ensure that only *E. coli* carrying plasmid comprise the culture. The United States FDA currently prefers plasmids that will be used as molecular medicines to avoid using ampicillin resistance, since ampicillin is prescribed to patients for use as an antibiotic. Another important feature is copy number, which is the number of plasmids per *E. coli* cell. High copy number plasmids are preferable since they provide a better yield of plasmid DNA from the same culture volume.

Plasmids are transformed into an optimal strain of *E. coli*. A culture is grown from a single colony, which may be cell banked. *E. coli* are obtained from liquid culture by centrifugation and then lysed and neutralized, resulting in the precipitation of protein and genomic DNA. The plasmid DNA is purified from the supernatant by chromatography and isopropanol precipitation. If the DNA will be used *in vivo*, endotoxins must also be removed to avoid serious side effects such as fever and shock. For use in the clinic, plasmids must be validated and then carefully monitored for purity and integrity during storage.[21]

3. Gene Transfer Protocols and Potential Applications

Mammalian cells do not normally take up plasmid DNA efficiently. Therefore, DNA integrating systems are limited by the availability of transfection techniques. We will provide a brief overview of several transfection methods and their gene therapy applications with integrating vectors.

3.1 *Hepatocyte Transfection via Hydrodynamic Injection*

Hydrodynamic injection is a fast, straightforward way to introduce DNA into the liver cells of mice.[22,23] This method uses a large volume of dilute DNA solution, injected quickly through the tail vein, to introduce the DNA into the hepatocytes. The human alpha-1 antitrypsin (hAAT) and human Factor IX (hFIX) transgenes have been delivered to normal mice using both SB and ϕC31 integrase, resulting in higher long-term transgene expression when recombinase was present.[24,25] SB and ϕC31 integrase have been compared directly for hFIX expression. Long-term transgene levels when wild-type integrase was used were about three-fold higher than those obtained using an optimized SB system.[26] Both systems have been shown to correct several mouse disease models using hydrodynamic injection and may be tested in the clinic by using balloon catheters to achieve hydrodynamic delivery.

3.2 *Lipophilic Complexes to Transfect Endothelial Cells and Glioblastoma*

Because DNA is negatively charged and must traverse the hydrophobic cell membrane, DNA may be complexed in solution with lipophilic reagents that have a positively charged head and a hydrocarbon tail. These reagents then encapsulate the DNA in a fatty ball, allowing it to be endocytosed and thus enter the cell. This strategy is extremely effective for cell culture and is also a viable *in vivo* transfection method. Complexes have been used to deliver the SB system to type II pneumocyte cells in the lung and to specifically express in endothelial cells by way of a tissue-specific promoter.[12] Lipophilic complexes were also used to target glioblastoma brain cancer cells with success, shrinking the tumors and increasing survival times of the mice when the SB system integrated cancer-killing genes.[27]

3.3 *Direct DNA Injection and Electroporation to Target Muscle, Retina, and Joints*

An effective way to transfect muscle fibers is to use electroporation. The muscle fiber is injected with plasmid DNA, then electrodes are placed on either side of the injection site and a current is passed through the tissue to disrupt the plasma membrane. This method was used to treat disease-model mice with a dystrophin gene using ϕC31 integrase.[28] ϕC31 integrase was also used in conjunction with electroporation to deliver the VEGF gene for hindlimb ischemia.[29] Both of these strategies were carried out in juvenile (12 day old) mice.

Electroporation has also been used to deliver DNA to the rat retina. This transfection method was combined with ϕC31 integrase to achieve long-term marker gene expression.[30] Additionally, in a rabbit model of arthritis, joints were injected intra-articularly with naked DNA, including ϕC31 integrase.[31] Transgene levels were higher when the integrase was present, but overall transgene levels were still relatively low, reflecting inefficient delivery.

3.4 *Integration into Cultured Cells for* ex vivo *Gene Therapy*

Depending on the ability of progenitor cells to be grown under tissue culture conditions and transplanted, *ex vivo* approaches may be suitable for treatment of diseases using a combined gene and cell therapy. The therapeutic cells are propagated, transfected with the integrating vector system, and used to treat the disease model, where the stem cells may contribute to the target tissue. Another use of this strategy is to improve cells so that they are better at attacking a tumor. SB has been used to modify peripheral blood leukocytes to stably express a suicide gene for cancer gene therapy.[32]

Both SB and ϕC31 integrase have been used effectively to integrate therapeutic genes into primary human keratinocytes isolated from patients with epidermolysis bullosa.[33,34] When these keratinocytes were grafted onto immune-deficient mice, healthy skin grew, indicating a viable gene therapy strategy. Muscle precursor cells have also been subjected to gene addition with ϕC31 integrase, in another example of a possible ϕC31 integrase-mediated treatment for dystrophin deficiency.[35] Additionally, a human T-cell line derived from a patient with X-linked severe combined immune deficiency was successfully corrected using ϕC31 integrase.[36] Other hematopoietic cells such as immortalized K-562 and CD34+ hematopoietic stem cells have been given SB carrying the reporter gene GFP.[37] While transposition was successful, engraftment was poor. However, blood outgrowth endothelial cells have been engineered with SB to produce hAAT, hFIX and hFVIII.[38] Embryonic stem cells and neural progenitor cells have been modified with ϕC31 integrase and were differentiated after modification into several lineages.[39,40]

4. Immunology

One of the major challenges facing gene therapy is the immune response of patients to the vectors used to deliver helpful genes. Systems utilizing plasmid DNA should be less immunogenic than viral vectors. No immune response has been observed in response to either SB or ϕC31 integrase in animals, and there is no reason to suspect that humans would be different in this respect. The duration of SB and ϕC31 integrase protein expression after hydrodynamic injection have been investigated. Both systems exhibit a short expression time of the recombinase, on the order of a couple of days.[41,42] This brief expression time may be due to the short period of transcription from the viral promoter (CMV) used to express the recombinase and is favorable to limit adverse effects.

5. Safety and Regulatory Issues

There is a risk associated with genomic modification because insertion events could cause oncogene activation or tumor suppressor loss, thereby increasing the likelihood of cancer.

5.1 *Integration Profiles and Associated Hazards*

SB transposase requires a TA dinucleotide to integrate and otherwise has few statistically significant sequence preferences. It prefers highly deformable sequences, probably because the DNA must be bent in order for SB to recombine it.[43] SB may be less harmful than retroviruses, because SB integration is essentially random, rather than favoring transcriptionally active regions. However, a randomly integrating system will still integrate in areas that are problematic if enough cells are subjected to transposition.

ϕC31 integrase, on the other hand, is naturally sequence-specific. By evaluating integration sites rescued from human cell lines a 28bp consensus pseudo *attP* sequence was identified.[44] While 60% of the 123 rescued integration events occurred at only 19 sites, many sites were rescued only once. Using a statistical method, it was estimated that approximately 370 potential integration sites may exist in the human genome. Of the rescued sites, none were found in regions of the genome that could be expected to contribute to malignant transformation of a cell, such as near an oncogene or disrupting a tumor suppressor. However, it was discovered that about 10% of integrations occurred across chromosomes, meaning that the plasmid integrated between pseudo *attP* sites located on two different chromosomes.[44]

5.2 *Efforts to Enhance Integration Specificity*

Integrating gene therapies may be improved by placing the therapeutic gene at a known, safe site in every cell. Strides have been made toward this goal. SB has been targeted to a particular genomic location in cultured human cells 10% of the time, which is a 10^7-fold enrichment over random integration.[45] This approach used expression of another protein that bound to both the transposase and the desired genomic sequence, thus bringing the transposase closer to the desired location. If this system could be further refined to eliminate random integration and were demonstrated *in vivo*, it would be a major advance for the SB system.

ϕC31 integrase, while many orders of magnitude more specific than SB, may also be modified to make integration more specific. Using a bacterial screen for directed evolution of the integrase gene, ϕC31 was mutated to integrate preferentially at a human pseudo *attP* site called psA. The best mutant was ~6-fold improved over the wild-type enzyme at integration into the genomic location psA.[46] ϕC31 integrase has also been mutagenized and screened for increased recombination at the native *attP* site. Mutants with ~8-fold higher specificity for *attP* were found.[46] This higher specificity acts as a proof of principle that the enzyme can be changed to recognize specific sequences better.

5.3 *Effects on Tumor Latency in Mouse Models of Cancer*

SB is so effective at random integration that it has been optimized and used for gene discovery to find cancer-causing genes in mouse and rat. Transgenic mice expressing the transposase under tissue-specific promoters have proven to be an advance over the traditional method of cancer gene discovery using retroviruses.[47] Although in these instances SB has been optimized to find the cancer genes, this application highlights the risk that even the more benign version of SB used in gene therapy could also cause problems by integrating near cancer genes.

ϕC31 integrase is currently being tested in a transgenic mouse model of liver cancer using hydrodynamic injection to deliver a donor plasmid. Preliminary evidence suggests that the presence of ϕC31 integrase does not cause acceleration of tumorigenesis in the liver, reducing concerns about the carcinogenic potential of ϕC31 integrase.[19] Upon completion, this study will provide rigorous data to augment prior observations that cancers have not been seen in treated mice.

In the future, integrating systems are expected to play an important role in the formulation of gene therapies for the lifelong treatment of genetic disease. Further improvements in these systems, as well as upcoming clinical trials using them for novel gene therapies, are eagerly awaited.

References

1. Ivics Z, Hackett PB, Plasterk RH, Izsvak Z (1997). Molecular reconstruction of Sleeping Beauty, a Tc1-like transposon from fish, and its transposition in human cells. *Cell* **91**: 501–510.
2. Groth AC, Olivares EC, Thyagarajan B, Calos MP (2000). A phage integrase directs efficient site-specific integration in human cells. *Proc Natl Acad Sci U S A* **97**: 5995–6000.
3. Thorpe HM, Wilson SE, Smith MC (2000). Control of directionality in the site-specific recombination system of the Streptomyces phage phiC31. *Mol Microbiol* **38**: 232–241.
4. Thyagarajan B, Olivares EC, Hollis RP, Ginsburg DS, Calos MP (2001). Site-specific genomic integration in mammalian cells mediated by phage phiC31 integrase. *Mol Cell Biol* **21**: 3926–3934.
5. Hackett PB, Ekker SC, Largaespada DA, McIvor RS (2005). Sleeping beauty transposon-mediated gene therapy for prolonged expression. *Adv Genet* **54**: 189–232.
6. Karsi A, Moav B, Hackett P, Liu Z (2001). Effects of insert size on transposition efficiency of the sleeping beauty transposon in mouse cells. *Mar Biotechnol (NY)* **3**: 241–245.
7. Izsvak Z, Ivics Z, Plasterk RH (2000). Sleeping Beauty, a wide host-range transposon vector for genetic transformation in vertebrates. *J Mol Biol* **302**: 93–102.
8. Yusa K, Takeda J, Horie K (2004). Enhancement of Sleeping Beauty transposition by CpG methylation: possible role of heterochromatin formation. *Mol Cell Biol* **24**: 4004–4018.
9. Yant SR, Park J, Huang Y, Mikkelsen JG, Kay MA (2004). Mutational analysis of the N-terminal DNA-binding domain of sleeping beauty transposase: critical residues for DNA binding and hyperactivity in mammalian cells. *Mol Cell Biol* **24**: 9239–9247.
10. Geurts AM, *et al.* (2003). Gene transfer into genomes of human cells by the sleeping beauty transposon system. *Mol Ther* **8**: 108–117.
11. Cui Z, Geurts AM, Liu G, Kaufman CD, Hackett PB (2002). Structure-function analysis of the inverted terminal repeats of the sleeping beauty transposon. *J Mol Biol* **318**: 1221–1235.
12. Liu L, Sanz S, Heggestad AD, Antharam V, Notterpek L, Fletcher BS (2004). Endothelial targeting of the Sleeping Beauty transposon within lung. *Mol Ther* **10**: 97–105.
13. Baus J, Liu L, Heggestad AD, Sanz S, Fletcher, BS (2005). Hyperactive transposase mutants of the Sleeping Beauty transposon. *Mol Ther* **12**: 1148–1156.
14. Woodard LE, Hillman RT, Keravala A, Lee S, Calos MP (2010). Effect of nuclear localization and hydrodynamic delivery-induced cell division on phiC31 integrase activity. *Gene Ther* **17**: 217–226.
15. Keravala A, *et al.* (2009). Mutational Derivatives of PhiC31 Integrase With Increased Efficiency and Specificity. *Mol Ther* **17**: 112–120.

16. Andreas S, Schwenk F, Kuter-Luks B, Faust N, Kuhn R (2002). Enhanced efficiency through nuclear localization signal fusion on phage PhiC31-integrase: activity comparison with Cre and FLPe recombinase in mammalian cells. *Nucleic Acids Res* **30**: 2299–2306.

17. Raymond CS, Soriano P (2007). High-efficiency FLP and PhiC31 site-specific recombination in mammalian cells. *PLoS ONE* **2**: e162.

18. Sharma N, Moldt B, Dalsgaard T, Jensen TG, Mikkelsen JG (2008). Regulated gene insertion by steroid-induced PhiC31 integrase. *Nucleic Acids Res* **36**: e67.

19. Woodard LE, Calos MP (Unpublished results).

20. Venken KJ, He Y, Hoskins RA, Bellen HJ (2006). P[acman]: a BAC transgenic platform for targeted insertion of large DNA fragments in *D. melanogaster. Science* **314**: 1747–1751.

21. Voss C (2007). Production of plasmid DNA for pharmaceutical use. *Biotechnol Annu Rev* **13**: 201–222.

22. Liu F, Song Y, Liu D (1999). Hydrodynamics-based transfection in animals by systemic administration of plasmid DNA. *Gene Ther* **6**: 1258–1266.

23. Zhang G, Budker V, Wolff JA (1999). High levels of foreign gene expression in hepatocytes after tail vein injections of naked plasmid DNA. *Hum Gene Ther* **10**: 1735–1737.

24. Yant SR, Meuse L, Chiu W, Ivics Z, Izsvak Z, Kay MA (2000). Somatic integration and long-term transgene expression in normal and haemophilic mice using a DNA transposon system. *Nat Genet* **25**: 35–41.

25. Olivares EC, Hollis RP, Chalberg TW, Meuse L, Kay MA, Calos MP (2002). Site-specific genomic integration produces therapeutic Factor IX levels in mice. *Nat Biotechnol* **20**: 1124–1128.

26. Ehrhardt A, Xu H, Huang Z, Engler JA, Kay MA (2005). A direct comparison of two nonviral gene therapy vectors for somatic integration: *in vivo* evaluation of the bacteriophage integrase phiC31 and the Sleeping Beauty transposase. *Mol Ther* **11**: 695–706.

27. Ohlfest JR, *et al.* (2005). Combinatorial antiangiogenic gene therapy by nonviral gene transfer using the sleeping beauty transposon causes tumor regression and improves survival in mice bearing intracranial human glioblastoma. *Mol Ther* **12**: 778–788.

28. Bertoni, C, *et al.* (2006). Enhancement of plasmid-mediated gene therapy for muscular dystrophy by directed plasmid integration. *Proc Natl Acad Sci U S A* **103**: 419–424.

29. Portlock JL, Keravala A, Bertoni C, Lee S, Rando TA, Calos MP (2006). Long-term increase in mVEGF164 in mouse hindlimb muscle mediated by phage phiC31 integrase after nonviral DNA delivery. *Hum Gene Ther* **17**: 871–876.

30. Chalberg TW, Genise HL, Vollrath D, Calos MP (2005). PhiC31 integrase confers genomic integration and long-term transgene expression in rat retina. *Invest Ophthalmol Vis Sci* **46**: 2140–2146.

31. Keravala A, Portlock JL, Nash JA, Vitrant DG, Robbins PD, Calos MP (2006). PhiC31 integrase mediates integration in cultured synovial cells and enhances gene expression in rabbit joints. *J Gene Med* **8**: 1008–1017.

32. Huang X, *et al.* (2006). Stable gene transfer and expression in human primary T cells by the Sleeping Beauty transposon system. *Blood* **107**: 483–491.

33. Ortiz-Urda S, Lin Q, Yant SR, Keene D, Kay MA, Khavari PA (2003). Sustainable correction of junctional epidermolysis bullosa via transposon-mediated nonviral gene transfer. *Gene Ther* **10**: 1099–1104.

34. Ortiz-Urda S, *et al.* (2002). Stable nonviral genetic correction of inherited human skin disease. *Nat Med* **8**: 1166–1170.

35. Quenneville SP, Chapdelaine P, Rousseau J, Tremblay JP (2007). Dystrophin expression in host muscle following transplantation of muscle precursor cells modified with the phiC31 integrase. *Gene Ther* **14**: 514–522.

36. Ishikawa Y, *et al.* (2006). Phage phiC31 integrase-mediated genomic integration of the common cytokine receptor gamma chain in human T-cell lines. *J Gene Med* **8**: 646–653.

37. Hollis RP, *et al.* (2006). Stable gene transfer to human CD34(+) hematopoietic cells using the Sleeping Beauty transposon. *Exp Hematol* **34**: 1333–1343.

38. Kren BT, Yin W, Key NS, Hebbel RP, Steer, CJ (2007). Blood outgrowth endothelial cells as a vehicle for transgene expression of hepatocyte-secreted proteins via Sleeping Beauty. *Endothelium* **14**: 97–104.

39. Thyagarajan B, *et al.* (2008). Creation of engineered human embryonic stem cell lines using phiC31 integrase. *Stem Cells* **26**: 119–126.

40. Keravala A, Ormerod BK, Palmer TD, Calos MP (2008). Long-term transgene expression in mouse neural progenitor cells modified with phiC31 integrase. *J Neurosci Methods* **173**: 299–305.

41. Chavez CL, Keravala A, Woodard LE, Calos MP (2009). Unpublished data.

42. Bell JB, *et al.* (2004). Duration of Expression of *Sleeping Beauty* Transposase by Hydrodynamic Injection of C57/BL6 Mice. *Molecular Therapy* **9**: S312.

43. Yant SR, Wu X, Huang Y, Garrison B, Burgess SM, Kay MA (2005). High-resolution genome-wide mapping of transposon integration in mammals. *Mol Cell Biol* **25**: 2085–2094.

44. Chalberg TW, *et al.* (2006). Integration specificity of phage phiC31 integrase in the human genome. *J Mol Biol* **357**: 28–48.

45. Ivics Z, Katzer A, Stuwe EE, Fiedler D, Knespel S, Izsvak Z (2007). Targeted sleeping beauty transposition in human cells. *Mol Ther* **15**: 1137–1144.

46. Sclimenti CR, Thyagarajan B, Calos MP (2001). Directed evolution of a recombinase for improved genomic integration at a native human sequence. *Nucleic Acids Res* **29**: 5044–5051.

47. Dupuy AJ, Jenkins NA, Copeland NG (2006). Sleeping beauty: a novel cancer gene discovery tool. *Hum Mol Genet* **15 Spec No 1**: R75–79.

Chapter 9

Homologous Recombination and Targeted Gene Modification for Gene Therapy

Matthew Porteus*

The ability to manipulate the genome in a site-specific manner would increase the specificity and perhaps the efficacy of gene therapy. Gene targeting by homologous recombination, whereby an endogenous segment of the genome is replaced by an introduced segment of DNA, can be used to both correct disease-causing mutations (gene correction) and to target the insertion site of transgenes (targeted transgene insertion). Gene targeting can be mediated by adeno-associated virus and stimulated by the induction of gene-specific double-strand breaks. Gene specific double-strand breaks can be made by either designed zinc finger nucleases or modified homing endonucleases. In addition, zinc finger nucleases can create gene specific mutations when the cell repairs an induced double-strand break in a mutagenic fashion. In this way, permanent gene-specific mutations in human somatic cells can be created with therapeutic intent.

1. Introduction

Gene therapy is based on the principle that we can manipulate the nucleic acid content of cells, usually the genome directly, for therapeutic purposes. The field was born from the recognition of the genetic origin of thousands of diseases and the discovery of the tools of molecular biology. The ability to create new DNA molecules and introduce DNA molecules into cells seemed like a direct way to cure or ameliorate genetic diseases. In the

---

*Correspondence: Department of Pediatrics and Biochemistry, UT Southwestern Medical Center.
E-mail: Matthew.Porteus@utsouthwestern.edu

laboratory, the modification of cellular phenotype by molecular biological techniques forms the backbone of contemporary biomedical research. The translation of these routine techniques from the laboratory into improvements in care for patients, however, has been more difficult. One of the most efficient ways to alter the genome of cells is to use vectors, such as retroviruses, lentiviruses, and transposons, that integrate new DNA into the genome. A current characteristic of these vectors is that they integrate in an uncontrolled fashion. These uncontrolled integrations have two fundamental problems. The first is that the integrations can occur in regions of the genome that cause the gene of interest to be expressed inappropriately secondary to "position-effects." Position effect variegation can cause both problems in safety (the gene of interest is expressed at a dangerous time, place or amount) or efficacy (the gene of interest is not expressed at the correct time, place or amount). The second problem is that uncontrolled integrations cause uncontrolled mutations. That is, each integration creates a new mutation. These mutations can result in the activation of a nearby onco-gene, thus causing cancer in a relatively short-time frame (several years). They can also result in the inactivation of potentially important genes that can cause the cell to behave inappropriately. If a tumor suppressor was inactivated by an insertion, for example, one might contribute to the oncogenic process, the result of which might not become apparent for decades. A solution to the problem of uncontrolled integrations is to use controlled genome modification (Fig. 9.1).

2. Problems with Using Gene Targeting by Homologous Recombination

In gene targeting by homologous recombination (hereafter referred to as "gene targeting" or "homologous recombination"), a fragment of DNA is introduced into a cell and that fragment replaces a similar but not identical region of the endogenous genome (Fig. 9.1). Through homologous recombination both large and small sequence changes can be introduced into the genome in a site-specific fashion.

The use of homologous recombination to modify genomes has been widely used in many experimental systems including bacteria, yeast, murine embryonic stem cells, and a few vertebrate cell lines. The major limitation

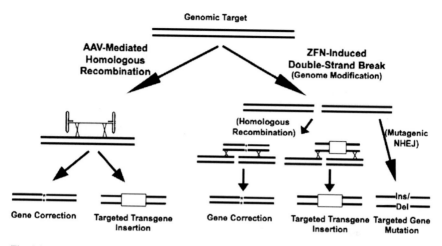

Fig. 9.1 Schematic Representation of Strategies to Site-Specifically Modify a Genomic Target by Either Homologous Recombination or Mutagenic Non-Homologous End-Joining (NHEJ). The "*" represents a correction of a small mutation, the "box" represents the targeted insertion of a transgene including new regulatory elements, and "Ins/Del" represents the insertion or deletion of a small number of nucleotides at the site of the ZFN induced double-strand break.

to using homologous recombination in mammalian cells has been two-fold. First, the spontaneous rate of gene targeting (10^{-6}) is too low to be clinically useful.[1,2] Second the relative rate of gene targeting, the ratio of targeted to non-targeted integrations, is also too low to be usable. In vertebrate cell types in which homologous recombination has been used experimentally, the relative rate of targeted vs. non-targeted integrations is higher than usual and the targeted events can be identified using simple screening strategies.[3] In this chapter, three approaches to using homologous recombination for gene therapy are discussed: 1) Using homologous recombination in embryonic stem cells and then differentiating embryonic stem cells into tissue specific stem cells for infusion back into a patient; 2) the use of adeno-associated virus; 3) the use of nucleases to create gene-specific DSBs to stimulate the rate of gene targeting.

3. Homologous Recombination in Embryonic Stem Cells

A standard paradigm for gene therapy for monogenic diseases is to remove tissue specific stem cells from a patient, use gene transfer technologies to

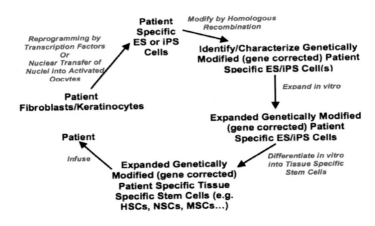

Fig. 9.2 Using Homologous recombination in ES/iPS cells for gene therapy.

manipulate the genome of those cells, and then re-infuse those cells back into the patient. The clinical trials for the immunodeficiencies have followed this paradigm.[4,5] An alternative paradigm is based on using embryonic stem (ES) cells (Fig. 9.2).

ES cells are immortal cells derived from the early blastocyst embryo and are capable of giving rise to all three germ layers (ectoderm, mesoderm, and endoderm). When implanted back into the early blastocyst, embryonic stem cells can give rise to every cell in the body. In addition, ES cells can be induced to form multiple different somatic cell types, including tissue specific stem cells, *in vitro*. In a gene therapy paradigm using ES cells, patient specific ES cells would be generated. Patient specific ES cells can currently be generated two possible ways. First, by nuclear transfer in which the nucleus of a patient cell is used to replace the nucleus of an activated oocyte thus creating a cell that is genetically identical to the patient.[6] This modified oocyte can then be converted *in vitro* into an ES cell. Second, patient specific pluripotent cells can be created by the use of re-programming by transcription factors. In this strategy defined transcription factors (such as Oct4, Nanog, Klf4, Sox2, and c-Myc) are introduced into patient somatic cells.[7] These transcription factors are able to reprogram a small fraction of the differentiated somatic cells, by a mechanism that is currently not understood, into cells with ES cell like properties (called "induced pluripotent stem cells" or "iPS" cells). Patient specific ES cells would be modified *in vitro* by homologous recombination. The manipulation of murine ES cells by homologous recombination has become a backbone of biomedical research, the importance of which was recognized by the awarding of the Nobel Prize in Medicine in 2007 to Capecchi, Smithies, and Evans for their development of the technology. The ES cells having undergone homologous recombination are identified and expanded. Finally, the modified ES cells can then be induced into tissue specific stem cells *in vitro* and re-infused to correct a disease. In the last decade, proof-of-principle experiments have been published in mice demonstrating the feasibility of this approach in correcting mouse models of immunodeficiencies and sickle cell anemia.[8,9] These mouse pre-clinical experiments highlight the promise of this strategy. There remain several important obstacles, however, before this strategy can be adopted in humans. First, a technique for the creation of patient specific iPS cells must be established such that the iPS cells have no increased propensity for cancer and are fully capable of being converted into tissue specific stem cells. Recent studies suggest that this problem may soon be solved for human cells using non-integrating gene transfer techniques or other strategies.[10] Second, while iPS cells seem to behave like true ES cells, there remains no definitive way to establish this for human cells. The ability to form teratomas in mice and differentiate into various tissue types *in vitro* are surrogates that may or may not accurately reflect the true potentiality of any given iPS line. Finally, reliable and non-oncogenic methods of converting ES cells into transplantable tissue specific stem cells need to be established. The simple injection of ES cells without such differentiation results in teratomas. The differentiation of ES cells into transplantable hematopoietic stem cells, for example, while possible in the murine model using proto-oncogenes such as Cdx2 and HoxB4, still remains relatively inefficient and requires the use of oncogenes.[11]

Although critical problems remain, overall, the paradigm of creating patient specific stem cells using the iPS technology, correcting those cells using homologous recombination or site-specific modification techniques, expansion of those modified cells and then transplanting tissue specific stem cells derived from the modified ES cells remains of high interest and great potential.

4. Homologous Recombination using Adeno-Associated Virus

The major limitation to using homologous recombination for gene therapy is the low spontaneous rate in human somatic cells. When naked DNA is introduced into a human somatic cell, the rate of gene targeting is on the order of 10^{-6}.[1,2] An intriguing property of adeno-associated virus (AAV), is that it stimulates gene targeting in a wide variety of human cell types.[12,13] Using high multiplicity of infections, targeting rates of greater than 1% have been obtained.[14] Using the AAV system, the entire range of genome modifications (correction of point mutations, the insertion of small or large segments of DNA, the creation of small deletions) have been created by homologous recombination. The use of AAV to create human somatic cell lines with genomes modified by homologous recombination has been adopted by a number of laboratories. The potential of using AAV-mediated homologous recombination for therapeutic purposes was highlighted by the work of Chamberlain, Russell and their colleagues.[15] They demonstrated that they could use AAV mediated gene targeting to selectively knockout the dominant negative allele in mesenchymal stem cells from patients with osteogenesis imperfecta.

5. Site-Specific Modification of the Genome using Double-Strand Breaks

A third method to site-specifically modify the genome is to use enzymes that create site-specific DNA double-strand breaks (DSBs). The way the DSB is repaired determines the type of genomic modification that occurs. Gene targeting occurs when the DSB is repaired using homologous recombination while insertions/deletions can occur at the site if the DSB is repaired by non-homologous end-joining (Fig. 9.1).

6. Double-Strand Break Repair

Double-strand breaks (DSBs) are among the most dangerous types of DNA damage because they can lead to chromosomal translocations, chromosomal loss and aneuploidy. The creation of a DSB activates a complex cell signaling pathway resulting in changes in chromatin structure, cell cycle arrest, the repair of the DSB or ultimately apoptosis if the DSB is not repaired in a timely fashion. Cells have redundant mechanisms to repair DSBs; the most prominent of which are homologous recombination and non-homologous end-joining (NHEJ).[16]

Conceptually the repair of a DSB by homologous recombination occurs by a "copy and paste" mechanism.[17] In homologous recombination, the DSB is processed to form long single-stranded 3' tails. These tails serve as a major signal to the cell that damage has occurred and initiate the cellular response to a DSB. In the repair of a DSB by homologous recombination, the repair machinery uses the 3' tails to initiate strand invasion into an undamaged, homologous repair template. The machinery then uses the undamaged template to synthesize new DNA that spans the site of the DSB. The amount of new DNA can vary from tens of bases (short-tract repair) to thousands of bases (long-tract repair).[18] The usual repair template is the sister chromatid (an identical copy to the damaged DNA). The repair of a DSB by homologous recombination using the sister chromatid will result in a perfect restoration of the damaged DNA and is considered a high-fidelity form of DSB repair. If the cell chooses a repair template that is homologous, but not identical, to the damaged DNA, then the cell will incorporate the sequence differences from the template into the damaged site, thereby creating a "gene conversion" event. Gene targeting occurs when the cell uses an experimentally introduced homologous, but not identical, DNA fragment as the template for repair. Thus, the stimulation of gene targeting by the induction of a site-specific DNA DSB is the result of harnessing the cell's own DNA repair machinery to repair the DSB by homologous recombination.

In contrast, the repair of a DSB by NHEJ is conceptually like gluing the two ends together.[19] In NHEJ, the ends are recognized, cell signaling is activated, the ends are processed and ultimately ligated together. If the ends of the break are "ragged," that is, not easily ligated back together, the repair by NHEJ will result in small insertions and deletions at the site of the break. Even if the ends are "clean," where the ends can be easily joined, the repair of a DSB by NHEJ can sometimes lead to small insertions and deletions at the site of the break. Thus, repeated cutting of a single site by an endonuclease can lead to mutations at the site of the break. This mutagenic effect of NHEJ is utilized to create site-specific mutations using ZFNs (discussed below and Fig. 9.1).

7. Double-Strand Break Induced Homologous Recombination

Double-strand breaks are powerful inducers of homologous recombination. There are multiple examples in nature where a nuclease induced DSB is made to stimulate genome rearrangement by homologous recombination including meiotic recombination, yeast mating type switching, and the genetic transfer of homing endonucleases. In addition, homologous recombination is a major mechanism by which cells repair spontaneous DSBs.[20] In the 1990's several labs determined that DSBs could stimulate gene targeting in mammalian cells.[2,21–26] In these studies, a recognition site for the I-SceI homing endonuclease (an 18 basepair sequence) was introduced into a reporter gene thereby inactivating the reporter gene. The reporter gene was then chromosomally integrated as a single copy into the genome to make a reporter cell line. The reporter cell line was co-transfected with an expression plasmid for I-SceI and a fragment of DNA (called the repair/donor) to fix the mutation in the reporter gene. I-SceI would create a DSB in the reporter gene, this DSB would be repaired by the cell by homologous recombination using the repair/donor DNA fragment and 1000-50,000 fold stimulation in gene targeting was achieved when compared to when the DSB was not induced. Thus, a DSB stimulated the process of gene targeting by 3-5 orders of magnitude in human somatic cells.[2] In principle, therefore, a path to using homologous recombination for gene therapy was established if a mechanism to create gene specific DSBs could be found.

8. Re-design of Homing Endonucleases to Recognize New Target Sites

Homing endonucleases, of which I-SceI is a member, are genetic parasites that catalyze their duplication by DSB mediated homologous recombination.[27] There are now hundreds of such proteins and the important characteristic of this family is that the endonucleases all have long recognition sites giving them excellent specificity. The widespread use of

I-SceI as a tool to explore DSB repair with no reports of off-target geno-toxicity, suggest that homing endonucleases could be powerful tools in stimulating homologous recombination for gene therapy. The major obstacle, however, is that the homing endonucleases have to be re-engineered to recognize endogenous gene target sites. In the last several years, progress using computational and selection approaches have been used to modify homing endonucleases to recognize novel target sites.[28–32] These homing endonucleases have been shown to stimulate gene targeting using reporter assays but have not yet been shown to be able to modify an endogenous gene target. It is likely that with improvements in the re-design of homing endonucleases to new target sites, however, that milestone will be reached.

9. Development of Zinc Finger Nucleases

An alternative approach to modifying homing endonucleases to recognize new target sites is the use of zinc finger nucleases (ZFNs) to generate sequence-specific genomic DSBs. ZFNs are artificial proteins that fuse a zinc finger DNA binding domain to the nuclease domain from FokI, a type IIS restriction endonuclease (Fig. 9.3). To create a DSB, two ZFNs need to bind their cognate binding sites in the proper orientation. In the proper orientation, the nuclease domains dimerize and generate a DSB between the two binding sites. ZFNs were originally developed by Chandrasegaran and his colleagues and further developed in collaboration with Carroll and his colleagues.[33–36] Those two groups first showed that purified ZFNs could create a DSB in a DNA fragment in a living cell and prompt the cell to repair the DSB in a recombinogenic fashion to create a new DNA molecule.[37] These results set the stage for using ZFNs to modify genomic DNA.

10. Using Zinc Finger Nucleases to Stimulate Gene Targeting

The stimulation of gene targeting by I-SceI induced DSBs in reporter assays in mammalian cells, suggested that any method that creates a gene-specific DSB could stimulate gene targeting by several orders of magnitude. This hypothesis was confirmed using a chromosomally integrated GFP based

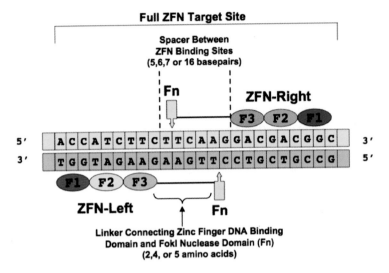

Fig. 9.3 Schematic depiction of a pair of ZFNs binding to a full target site. In this figure, three-finger ZFNs are shown but ZFNs with more fingers have also been used. The sequence of the spacer region (shown in purple text) has not been shown to affect the ability of ZFNs to cut DNA. If two 3-finger ZFNs are used, the full recognition site consists of 18 basepairs, a length that is statistically unique in the mammalian genome.

reporter assay in which recognition sites for model ZFNs were inserted adjacent to the recognition site for I-SceI.[2] The model ZFNs stimulated gene targeting as efficiently as I-SceI. In these experiments, the ZFNs were designed to target arbitrary sequences and just like I-SceI did not target sequences already present in a gene. The power of ZFNs, however, is that there are multiple ways to re-design the zinc finger DNA binding domain to recognize target sequences from natural genes. First, using modular-assembly, ZFNs were designed to stimulate gene targeting in GFP directly, rather than at sequences inserted into the GFP gene.[38] These ZFNs stimulated targeting at approximately 10-20% of the efficiency of I-SceI. More importantly, using a combination of selection based approaches, modular-assembly, and empiric design, a pair of ZFNs were made to target exon 5 of the IL2RG gene (mutations in which are the cause of X-linked severe combined immunodeficiency, SCID).[39] In these experiments, the gene specific ZFN pair were transfected into cells along with a repair/donor plasmid that contained a single basepair change from the chromosomal target creating a restriction site polymorphism. Gene targeting was measured by determining the presence or absence of this polymorphism in the gene. Using the IL2RG-ZFNs in K562 cells, a human erythroleukemia cell line, 11% of cells underwent targeting at a single allele and 6% of cells underwent targeting at both alleles. In primary T-cells, 4% of the IL2RG alleles could be targeted. Moreover, using these ZFNs one could create bi-allelic modification of the IL2RG gene and then subsequently re-modify both alleles. This established a proof-of-principle on the potential use of ZFNs to stimulate gene targeting by homologous recombination to correct small mutations in disease-associated genes.

The repertoire of genomic changes that can be created by ZFN-mediated homologous recombination, however, is not limited to small point mutations. In addition, ZFNs can induce both small and large insertions into a target locus by gene targeting.[38,40,41] In this way, full genes including regulatory elements such as promoters and enhancers, can be specifically targeted to integrate into specific genomic locations by homologous recombination. By targeting transgene insertion to a site that has no oncogenic potential and is not subject to chromatin silencing, the use of ZFN-mediated gene targeting could increase both the safety and efficacy of gene therapy based on the integration of transgenes.

In summary, the use of ZFN mediated gene targeting by homologous recombination has been shown in pre-clinical models to be a method that can directly correct disease causing mutations and a way of targeting transgene insertion in situations where direct correction of a disease causing mutation is neither desirable nor feasible.

11. Using Zinc Finger Nucleases to Site-Specifically Modify Genes by Mutagenic Non-Homologous End-Joining

Carroll and his colleagues established in flies an alternative use of ZFNs to perform site-specific modifications of the genome. In addition to showing that ZFNs could stimulate gene targeting by homologous recombination, they demonstrated that gene specific mutations could be created by the mutagenic repair of a ZFN induced DSB by NHEJ.[42] In their seminal work, they used ZFNs to recognize the *yellow* locus, a gene that controls eye color. They demonstrated that their ZFNs could induce a high rate of germline mutations in the *yellow* gene. When they sequenced the gene, they found small insertions and deletions that were created by the mutagenic repair of a DSB by NHEJ. This strategy has been applied to create gene specific mutations in zebrafish and has also been applied to mammalian cells where ZFNs have been used to create gene-specific inactivating mutations in the DHFR gene, the CCR5 gene, and the glucocorticoid receptor gene.[43-46] This strategy of using ZFNs to create gene specific mutations by the mutagenic repair of a DSB is one that has great potential in creating experimental cell lines, particularly human lines, that could not be easily generated in other ways. The ability to create such specific cell lines should be a powerful tool in biomedical research. In addition, the possibility of using this strategy to knock-out dominant disease causing alleles is a potential use in gene therapy.

The inactivation of the CCR5 gene by CCR5 specific ZFNs deserves special attention because it is the first use of ZFNs in a clinical trial.[46] In HIV infections, patients who have a bi-allelic deletion in the CCR5 gene are resistant to HIV infection and AIDS.[47] While these patients do seem to be more susceptible to other infections, such as Listeria and West Nile Virus, their immune systems are otherwise functional. The strategy is that by harvesting T cells from HIV infected patients, treating them with CCR5 specific ZFNs to inactivate both alleles of the CCR5 gene by mutagenic NHEJ, a pool of patient specific T cells would be created that would be relatively resistant to HIV infection and thus might protect the patient from developing AIDS or even help limit the proliferation of HIV itself. In contrast to RNAi or ribozyme based strategies to target CCR5, the ZFN strategy has the advantage of permanently modifying the CCR gene itself and thus is not dependent on the continued trans-repression of the gene. In pre-clinical models, CCR5-ZFNs have been designed that after delivery into primary T cells by adenovirus can bi-allelically inactivate the CCR5 gene in approximately 1-3% of cells. When the ZFN modified cells were injected into an immunodeficient mouse (NOG mice) the modified T cells had a selective advantage compared to un-modified cells when the animals are inoculated with HIV. Based on these results, a clinical trial to test the safety of this approach in humans has opened. One caveat is that the in the mouse pre-clinical studies, the ZFN modified T cells consisted of 10% of the T cells in the animal prior to HIV challenge. In contrast, the ZFN modified T cells will likely represent less than 0.1% of the total number of T cells in the body in patients. Furthermore, in the mouse studies a single CCR-tropic virus was used to challenge the animal. In contrast, human HIV infected patients harbor a mixture of both CCR5 and CXCR4 tropic viruses. The knockout of the CCR5 gene would not be expected to have any effect on the kinetics of the CXCR4 tropic viruses. These quantitative and qualitative factors may be of such magnitude to prevent any clinical efficacy from being demonstrated. Nonetheless, this application is a logical first use of ZFNs in a human clinical trial.

12. Strategies of Zinc Finger Nuclease Design

ZFNs have been used to stimulate homologous recombination and to create gene specific mutations in mammalian cells; uses that have direct application to gene therapy. The key factor to ZFNs is to design the zinc finger DNA binding domain such that it has high affinity and high specificity for its target binding site, as that will determine the quality of the ZFN. A zinc finger DNA binding domain consists of an array of individual zinc finger units. Each unit binds to a 3-4 basepair sequence of DNA. To create a zinc finger DNA binding domain with sufficient affinity for a specific sequence, at least three individual fingers must be assembled in an array and this 3-finger domain recognizes a 9-10 basepair site. Proteins with more fingers, including 4, 5, or 6 fingers, which will have longer target sites, can be made. Proteins with more fingers can have increased affinity for their target site but may not necessarily have greater specificity. There are three general methods to designing the zinc finger DNA binding domain: 1) Modular-Assembly; 2) The Sangamo Biosciences Platform; 3) Selection.

In modular-assembly, new zinc finger proteins are assembled from previously identified individual zinc finger units to create a new zinc finger protein.[48] These individual units have been identified for most of the 64 possible 3-basepair binding sites.[49-52] These units have been identified using a variety of methods, including from naturally occurring zinc finger proteins, phage display and empiric design. Using standard molecular biology techniques individual units can be assembled into a multi-finger protein to target a specific DNA sequence. This strategy has been used with success in creating artificial zinc finger transcription factors and several ZFNs (including those that target the GFP gene and the fly *yellow* genes).[38,42,48] A broad study of three-finger proteins assembled by modular-assembly, highlighted the limitations of this approach.[53] Modular-assembly has a relatively low probability of being a successful design strategy for a large proportion of potential target sequences.[54] In addition, in comparison with ZFNs designed by the approaches described below, ZFNs designed by modular-assembly have less specificity and more toxicity in cells.[55] Thus, the ease of creating a ZFN by modular-assembly must be balanced against the possibility that the ZFN will not have sufficient affinity to be active and may be more toxic than desired.

A second approach to ZFN design is a proprietary strategy from Sangamo Biosciences and now licensed for research use to Sigma-Aldrich. In this strategy, two-finger cassettes are selected by phage display against a wide range of 6 basepair binding sites.[56] To create a ZFN, a target sequence is compared against a private database of the two finger cassettes to identify potential ZFN target sites. In this strategy, two two-finger cassettes are assembled to create a four-finger protein to recognize a specific twelve-basepair target site. A series of four-finger proteins are created and these are screened using both *in vitro* binding assays and cell-based nuclease assays to identify a pair of ZFNs that have the best activity at a target site within a given locus.[44] This strategy was used to design the IL2RG and CCR5 ZFNs described above.[39,46] The major limitation of this ZFN design strategy is that it depends on a proprietary algorithm and database that is not available to academic researchers. For investigators interested in using ZFNs for research purposes, they can buy ZFNs made by this strategy from Sigma-Aldrich. Investigators interested in using ZFNs made with this platform for therapeutic use, must establish a collaborative agreement with Sangamo Biosciences.

A third approach to designing ZFNs is to use a selection-based strategy.[55,57–59] In this approach, a large library of three-finger proteins is created and then using either iterative phage display or serial bacterial based selections, proteins with high affinity for the desired target site are identified. While phage display was the first type of selection strategy to be used, this selection strategy has been replaced by the simpler, but still relatively complex, bacterial based systems.[43,55,58,60] Using the bacterial based systems, active ZFNs have been made to stimulate gene targeting by homologous recombination in human cells, to create site-specific modifications in endogenous genes in human cells and, to create heritable site-specific gene mutations in zebrafish.[43,55,58] ZFNs made by bacterial selection strategies have been shown to have greater specificity and consequently less cellular toxicity than ZFNs made by modular-assembly and to have similar, if not better, activity and toxicity characteristics made by the proprietary Sangamo Biosciences approach.[55,58] While this approach is available to all researchers, the major limitation to this strategy is that it remains relatively complex, currently requires substantial molecular biology expertise to perform and still can take weeks to months to generate the desired proteins.

There are advantages and disadvantages to each of the ZFN design techniques. With the development of ZFNs, there has been a renewed interest in developing bioinformatic and computational design approaches to making zinc finger proteins. It is likely that as the database of ZFNs increases, these approaches will become increasingly useful. Even if they do not absolutely determine the best ZFN in terms of activity and toxicity, if computational approaches can provide guidance to selecting a relatively small number of potential ZFNs to screen, they will be a significant advance in the field and dramatically shorten the time from experimental use to concrete results.

13. Aspects of Zinc Finger Binding Sites and Structure of Zinc Finger Nucleases

Figure 9.3 shows the general structure of a pair of ZFNs binding to their cognate binding sites. In this figure, the separation of the ZFN binding sites (the "spacer") is 6 basepairs. In practice, the ZFN binding sites can be separated by 5, 6, 7, or 16 basepairs.[61] Spacer distances of 3 and 4 basepairs have been shown not to work and a spacer length of 8 basepairs has been shown to be sup-optimal in comparison to a 6 basepair spacer.[2]

Figure 9.3 also shows a pair of three finger ZFNs binding to their 9-basepair cognate binding sites. In a 3-finger zinc finger protein, each finger is linked to its neighbor by a 5 amino acid linker. In 4-finger ZFNs there is an extra serine in the linker region between fingers 2 and 3. This extra amino acid allows the zinc finger DNA binding domain and the major groove of DNA in which the zinc finger binds to remain in register. Finally,

Fig. 9.3 depicts an amino acid linker that connects the zinc finger DNA binding domain to the nuclease domain. This linker can be 2, 4, or 5 amino acids long without affecting the activity of the ZFN. The linker length between the zinc finger DNA binding domain and the nuclease domain does affect the structure of the ZFN target site. For example, if the linker length between the two domains is 5 amino acids, the ZFNs effectively cut target sites with a 6 or 7 basepair spacer but do not efficiently cut sites separated by 5 basepairs. In contrast, if the linker length is 2 or 4 amino acids, the ZFNs effectively cut sites separated by 5 or 6 basepairs, but do not seem to cut sites separated by 7 basepairs. Thus, the target site guides what linker needs to be used to connect the zinc finger DNA binding domain and the nuclease domain.

14. Zinc Finger Nuclease Toxicity: Measuring and Minimizing

In addition to the challenges in designing active ZFNs, the other major problem in the use of ZFNs is limiting their cellular toxicity. When ZFNs were compared to I-SceI in gene targeting in mammalian cells, they were found to have significantly more cellular toxicity.[2] The toxicity of ZFNs was also observed in flies.[62] The toxicity in flies was abrogated when the nuclease activity of the ZFNs was mutated suggesting that the toxicity of ZFNs was due in major part from off-target DSBs.[62] The translation of ZFN technology to use in humans for gene therapy requires that these off-target effects be quantitatively determined, better understood, and minimized to a substantial degree because the creation of off-target DSBs would be predicted to lead to oncogenic mutations.

ZFN toxicity has been measured in a number of different ways including the use of assays that measure cell survival after exposure to ZFNs, the frequency with which ZFNs create extra DSBs, and by deep sequencing of potential off-target break sites.[46,55] Using these assays, several different approaches have been used to minimize the off-target effects. The first approach was to use more sophisticated design methods for the zinc finger DNA binding domain.[39,55,58] These more sophisticated design methods have led to improvements in ZFN cellular toxicity but as more sensitive measures of cellular toxicity have been developed, they show that improved zinc finger design does not solve the problem entirely.[63] Further decreases in toxicity were achieved by modifying the nuclease domain.[55,64,65] As Fig. 9.3 depicts, cutting of the target site requires dimerization of the nuclease domains from two different ZFNs. It is possible, however, that off-target cutting can arise from the homodimerization of a single ZFN. To prevent such homodimer cutting at off-target sites, the nuclease domain has been re-engineered using a combination of structure based approaches and empirical design to create domains that are only permissive for heterodimerization. That is, the nuclease domain from ZFN1 can only dimerize with the nuclease domain of ZFN2, not with itself, thus allowing cutting at the target site but reducing cutting at off-target sites. This modification should theoretically result in a 50% reduction of off-target cutting, but empirically it is observed that a much greater reduction in toxicity is observed with these "obligate heterodimer" ZFNs. Better understanding of the mechanism by which these have reduced cellular toxicity and reduced creation of extra DSBs should help unravel the different ways that ZFNs can create off-target damage. Finally, a third approach to minimizing cellular toxicity is to use small molecule regulation of ZFN expression level.[63] ZFN based genome modification is based on a "hit and run" strategy and continued expression of the ZFN is neither needed nor desired. By attaching a domain to the N-terminus of the ZFN that destabilizes the ZFN except in the presence of a small molecule, the expression level of the ZFN can be tightly regulated so it is only expressed for a short window of time. This tight regulation of ZFN expression results in reduced cellular toxicity without compromising ZFN efficacy.[63]

In summary, there are now multiple strategies that can reduce the off-target effects of ZFNs and increasingly sensitive assays to measure such effects. One of the major lessons from the development of retroviral vectors for gene therapy for the immunodeficiencies, however, is to push the envelope on developing pre-clinical models to quantitatively assess the oncogenic potential of any gene therapy approach. This lesson applies to the use of ZFNs — the potential safety of using site-specific genome modification needs to be measured against the potential oncogenic risk of off-target DSBs. Until better pre-clinical assays are developed, it remains unclear whether the oncogenic risk of ZFNs falls in the tolerable range, such as that created by radiation therapy for life threatening malignancies, or falls in a range that would be considered too high for the disease processes the ZFNs are designed to treat.

15. The Challenge of Delivery

In addition to the challenge of designing highly active and specific ZFNs and minimizing the off-target toxicity of ZFNs, the challenge of how to best deliver the ZFN based components to therapeutically important cell types, such as tissue specific stem cells, needs to be determined. While ZFN manipulation of the genome can be classified as a "non-viral" strategy, the problem of how to deliver the components for ZFN manipulations into the cell still remains. The upcoming clinical trial with the CCR5 directed ZFNs for treatment of HIV uses *ex vivo* adenoviral delivery of ZFN expression constructs followed by expansion of transduced cells.[46] The expansion of T cells after manipulation should reduce the probability of a host immune response to either the viral vector or the foreign ZFN. But this must be carefully monitored given the experience with the *in vivo* use of adenovirus for the treatment of OTC deficiency.[66] The use of integration deficient lentivirus has also shown promise in pre-clinical studies, but the low rate of gene targeting in human CD34+ using this delivery method demonstrates that further improvements are needed.[41] Since ZFN based genome manipulation is a hit and run strategy, delivery methods that introduce either protein or mRNA for the ZFNs rather than DNA ZFN expression constructs may provide an added degree of safety. In fact, the delivery of mRNA to zebrafish fertilized eggs was an effective method in that organism,[43,44] but scaling that process up for human clinical use, where millions if not hundreds of millions of cells might need to be manipulated remains a technological hurdle. Another possible delivery method is the use of AAV. As discussed above, AAV, through mechanisms not entirely clear, can stimulate targeting on its own. But AAV can be combined with the creation of a gene-specific DSB to stimulate gene targeting.[67,68] While the packaging size of AAV can limit certain gene therapy applications, it should not pose as great a limitation in a gene targeting approach. Finding the optimal serotype to transduce human embryonic or tissue specific stem cells remains an important problem for either the *ex vivo* or *in vivo* modification of cells for therapeutic use. In summary, once again, gene transfer remains an important problem in translating an exciting laboratory based approach into human therapy.

16. Future Directions and Promise of Homologous Recombination as a Gene Correction Approach to Gene Therapy

While the gene therapy field has worked hard to translate laboratory based techniques into therapy for humans and the progress has been slower than either expected or desired, the promise of being able to cure genetic diseases with genetic techniques remains great. Moreover, while short and intermediate term successes are likely to continue using vectors that integrate in an uncontrolled fashion, the ultimate long-term broad success of gene therapy is likely to depend on finding ways to either directly correct disease causing mutations or to control the target site of transgene integration. The use of homologous recombination by AAV, the potential of embryonic stem cell therapy, and the progress with gene specific modifications using modified homing endonucleases and zinc finger nucleases all suggest that the specific targeting of the mammalian genome for therapeutic purposes may one day be successful.

References

1. Sedivy JM, Dutriaux A (1999). Gene targeting and somatic cell genetics–a rebirth or a coming of age? *Trends Genet* **15**: 88–90.
2. Porteus MH, Baltimore D (2003). Chimeric nucleases stimulate gene targeting in human cells. *Science* **300**: 763.
3. Capecchi M (1989). Altering the genome by homologous recombination. *Science* **244**: 1288-1292.
4. Aiuti A, *et al.* (2009). Gene therapy for immunodeficiency due to adenosine deaminase deficiency. *N Engl J Med* **360**: 447–458.
5. Cavazzana-Calvo M, Lagresle C, Hacein-Bey-Abina S, Fischer A (2005). Gene therapy for severe combined immunodeficiency. *Annu Rev Med* **56**: 585–602.
6. Hochedlinger K, Jaenisch R (2006). Nuclear reprogramming and pluripotency. *Nature* **441**: 1061–1067.
7. Yamanaka S (2007). Strategies and new developments in the generation of patient-specific pluripotent stem cells. *Cell Stem Cell* **1**: 39–49.

8. Rideout WM, 3rd, Hochedlinger K, Kyba M, Daley GQ, Jaenisch R (2002). Correction of a genetic defect by nuclear transplantation and combined cell and gene therapy. *Cell* **109**: 17–27.

9. Hanna J, *et al.* (2007). Treatment of sickle cell anemia mouse model with iPS cells generated from autologous skin. *Science* **318**: 1920–1923.

10. Maherali N, Hochedlinger K (2008). Guidelines and techniques for the generation of induced pluripotent stem cells. *Cell Stem Cell* **3**: 595–605.

11. Lengerke C, *et al.* (2007). The cdx-hox pathway in hematopoietic stem cell formation from embryonic stem cells. *Ann N Y Acad Sci* **1106**: 197–208.

12. Russell DW, Hirata RK, Inoue N (2002). Validation of AAV-mediated gene targeting. *Nat Biotechnol* **20**: 658.

13. Hirata R, Chamberlain J, Dong R, Russell DW (2002). Targeted transgene insertion into human chromosomes by adeno-associated virus vectors. *Nat Biotechnol* **20**: 735–738.

14. Russell DW, Hirata RK (1998). Human gene targeting by viral vectors. *Nat Genet* **18**: 325–330.

15. Chamberlain JR, *et al.* (2004). Gene targeting in stem cells from individuals with osteogenesis imperfecta. *Science* **303**: 1198–1201.

16. Wyman C, Kanaar R (2006). DNA double-strand break repair: all's well that ends well. *Annu Rev Genet* **40**: 363–383.

17. West SC, Chappell C, Hanakahi LA, Masson JY, McIlwraith MJ, Van Dyck E (2000). Double-strand break repair in human cells. *Cold Spring Harb Symp Quant Biol* **65**: 315–321.

18. Elliott B, Richardson C, Winderbaum J, Nickoloff JA, Jasin M (1998). Gene conversion tracts from double-strand break repair in mammalian cells. *Mol Cell Biol* **18**: 93–101.

19. Burma S, Chen BP, Chen DJ (2006). Role of non-homologous end joining (NHEJ) in maintaining genomic integrity. *DNA Repair (Amst)* **5**: 1042–1048.

20. Wyman C, Ristic D, Kanaar R (2004). Homologous recombination-mediated double-strand break repair. *DNA Repair (Amst)* **3**: 827–833.

21. Smih F, Rouet P, Romanienko PJ, Jasin M (1995). Double-strand breaks at the target locus stimulate gene targeting in embryonic stem cells. *Nucleic Acids Res* **23**: 5012–5019.

22. Brenneman M, Gimble FS, Wilson JH (1996). Stimulation of intrachromosomal homologous recombination in human cells by electroporation with site-specific endonucleases. *Proc Natl Acad Sci USA* **93**: 3608–3612.

23. Sargent RG, Brenneman MA, Wilson JH (1997). Repair of site-specific double-strand breaks in a mammalian chromosome by homologous and illegitimate recombination. *Mol Cell Biol* **17**: 267–277.

24. Choulika A, Perrin A, Dujon B, Nicolas J-F (1995). Induction of homologous recombination in mammalian chromosomes by using the I-SceI system of *Saccaromyces cerevisiae*. *Molecular and Cellular Biology* **15**: 1968–1973.

25. Taghian DG, Nickoloff JA (1997). Chromosomal double-strand breaks induce gene conversion at high frequency in mammalian cells. *Mol Cell Biol* **17**: 6386–6393.

26. Donoho G, Jasin M, Berg P (1998). Analysis of gene targeting and intrachromosomal homologous recombination stimulated by genomic double-strand breaks in mouse embryonic stem cells. *Mol Cell Biol* **18**: 4070–4078.

27. Chevalier BS, Stoddard BL (2001). Homing endonucleases: structural and functional insight into the catalysts of intron/intein mobility. *Nucleic Acids Res* **29**: 3757–3774.

28. Ashworth J, *et al.* (2006). Computational redesign of endonuclease DNA binding and cleavage specificity. *Nature* **441**: 656–659.

29. Sussman D, *et al.* (2004). Isolation and characterization of new homing endonuclease specificities at individual target site positions. *J Mol Biol* **342**: 31–41.

30. Arnould S, *et al.* (2007). Engineered I-CreI derivatives cleaving sequences from the human XPC gene can induce highly efficient gene correction in mammalian cells. *J Mol Biol* **371**: 49–65.

31. Redondo P, *et al.* (2008). Molecular basis of xeroderma pigmentosum group C DNA recognition by engineered meganucleases. *Nature* **456**: 107–111.

32. Paques F, Duchateau P (2007). Meganucleases and DNA double-strand break-induced recombination: perspectives for gene therapy. *Curr Gene Ther* **7**: 49–66.

33. Kim YG, Cha J, Chandrasegaran S (1996). Hybrid restriction enzymes: zinc finger fusions to Fok I cleavage domain. *Proc Natl Acad Sci USA* **93**: 1156–1160.

34. Smith J, Berg JM, Chandrasegaran S (1999). A detailed study of the substrate specificity of a chimeric restriction enzyme. *Nucleic Acids Res* **27**: 674–681.

35. Smith J, Bibikova M, Whitby FG, Reddy AR, Chandrasegaran S, Carroll D (2000). Requirements for double-strand cleavage by chimeric restriction enzymes with zinc finger DNA-recognition domains. *Nucleic Acids Res* **28**: 3361–3369.

36. Durai S, Mani M, Kandavelou K, Wu J, Porteus MH, Chandrasegaran S (2005). Zinc finger nucleases: custom-designed molecular scissors for genome engineering of plant and mammalian cells. *Nucleic Acids Res* **33**: 5978–5990.

37. Bibikova M, *et al.* (2001). Stimulation of homologous recombination through targeted cleavage by chimeric nucleases. *Mol Cell Biol* **21**: 289–297.

38. Porteus MH (2006). Mammalian gene targeting with designed zinc finger nucleases. *Mol Ther* **13**: 438–446.

39. Urnov FD, *et al.* (2005). Highly efficient endogenous human gene correction using designed zinc-finger nucleases. *Nature* **435**: 646–651.

40. Moehle EA, *et al.* (2007). Targeted gene addition into a specified location in the human genome using designed zinc finger nucleases. *Proc Natl Acad Sci USA* **104**: 3055–3060.

41. Lombardo A, *et al.* (2007). Gene editing in human stem cells using zinc finger nucleases and integrase-defective lentiviral vector delivery. *Nat Biotechnol* **25**: 1298–1306.

42. Bibikova M, Beumer K, Trautman JK, Carroll D (2003). Enhancing gene targeting with designed zinc finger nucleases. *Science* **300**: 764.

43. Meng X, Noyes MB, Zhu LJ, Lawson ND, Wolfe SA (2008). Targeted gene inactivation in zebrafish using engineered zinc-finger nucleases. *Nat Biotechnol* **26**: 695–701.

44. Doyon Y, *et al.* (2008). Heritable targeted gene disruption in zebrafish using designed zinc-finger nucleases. *Nat Biotechnol* **26**: 702–708.

45. Santiago Y, *et al.* (2008). Targeted gene knockout in mammalian cells by using engineered zinc-finger nucleases. *Proc Natl Acad Sci USA* **105**: 5809–5814.

46. Perez EE, *et al.* (2008). Establishment of HIV-1 resistance in CD4+ T cells by genome editing using zinc-finger nucleases. *Nat Biotechnol* **26**: 808–816.

47. Stewart GJ, *et al.* (1997). Increased frequency of CCR-5 delta 32 heterozygotes among long-term non-progressors with HIV-1 infection. The Australian Long-Term Non-Progressor Study Group. *AIDS* **11**: 1833–1838.

48. Beerli RR, Barbas CF, 3rd (2002). Engineering polydactyl zinc-finger transcription factors. *Nat Biotechnol* **20**: 135–141.
49. Segal DJ, Dreier B, Beerli RR, Barbas CF, 3rd (1999). Toward controlling gene expression at will: selection and design of zinc finger domains recognizing each of the 5′-GNN-3′ DNA target sequences. *Proc Natl Acad Sci USA* **96**: 2758–2763.
50. Dreier B, Beerli RR, Segal DJ, Flippin JD, Barbas CF, 3rd (2001). Development of zinc finger domains for recognition of the 5′-ANN-3′ family of DNA sequences and their use in the construction of artificial transcription factors. *J Biol Chem* **276**: 29466–29478.
51. Dreier B, *et al.* (2005). Development of zinc finger domains for recognition of the 5′-CNN-3′ family DNA sequences and their use in the construction of artificial transcription factors. *J Biol Chem* **280**: 35588–35597.
52. Liu Q, Xia Z, Zhong X, Case CC (2002). Validated zinc finger protein designs for all 16 GNN DNA triplet targets. *J Biol Chem* **277**: 3850–3856.
53. Ramirez CL, *et al.* (2008). Unexpected failure rates for modular assembly of engineered zinc fingers. *Nat Methods* **5**: 374–375.
54. Sander JD, Zaback P, Joung JK, Voytas DF, Dobbs D (2009). An affinity-based scoring scheme for predicting DNA-binding activities of modularly assembled zinc-finger proteins. *Nucleic Acids Res* **37**: 506–515.
55. Pruett-Miller SM, Connelly JP, Maeder ML, Joung JK, Porteus MH (2008). Comparison of zinc finger nucleases for use in gene targeting in mammalian cells. *Mol Ther* **16**: 707–717.
56. Isalan M, Klug A, Choo Y (2001). A rapid, generally applicable method to engineer zinc fingers illustrated by targeting the HIV-1 promoter. *Nat Biotechnol* **19**: 656–660.
57. Hurt JA, Thibodeau SA, Hirsh AS, Pabo CO, Joung JK (2003). Highly specific zinc finger proteins obtained by directed domain shuffling and cell-based selection. *Proc Natl Acad Sci USA* **100**: 12271–12276.
58. Maeder ML, *et al.* (2008). Rapid "open-source" engineering of customized zinc-finger nucleases for highly efficient gene modification. *Mol Cell* **31**: 294–301.
59. Greisman HA, Pabo CO (1997). A general strategy for selecting high-affinity zinc finger proteins for diverse DNA target sites. *Science* **275**: 657–661.
60. Durai S, Bosley A, Abulencia AB, Chandrasegaran S, Ostermeier M (2006). A bacterial one-hybrid selection system for interrogating zinc finger-DNA interactions. *Comb Chem High Throughput Screen* **9**: 301–311.
61. Handel EM, Alwin S, Cathomen T (2009). Expanding or restricting the target site repertoire of zinc-finger nucleases: the inter-domain linker as a major determinant of target site selectivity. *Mol Ther* **17**: 104–111.
62. Beumer K, Bhattacharyya G, Bibikova M, Trautman JK, Carroll D (2006). Efficient gene targeting in Drosophila with zinc-finger nucleases. *Genetics* **172**: 2391–2403.
63. Pruett-Miller SM, Reading DW, Porter SN, Porteus MH (2009). Attenuation of Zinc Finger Nuclease Toxicity by Small Molecule Regulation of Protein Levels. *PLOS Genetics* **in press**.
64. Szczepek M, Brondani V, Buchel J, Serrano L, Segal DJ, Cathomen T (2007). Structure-based redesign of the dimerization interface reduces the toxicity of zinc-finger nucleases. *Nat Biotechnol* **25**: 786–793.

65. Miller JC, *et al.* (2007). An improved zinc-finger nuclease architecture for highly specific genome editing. *Nat Biotechnol* **25**: 778–785.
66. Raper SE, *et al.* (2003). Fatal systemic inflammatory response syndrome in a ornithine transcarbamylase deficient patient following adenoviral gene transfer. *Mol Genet Metab* **80**: 148–158.
67. Porteus MH, Cathomen T, Weitzman MD, Baltimore D (2003). Efficient gene targeting mediated by adeno-associated virus and DNA double-strand breaks. *Mol Cell Biol* **23**: 3558–3565.
68. Miller DG, Petek LM, Russell DW (2003). Human gene targeting by adeno-associated virus vectors is enhanced by DNA double-strand breaks. *Mol Cell Biol* **23**: 3550–3557.

Chapter 10
Gene Switches for Pre-Clinical Studies in Gene Therapy

Caroline Le Guiner*, Knut Stieger, Alice Toromanoff*,
Fabienne Rolling*, Philippe Moullier*
and Oumeya Adjali*,†

Numerous preclinical studies have demonstrated the efficacy of viral gene delivery vectors. However, the tight control of transgene expression is likely to be required for therapeutic applications and/or safety reasons. For this purpose, several ligand-dependent transcription regulatory systems have been developed. Among these, the tetracycline and the rapamycin dependent systems have been largely used. However, if long-term regulation of the transgene has been obtained in small animal models using these inducible systems, when translational studies were initiated in larger animals, especially in nonhuman primates (NHP) the development of an immune response against proteins involved in transgene regulation were often observed. Such immune response was especially documented when using the TetOn tetracycline (Tet) regulatable system in NHP especially after intramuscular injection of the vector encoding for the Tet-regulated transgene. This chapter will emphasize on the tetracycline and rapamycin gene switch systems, which we believe are the more relevant in a preclinical setting and potentially in future clinical trials.

1. Introduction

The clinical efficacy and safety as well as the application range of gene therapy will be broadened by developing systems capable of fine-tuning the expression of therapeutic genes. A regulatable system that can turn

*Correspondence: INSERM UMR 649, Institut de Recherche Therapeutique IRT UN, Nantes, Fance.
E-mail: †oumeya.adjali@univ-nantes.fr

"on" and "off" therapeutic gene expression will not only be crucial for maintaining appropriate levels of a gene product within the therapeutic range, thus preventing toxicity, but also allow the turning "off" of therapeutic gene expression to avoid harmful side effects. The development of ligand-dependent transcription regulatory systems is thus of great importance.

In human gene therapy, a pharmacologically regulated system should fulfill several criteria: (1) the ligand should activate (On-switch) rather than silence transcription (Off-switch). Indeed, Off-systems suffer from 2 drawbacks: a prolonged exposure to the drug is required to silence the system and the induction kinetic is mainly determined by the rate of drug clearance; (2) the inducer drug should be orally bioavailable and penetrate the target tissue; (3) the system should not interfere with endogenous metabolic pathways and the inducer compound should have a drug metabolism profile compatible with prolonged therapeutic uses; (4) transcription should be fully and rapidly reversible to enable prompt changing of the dosing regimen; (5) the system should be inactive in the absence of the inducer drug but strongly stimulated by the drug administration; (6) a precise correlation must exist between drug dosage and target gene expression level; (7) the transcriptional activator should not elicit an immune response in human.

Four drug-dependent regulatable systems were developed for the *in vivo* control of transgene expression: the tetracycline (Tet) and the rapamycin regulatable systems, the mammalian steroid receptor (mifepristone and tamoxifen) and the insect steroid receptor (ecdysteroid)-based systems. The molecular mechanisms underlying the regulatory effect and the pharmacologic characteristics of the inducer drug are distinct in these different regulatory systems, but transgene regulation has been documented *in vitro* and in rodent models with all of them. Only the Tet and the rapamycin regulatable systems were tested in large animal models.[1−7] However, when these translational studies were initiated in larger animals, the development of an immune response against the proteins involved in transgene regulation was observed,[1−3,6] whereas in the murine model both were able to permanently sustain transgene regulation.[8,9]

Because the Tet and the rapamycin-based systems are the most potent and clinically relevant for future trials, this chapter will focus on the molecular mechanisms involved in transgene regulation, the pharmacology of the inducer drug and the immunotoxicity observed (or not) for each of the two systems.

2. Rapamycin-Dependent Regulatable System

2.1 *Molecular Mechanisms Involved in Transgene Regulation*

The basic concept of the rapamycin regulatory system is to express the two critical domains of a transcription factor as separate polypeptides that can be reversibly crosslinked only in the presence of the inducer drug, the bivalent "dimerizer" drug (Fig. 10.1). In cells, the natural product rapamycin or its analogs, binds to two peptides, FKBP12 (FK506 binding protein of 12kDa) and FRAP (FKBP rapamycin associated protein), brings them into close proximity, thus, activating their transcriptional specificity as inhibitor of T-cell proliferation and immunosuppression. The fusion of the FKBP12 domain to a DNA binding domain, called ZFHD1 (zinc finger homeo domain 1) and of one portion of FRAP, called FRB (FKBP rapamycin binding), to a transcriptional activation domain from the p65 subunit of the transcription factor NF-kB, results, when dimerization of FKBP12 and FRB occurs, in the formation of a functional transactivator. In the presence of rapamycin, the complex FKBP12-ZFHD1/FRB-p65 is able to associate to the ZFHD1 corresponding DNA sequence upstream of a minimal Interleukin-2 promoter (pIL2min), and the transactivator allows

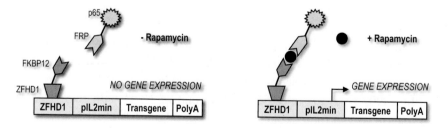

Fig. 10.1   Schematic representation of the rapamycin dependent regulation system. This system is composed of two chimeric proteins, FRB-p65 and FKBP12-ZFHD1. The FKBP12-ZFHD1 protein is able to interact with the corresponding DNA sequence (ZFHD1), which is fused to a minimal IL2 promoter. The second protein is soluble. Rapamycin allows dimerization of FKBP12 and FRB, resulting in the formation of a functional transactivator, which allows transgene expression. In the absence of rapamycin, the FRB and FKBP12 domains cannot interact, and no transgene expression is observed.

the expression of the transgene. In the absence of rapamycin, the two peptides FKBP12-ZFHD1 and FRB-p65 are unable to interact, and are then transcriptionally inert (Fig. 10.1).[9]

2.2 *Pharmacology of Rapamycin*

The pharmacologic profile of rapamycin is well known, but one major obstacle for its application in gene therapy clinical trial is that rapamycin is a powerful immunosuppressive agent used in humans for many years. Since a long-term treatment with rapamycin could have significant deleterious effects, efforts were made to develop non-immunosuppressive analogs. Analysis of the rapamycin-FRB complex led to the development of the so called "rapalogs" carrying punctual substitutions.[10,11] Several of these rapalogs are considerably less immunosuppressive and, yet, were shown to efficiently induce transgene expression in nonhuman primate (NHP) muscle.[6] However, one analog was tested in the retina and failed to be as efficient as the original drug.[5]

2.3 *Translation Development of the Rapamycin Dependent Regulation System*

This system employs exclusively human peptides, thus rendering the possibility for immune reactions against the chimeric transactivator minimal when tested in other species. Since its first description in 1996, the rapamycin regulatable system was used *in vivo* in rodents, using different viral vector platforms.[1,12−14] The long-term regulation of the erythropoietin (Epo) transgene was also obtained in NHP eye and muscle, after subretinal, intravitreal and intramuscular (IM) injection of recombinant Adeno-Associated Virus (rAAV) vectors.[5,6] The NHP that received rapamycin-regulated expression cassettes are summarized in Table 10.1.

In 1999, Ye *et al.* injected a first group of NHP, with two rAAV-2 vectors (in grey in Table 10.1). If the first two administrations of rapamycin resulted in the rapid induction of Epo, its expression was decreased 20 fold at the third induction, and became indistinguishable from endogenous Epo after the fourth induction.[1,6] Antibodies directed against the influenza virus hemagglutinin tag contained in the transcription factor were detected in serum of at least one animal after two rounds of rapamycin induction.

Table 10.1. Summary of NHP that received rapamycin-regulated vectors (*nd* = non determined; IM = intramuscular; SR = subretinal; IVit = intravitreous).

Study	Mode of delivery	NHP identification	Injected vector	Expression cassette(s)	Duration of the regulation	Host immune response?
Ref. 6	IM	95C002	rAAV-2	CMV-TF1 + rhEpo-3	2 months	Yes
	IM	EWP	rAAV-2	CMV-TF1 + rhEpo-3	1,5 month	*nd*
	IM	FJX	rAAV-2	CMV-TF1 + rhEpo-3	2,5 months	*nd*
	IM	GXF	rAAV-2	CMV-TF1 + rhEpo-2	>24 months	*nd*
	IM	FFT	rAAV-2	CMV-TF1 + rhEpo-2	>34 months	*nd*
	IM	94B091	rAAV-2	CMV-TF1 + rhEpo-2	>73 months	*nd*
	IM	FTF	rAAV-2	CMV-TF1Nc + rhEpo-2	>66 months	*nd*
	IM	AC7B	rAAV-2	CMV-TF1Nc + rhEpo-2	>66 months	*nd*
	IM	97E010	rAAV-2	CMV-TF1Nc + rhEpo-2	>12 months	*nd*
	IM	97E036	rAAV-2	CMV-TF1Nc + rhEpo-2	>58 months	*nd*
	IM	97E081	rAAV-2	CMV-TF1Nc + rhEpo-2	>12 months	*nd*
	IM	97E102	rAAV-2	CMV-TF1Nc + rhEpo-2	>6 months	*nd*
	IM	97E091	rAAV-2	CMV-TF1Nc + rhEpo-2	>54 months	*nd*
	IM	97E099	rAAV-2	CMV-TF1Nc + rhEpo-2	>54 months	*nd*
	IM	99E120	rAAV-2	CMV-TF1Nc + rhEpo-2	>20 months	*nd*
	IM	00E022	rAAV-2/1	CMV-TF-rhEpo-2.3	>18 months	*nd*
Ref. 14	SR	NHP	rAAV-2	CMV-TF1Nc + rhEpo-2	>3 months	*nd*
Ref. 5	SR	94B106	rAAV-2	CMV-TF1Nc + rhEpo-2	>27 months	*nd*
	SR	E009	rAAV-2	CMV-TF1Nc + rhEpo-2	>23 months	*nd*
	IVit	AC3H	rAAV-2	CMV-TF1Nc + rhEpo-2	>25 months	*nd*

However, the presence of a cellular mediated immune response against the transactivator was not clearly established.[1] In contrast, three NHP that were co-injected with a slightly modified target vector (rhEpo-2) and the same vector encoding the transactivator (CMV-TF1), showed long term (more than 6 years) regulation of Epo expression.[6] The only difference between the rhEpo-3 and the rhEpo-2 expression cassettes was the deletion of a chimeric intron, initially present after the pIL2min driving the Epo expression. Therefore, this result suggests that rather than the presence of heterologous sequences in the transactivator, the presence of this intron had a harmful effect on the ability to sustain long-term regulation. Unfortunately, if the study also showed that long-term regulation of Epo was achieved with a modified transactivator in which nonhuman sequences were removed (CMV-TF1Nc), or with one single rAAV vector encoding all three components of the expression/regulation system (CMV-TF-rhEpo-2, -3), the immune status against the transactivator in this NHP cohort was not reported (Table 10.1).

Finally, the retina of other NHP was injected with the same rAAV vectors encoding for the Epo under the control of the rapamycin-regulated transactivator. In these animals, tight and long-term regulation of Epo expression was observed for more than two years.[5,14] Again, if this result suggests the absence of an immune response against the transcription factor, there was no indication on the absence/presence of circulating antibodies against the transactivator.

In summary, it is now clearly demonstrated that an efficient rapamycin-regulation of transgene expression can be obtained after a single IM or intraocular administration of a rAAV vector in NHP. One major advantage of this regulation system would be that it employs human peptides exclusively, thus minimizing the possibility for an immune reaction against the transactivator. However, the presence of a humoral and/or a cellular immune response against this transcription factor has not yet been fully characterized.

3. Tetracycline-Dependent Regulatable Systems

3.1 *Molecular Mechanisms Involved in Transgene Regulation*

Tet-responsive systems are widely used because they are able to tightly control transgene expression in a wide range of cultured cells as well as plants, yeast, insects and mammals *in vivo*.[2,4,7,8,15–18] They are based on the Escherichia coli Tn10 Tet resistance operator, which consists of the Tet repressor protein (TetR) and the Tet operator DNA sequence (TetO). In the absence of Tet or its derivate doxycycline (Dox), the TetR protein interacts with the TetO DNA sequence, while in the presence of this drug, TetR changes its conformation, thus detaching from the DNA. Based on this bacterial operator system, three different Tet-responsive regulatory systems have been developed (Fig. 10.2).

The first two systems contain the TetR protein fused to the transactivator VP16 (viral protein 16) of the Herpes simplex virus (HSV).[19,20] If transgene expression is allowed only in the absence of Dox, the system is called TetOff. If transgene expression is allowed in the presence of Dox, the system is called TetOn. The third system contains the TetR protein fused to a human inhibitor protein: the krüppel associated box (KRAB) of zinc-finger

A. TetOff System

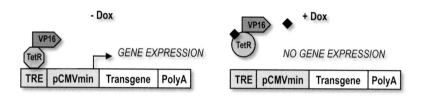

B. TetOn System

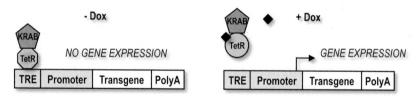

C. TetR-KRAB System

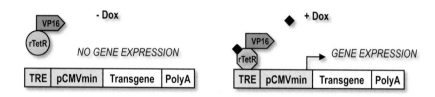

Fig. 10.2 Schematic representations of the three Tet dependent regulation systems. A. In the TetOff system, the chimeric protein tTa is composed of the Tet repressor protein TetR and the viral transactivator VP16. In the absence of Dox, the tTA protein binds to the TRE, which is fused to a minimal CMV promoter, and allows transgene expression. In the presence of Dox, the tTA protein cannot interact with the TRE, and no transgene expression is observed. B. In the TetOn system, the rtTA transactivator is also a chimeric protein composed of the TetR and the VP16 domains, but due to four mutations in the TetR domain, it fixes the TRE only in the presence of Dox, thus allowing expression of the transgene. In the absence of Dox, rtTA detaches from the DNA and no transgene expression is observed. C. The chimeric TetR-KRAB transactivator is a fusion protein of the original TetR domain with the KRAB box of the human zinc finger protein Kox1. In the absence of Dox, TetR-KRAB binds to the TRE and inhibits transgene expression driven by a normal promoter. In the presence of Dox, TetR-KRAB detaches from the TRE, thus disabling the inhibition of the promoter and allowing transgene expression.

proteins.[21] This system is designed to allow transgene expression only in the presence of Dox.

In the TetOff system, the transactivator protein, called tTA (for Tet-controlled transcriptional activator) is a chimera made by the fusion of the TetR protein and the transactivator VP16. In the absence of Dox, tTA is able to bind to seven repeats of the TetO sequence, also referred to as TRE, for Tet responsive element. Binding of tTA to the TRE activates the TRE associated promoter (usually a minimal CMV promoter), resulting in the expression of the adjacent gene. If Dox is present, tTA detaches from the TRE and gene expression is silenced (Fig. 10.2A).[19]

Four mutations in the TetR part of the transactivator tTA were shown to reverse its interaction with the TRE in the presence or absence of Dox[20]: rather than attaching to the TRE in the absence of Dox, this mutant, called rtTA (for reverse tTA), stays soluble. In the presence of Dox, the conformation of the rtTA changes, which in turn allows the transactivator to interact with TRE. Therefore, the system changes to a TetOn system, i.e. only the presence of Dox allows expression of the transgene (Fig. 10.2B). Because the original rtTA version was prone to high background leakage in the "Off" stage and poor inducibility in some instances, novel rtTA variants were generated after random and directed mutagenesis as well as codon optimization. These variants, called rtTA-S2 and rtTA-M2, contain mutations in the DNA binding domain that connects the DNA with the core of the protein. They showed an improved sensitivity to the inducer drug and a reduced background activity in the absence of Dox.[22]

In 1995, Deuschle *et al.* developed a system, in which the original TetR protein was fused to the repressing KRAB domain of the human zinc-finger protein Kox1, generating a transrepressor called TetR-KRAB.[21] Zinc-finger proteins are mammalian regulatory proteins and their KRAB box is known to inhibit all DNA Polymerases I, II or III within a range of 3 kilobases of its DNA attachment site.[23] In the absence of Dox, TetR-KRAB binds specifically to the TRE and suppresses the activity of the nearby promoter(s), while in the presence of Dox, TetR-KRAB detaches from the TRE, thus allowing transgene expression (Fig. 10.2C). The mechanism by which KRAB inhibits expression of genes is not entirely known. Three pathways seem to be involved: (i) local change in chromatin structure; (ii) local histone deacetylation, and; (iii) the indirect influence of the arrangement of the basal transcription machinery. Therefore, until recently the hypothesis was that only integrated promoters could be inhibited by the TetR-KRAB.[23,24] But, it was recently demonstrated that the TetR-KRAB system could also be functional *in vitro* and *in vivo* in the mouse model in a non-integrative context, using non-integrative lentiviral and AAV vectors.[25]

This KRAB-mediated inhibition system was also incorporated in the TetOn system, in order to minimize the background expression observed in the latter regulation system.[26] For this purpose, the TetR-KRAB gene, also called tTS (for Tet controlled transcriptional silencer), was cloned into a bicistronic expression cassette downstream of the transactivator separated by an IRES (internal ribosomal entry site). In the absence of Dox, the rtTA transactivator stays detached from the TRE while the tTS transrepressor binds and inhibit background expression of the transgene. In the presence of Dox, the two chimeric proteins change their conformation and reverse their respective activity, allowing transgene expression.

3.2 *Pharmacology of Doxycycline (Dox)*

The antibiotic Dox has been extensively documented over the past 40 years. It has a half-life of 14–22 hours after absorption and an excellent tissue penetration, including the brain. The bioavailability of Dox after oral administration is 100% and it is rapidly metabolized and eliminated by the gastrointestinal tract. Dox has been extensively used in patients and is well tolerated, at least for medium term (months) applications. Side effects during prolonged administration of Tet include infrequent rise in blood-urea nitrogen, which is apparently dose-related as well as a dose-dependant photosensitivity. A concern when using an antibiotic-driven regulatory switch remains the possibility of raising resistance to the antibiotic itself. The generation of a nonantibiotic Dox analog-system would improve biosafety in that respect.

3.3 *Translational Development of Tet-dependant Regulation Systems*

The different Tet-derived regulation systems were used to study transgene expression in several rodent models using recombinant retrovirus,

AAV, adenovirus or naked DNA as vehicles.[8,18,24,25,27–30] All these studies demonstrated the ability for the Tet-derived systems to sustain permanent transgene regulation with no adverse effect. Of note, however, and even with more advanced versions of the chimeric transactivators, background activity in the absence of the inducer remains an issue, especially when high copy numbers of the vector per cell are achieved.

To date, among the Tet-derived systems used to regulate gene transcription, the TetOn system is the most widely developed. Importantly, it was tested in large animal models particularly in NHP, in the liver,[31] in the skeletal muscle[2,4] and in the retina.[7,32]

The NHP that received Tet-regulated vectors are summarized in Table 10.2. Unlike in rodent models, translation of the Tet-based systems into NHP resulted in some studies in the rapid loss of transgene regulation with an immune response towards the chimeric transactivators (in grey in Table 10.2). It appears that the targeted organ, as well as the route of administration of the vector, are likely to impact the development (or not) of this immune response.

When rtTA and TetR-KRAB were used to regulate the expression of either macaque Epo or the secreted alkaline phosphatase (SEAP) from skeletal muscle, after rAAV, plasmid or recombinant adenovirus-mediated gene transfer, no immune response was observed against the transgene protein, but a large majority of individuals developed an immune response against the rtTA or the TetR-KRAB proteins. The mechanism involved included both humoral and cellular responses, and resulted in the destruction of the genetically modified cells.[2,3]

Such immune response even occurred when using the so-called muscle-specific desmin promoter instead of an ubiquitous promoter to drive the expression of the transactivator rtTA [Moullier *et al.*, unpublished data]. However, one should emphasize that when the NHP immune system tolerates the chimeric transactivator rtTA, transgene regulation is achieved for several years (more than 6 years) [Moullier *et al.*, unpublished data].

Importantly, Table 10.2 indicates that skeletal muscle transduction after Regional Intravenous delivery (RI) using an rAAV vector, leads to permanent TetOn-mediated transgene regulation (more than 7 Dox-inductions). In the fully immunocompetent injected animals, neither anti-rtTA nor lymphocyte infiltration were detected in the muscle [Toromanoff *et al.*, submitted].

Table 10.2. Summary of NHP that received tetracycline-regulated vectors (*nd* = non determined; IM = intramuscular; IV = intravenous; RI = Regional Intravenous; SR = subretinal).

Study	Mode of delivery	NHP identification	Injected vector	Expression cassette(s)	Duration of the regulation	Host immune response?
Ref. 3	IM	M113	Adenovirus	TK.TetOn.SEAP	0.5 month	Yes
	IM	M114	Adenovirus	TK.TetOn.SEAP	0.5 month	Yes
	IM	F128	Adenovirus	TK.TetOn.SEAP	0.5 month	Yes
	IM	F129	Adenovirus	TK.TetOn.SEAP	0.5 month	Yes
	IM	F132	Adenovirus	TK.TetOn.SEAP	0.5 month	Yes
	IM	M102	Plasmid	TK.TetOn.SEAP	0.7 month	Yes
	IM	M107	Plasmid	TK.TetOn.SEAP	0.7 month	Yes
	IM	M109	Plasmid	TK.TetOn.SEAP	0.7 month	Yes
	IM	F121	Plasmid	TK.TetOn.SEAP	0.7 month	Yes
	IM	F124	Plasmid	TK.TetOn.SEAP	0.7 month	Yes
(Refs. 2, 4, Moullier *et al.*, unpublished data and Toromanoff *et al.*, submitted (or the reference number if updated in the list))	IM	Mac 4	rAAV-2	LTR.TetOn.Epo	5 months	Yes
	IM	Mac 5	rAAV-2	CAG.TetOn.Epo	3 months	Yes
	IM	Mac 6	rAAV-2	CAG.TetOn.Epo	3 months	Yes
	IM	Mac 7	rAAV-2	CAG.TetOn.Epo	6 months	Yes
	IM	Mac 8	rAAV-2	CAG.TetOn.Epo	>13 months	No
	IM	Mac 9	rAAV-2	Desmine.TetOn.Epo	4 months	Yes
	IM	Mac 10	rAAV-2	Desmine.TetOn.Epo	3 months	Yes
	IM	Mac 11	rAAV-2	Desmine.TetOn.Epo	2 months	Yes
	IM	Mac 12	rAAV-2	Desmine.TetOn.Epo	2 months	Yes
	IM	Mac 13	rAAV-2/1	Desmine.TetOn.Epo	>72 months	Yes
	IM	Mac 14	rAAV-2/1	Desmine.TetOn.Epo	2 months	Yes
	IM	Mac 15	rAAV-2/1	Desmine.TetOn.Epo	4 months	Yes
	IM	Mac 16	rAAV-2/1	Desmine.TetOn.Epo	3 months	Yes
	IM	Mac 17	rAAV-2/5	Desmine.TetOn.Epo	5 months	Yes
Ref. 31	IV	95-14	Adenovirus	TTR-TetOn-tINFα	<3 months	*nd*
	IV	96-15	Adenovirus	TTR-TetOn-tINFα	<3 months	*nd*
	IV	96-20	Adenovirus	TTR-TetOn-tINFα	<3 months	*nd*
	IV	97-01	Adenovirus	TTR-TetOn-tINFα	0 month	*nd*
[Toromanoff *et al.*, submitted]	RI	Mac 19	rAAV-2/1	Desmine.TetOn.Epo	>22 months	No
	RI	Mac 20	rAAV-2/1	Desmine.TetOn.Epo	>22 months	No

(Continued)

Table 10.2. (*Continued*)

Study	Mode of delivery	NHP identification	Injected vector	Expression cassette(s)	Duration of the regulation	Host immune response?
	RI	Mac 21	rAAV-2/1	Desmine.TetOn.Epo	>22 months	No
(adjali *et al.*, in preparation)	IM	Mac 22	rAAV-2/1	CAG.TetR-KRAB.cmEpo	0 month	Yes
	IM	Mac 23	rAAV-2/1	CAG.TetR-KRAB.cmEpo	2 months	Yes
	IM	Mac 24	rAAV-2/1	CAG.TetR-KRAB.cmEpo	2 months	Yes
Refs. 7, 32	SR	P5.1	rAAV-2/5	CAG.TetOn.Epo	>48 months	No
	SR	P5.2	rAAV-2/5	CAG.TetOn.Epo	>32 months	No
	SR	P5.3	rAAV-2/5	CAG.TetOn.Epo	>32 months	No
	SR	P4.1	rAAV-2/4	CAG.TetOn.Epo	>31 months	No
	SR	P4.2	rAAV-2/4	CAG.TetOn.Epo	>31 months	No
	SR	P4.3	rAAV-2/4	hRPE65.TetOn.Epo	>12, 5 months	No
	SR	P4.4	rAAV-2/4	hRPE65.TetOn.Epo	>12, 5 months	No

What makes the classical IM route of delivery rather efficient to trigger the host immune system against the transgene is believed to be, at least in part, local factors including high vector concentration at the site of injection, poor diffusion in the muscle, and high transgene expression in focal areas of the tissue.[33] In contrast, the RI route results in homogeneous and uniform transgene distribution within the injected limb with no transgene overexpression in muscular focal areas.[33] While immune tolerance (i.e. Treg induction) or ignorance with respect to the rtTA transactivator, in this context still needs to be analyzed, this study suggested that the RI route of administration could be an attractive procedure to achieve safe and non-immunogenic regulated gene transfer to the skeletal muscle using Tet-regulatable systems.

Also, and likewise the rapamycin regulatable system, ability for the TetOn system to establish long-term transgene regulation was systematically demonstrated in the retina of NHP following subretinal injection of rAAV vectors (Table 10.2).

The fact that no antibody directed against the rtTA could be detected in the eye or in the serum of these NHP, and that all the animals displayed persistent regulation of Epo secretion (more than 4 year study period), is consistent with the hypothesis that, unlike skeletal muscle, the retina is likely an immune-privilege site. An important aspect for future clinical applications was that transgene regulation was achieved with equivalent efficiency, whether Dox was given intravenously or orally by pills.[32]

Another study suggested that the brain might also be an immunologically "silent" site for transgene regulation when using the TetOn system, even in the presence of a systemic immune response against the vector, a recombinant adenovirus.[34] However, these results were obtained in rats and have not yet been translated in a large animal model.

4. Other Regulatable Systems

Other regulatable systems were developed for the regulation of transgene expression, but have not yet been tested in large animal models. In the mifepristone (RU486) regulatable system, the original DNA binding domain of the human mifepristone receptor was replaced by a GAL4 DNA binding domain, and the promoter-activating domain by the Herpes Simplex Virus (HSV) VP16 transcriptional activation domain to construct a chimeric protein. In the presence of mifepristone, this protein is able to bind to the GAL4 corresponding DNA sequence and to activate the downstream promoter to express the transgene of interest. This system was commercialized with the registered trademark Geneswitch®, and was validated for transgene regulation in rodent models.[35] However, since it contains the GAL4 binding domain, which is a peptide derived from yeast, it is reasonable to expect an immune response when tested in higher mammals.

A fully humanized tamoxifen-dependent transactivator has been developed for exclusive use in muscle tissue.[36] The chimeric transactivator, called HEA-3, comprises: (i) a DNA binding domain from human hepatocyte nuclear factor 1a (HNF 1a), a protein not expressed in muscle cells; (ii) a mutant antiprogesterone receptor that binds tamoxifen specifically; and; (iii) the human p65 activating domain of NF-kB. Induction of transgene expression was observed in mouse muscle over a long time period.[36] To date, the system has not been tested in higher animals. As this system lacks any non-human peptides, induction of an immune response against the transactivator would, theoretically, be minimized. However, if tamoxifen is a clinically validated drug for the treatment of breast cancer, doses to induce transgene expression would exceed those tolerated in clinical practice. Therefore, before this system can be applied in clinical gene therapy, development of less toxic tamoxifen analogs is warranted.

Another inducible regulation system, which was never tested in large animal models, is the ecdysteroid receptor-based regulatable system. Ecdysteroid hormones, such as ecdysone, regulate metamorphosis and other developmental processes in insects. It binds and activates the ecdysteroid receptor EcR, which in turn forms a heterodimer with the ultraspiracle protein (USP), leading to the activation of gene expression. USP is the insect homolog to mammalian retinoid X receptor (RXR). By replacing the original DNA binding site of fruitfly *Drosophila melanogaster* (Dm) EcR by a GAL4 DNA binding domain, and the activation protein by human RXR fused to a VP16 activation domain, a dimerizer-based regulatory system was created and commercialized under the name Rheoswitch®. Another version of this system contains the silkworm *Bombyx mori* (Bm) EcR fused to the VP16 activation domain. This system was used to regulate transgene expression *in vitro* and in rodents.[37,38] However, as these transactivators contain insect sequences, the possibility of an immune response to the chimeric proteins can be expected. Here too, testing in higher mammals is warranted.

Finally, recently developed protein-free regulation systems, which control transgene translation, either by specific small molecule-responsive riboswitches,[39,40] would likely alleviate the host immune response in large animal models. These molecular switches consist solely of RNA molecules, which undergo restructuring on metabolite binding. This affects gene expression by either causing inhibition of translation initiation, or ribozyme mediated mRNA degradation. Efficient gene regulation has been demonstrated in yeasts and bacteria using tetracycline-responsive riboswitches.[39] However, no such regulation was yet reported with such elements in mammalian cells, and even less *in vivo*. One study reported the control of mammalian gene expression through modulation of RNA self-cleavage in mice, using a toyocamycin-responsive riboswitch.[40] Unfortunately at this time, the toxicity of the inducer drug is not compatible with preclinical evaluation in large animal models.

A recent approach described the exogenous control of gene expression relying on the control of translational termination.[41] To achieve gene regulation, the transgene-coding sequences were modified by the introduction of a translational termination codon just downstream from the initiator AUG codon. Translation of the resulting mRNA leads to significant reduction in expression of the desire gene product. Addition of small molecules such as aminoglycoside antibiotics capable of suppressing translational termination results in production of the desired full-length protein. This original system was successfully tested *in vitro* and *in vivo* in the mouse (lung and bone marrow).[41] Evaluation in NHP of such regulatable systems should be of great interest.

5. General Conclusions

Regulatable systems have now been used in a variety of platforms such as plasmids, AAV, adenovirus, lentivirus or HSV vectors. Most applications were experimental studies to demonstrate the feasibility of such an approach *in vitro* or *in vivo* in rodent models but only a few moved to the large animal models, i.e. the NHP. Among the different gene switch platforms, the tetracycline- and the rapamycin-regulatable systems are the most widely developed. In particular, Tet-based systems have been fairly developed in relevant and challenging preclinical models with respect to future gene therapy trials targeting different organs such as the skeletal muscle, the retina or the liver. Among the several information derived from such preclinical studies, the immune reactions against the chimeric transactivators emerged as a main limit for the clinical translation. Factors involved in this immunotoxicity are likely numerous, but depend at least in part on the vector system used, the targeted organ and the route of administration.

Finally, new developments in the field avoiding the use of chimeric drug-sensitive transactivators are emerging, with the potential to not trigger the host immune system. However, they need to be tested in higher species to appreciate their clinical potential.

References

1. Ye X, *et al.* (1999). Regulated delivery of therapeutic proteins after *in vivo* somatic cell gene transfer. *Science* **283**: 88–91.
2. Favre D, *et al.* (2002). Lack of an immune response against the tetracycline-dependent transactivator correlates with long-term doxycycline-regulated transgene expression

in nonhuman primates after intramuscular injection of recombinant adeno-associated virus. *J Virol* **76**: 11605–11611.

3. Latta-Mahieu M, *et al.* (2002). Gene transfer of a chimeric trans-activator is immunogenic and results in short-lived transgene expression. *Hum Gene Ther* **13**: 1611–1620.

4. Chenuaud P, *et al.* (2004). Optimal design of a single recombinant adeno-associated virus derived from serotypes 1 and 2 to achieve more tightly regulated transgene expression from nonhuman primate muscle. *Mol Ther* **9**: 410–418.

5. Lebherz C, *et al.* (2005). Long-term inducible gene expression in the eye via adeno-associated virus gene transfer in nonhuman primates. *Hum Gene Ther* **16**: 178–186.

6. Rivera VM, *et al.* (2005). Long-term pharmacologically regulated expression of erythropoietin in primates following AAV-mediated gene transfer. *Blood* **105**: 1424–1430.

7. Stieger K, *et al.* (2006). Long-term doxycycline-regulated transgene expression in the retina of nonhuman primates following subretinal injection of recombinant AAV vectors. *Mol Ther* **13**: 967–975.

8. Bohl D, Salvetti A, Moullier P, Heard JM (1998). Control of erythropoietin delivery by doxycycline in mice after intramuscular injection of adeno-associated vector. *Blood* **92**: 1512–1517.

9. Rivera VM, *et al.* (1996). A humanized system for pharmacologic control of gene expression. *Nat Med* **2**: 1028–1032.

10. Chong H, Ruchatz A, Clackson T, Rivera VM, Vile RG (2002). A system for small-molecule control of conditionally replication-competent adenoviral vectors. *Mol Ther* **5**: 195–203.

11. Pollock R, Clackson T (2002). Dimerizer-regulated gene expression. *Curr Opin Biotechnol* **13**: 459–467.

12. Pollock R, Issner R, Zoller K, Natesan S, Rivera VM, Clackson T (2000). Delivery of a stringent dimerizer-regulated gene expression system in a single retroviral vector. *Proc Natl Acad Sci USA* **97**: 13221–13226.

13. Auricchio A, *et al.* (2002). Constitutive and regulated expression of processed insulin following *in vivo* hepatic gene transfer. *Gene Ther* **9**: 963–971.

14. Auricchio A, *et al.* (2002). Pharmacological regulation of protein expression from adeno-associated viral vectors in the eye. *Mol Ther* **6**: 238–242.

15. Weinmann P, Gossen M, Hillen W, Bujard H, Gatz C (1994). A chimeric transactivator allows tetracycline-responsive gene expression in whole plants. *Plant J* **5**: 559–569.

16. Gari E, Piedrafita L, Aldea M, Herrero E (1997). A set of vectors with a tetracycline-regulatable promoter system for modulated gene expression in *Saccharomyces cerevisiae*. *Yeast* **13**: 837–848.

17. Bieschke ET, Wheeler JC, Tower J (1998). Doxycycline-induced transgene expression during Drosophila development and aging. *Mol Gen Genet* **258**: 571–579.

18. Folliot S, *et al.* (2003). Sustained tetracycline-regulated transgene expression *in vivo* in rat retinal ganglion cells using a single type 2 adeno-associated viral vector. *J Gene Med* **5**: 493–501.

19. Gossen M, Bujard H (1992). Tight control of gene expression in mammalian cells by tetracycline-responsive promoters. *Proc Natl Acad Sci USA* **89**: 5547–5551.

20. Gossen M, Freundlieb S, Bender G, Muller G, Hillen W, Bujard H (1995). Transcriptional activation by tetracyclines in mammalian cells. *Science* **268**: 1766–1769.

21. Deuschle U, Meyer WK, Thiesen HJ (1995). Tetracycline-reversible silencing of eukaryotic promoters. *Mol Cell Biol* **15**: 1907–1914.
22. Urlinger S, Baron U, Thellmann M, Hasan MT, Bujard H, Hillen W (2000). Exploring the sequence space for tetracycline-dependent transcriptional activators: novel mutations yield expanded range and sensitivity. *Proc Natl Acad Sci USA* **97**: 7963–7968.
23. Wiznerowicz M, Trono D (2003). Conditional suppression of cellular genes: lentivirus vector-mediated drug-inducible RNA interference. *J Virol* **77**: 8957–8961.
24. Szulc J, Wiznerowicz M, Sauvain MO, Trono D, Aebischer P (2006). A versatile tool for conditional gene expression and knockdown. *Nat Methods* **3**: 109–116.
25. Barde I, *et al.* (2009). Regulation of episomal gene expression by KRAB/KAP1-mediated histone modifications. *J Virol.* **83**: 5574-5580.
26. Freundlieb S, Schirra-Muller C, Bujard H (1999). A tetracycline controlled activation/repression system with increased potential for gene transfer into mammalian cells. *J Gene Med* **1**: 4–12.
27. Harding TC, Geddes BJ, Murphy D, Knight D, Uney JB (1998). Switching transgene expression in the brain using an adenoviral tetracycline-regulatable system. *Nat Biotechnol* **16**: 553–555.
28. Dejneka NS, *et al.* (2001). Pharmacologically regulated gene expression in the retina following transduction with viral vectors. *Gene Ther* **8**: 442–446.
29. Lamartina S, *et al.* (2003). Construction of an rtTA2(s)-m2/tts(kid)-based transcription regulatory switch that displays no basal activity, good inducibility, and high responsiveness to doxycycline in mice and non-human primates. *Mol Ther* **7**: 271–280.
30. Koponen JK, *et al.* (2003). Doxycycline-regulated lentiviral vector system with a novel reverse transactivator rtTA2S-M2 shows a tight control of gene expression *in vitro* and *in vivo*. *Gene Ther* **10**: 459–466.
31. Aurisicchio L, De Tomassi A, La Monica N, Ciliberto G, Traboni C, Palombo F (2005). Regulated and liver-specific tamarin alpha interferon gene delivery by a helper-dependent adenoviral vector. *J Virol* **79**: 6772–6780.
32. Stieger K, *et al.* (2007). Oral administration of doxycycline allows tight control of transgene expression: a key step towards gene therapy of retinal diseases. *Gene Ther* **14**: 1668–1673.
33. Toromanoff A, *et al.* (2008). Safety and efficacy of regional intravenous (r.i.) versus intramuscular (i.m.) delivery of rAAV1 and rAAV8 to nonhuman primate skeletal muscle. *Mol Ther* **16**: 1291–1299.
34. Xiong W, *et al.* (2006). Regulatable gutless adenovirus vectors sustain inducible transgene expression in the brain in the presence of an immune response against adenoviruses. *J Virol* **80**: 27–37.
35. Wang L, *et al.* (2004). Prolonged and inducible transgene expression in the liver using gutless adenovirus: a potential therapy for liver cancer. *Gastroenterology* **126**: 278–289.
36. Roscilli G, *et al.* (2002). Long-term and tight control of gene expression in mouse skeletal muscle by a new hybrid human transcription factor. *Mol Ther* **6**: 653–663.
37. Hoppe UC, Marban E, Johns DC (2000). Adenovirus-mediated inducible gene expression *in vivo* by a hybrid ecdysone receptor. *Mol Ther* **1**: 159–164.
38. Galimi F, *et al.* (2005). Development of ecdysone-regulated lentiviral vectors. *Mol Ther* **11**: 142–148.

39. Bayer TS, Smolke CD (2005). Programmable ligand-controlled riboregulators of eukaryotic gene expression. *Nat Biotechnol* **23**: 337–343.

40. Yen L, *et al.* (2004). Exogenous control of mammalian gene expression through modulation of RNA self-cleavage. *Nature* **431**: 471–476.

41. Streilein JW (2003). Ocular immune privilege: therapeutic opportunities from an experiment of nature. *Nat Rev Immunol* **3**: 879–889.

Chapter 11

Gene Therapy for Central Nervous System Disorders

Deborah Young* and Patricia A. Lawlor

The chronic nature of many diseases affecting the central nervous system makes gene therapy an attractive alternative treatment option. In this chapter, we provide an overview of the gene therapy strategies for the major brain diseases including Parkinson's disease, Alzheimer's disease and temporal lobe epilepsy and highlight some early clinical findings using gene therapy approaches that suggest a promising outlook for this technology.

1. Introduction

One of the greatest potential applications of gene therapy is for the treatment of diseases affecting the central nervous system (CNS). Debilitating diseases like Parkinson's disease (PD), Alzheimer's disease (AD), Huntington's disease (HD), epilepsy, stroke, and amyotrophic lateral sclerosis (ALS) still affect a large percentage of the world's population. Pharmacological agents that help alleviate disease symptoms or disease progression are introduced into the market each year, yet many only provide symptomatic relief and have modest effects on disease progression. Many do not target the disease process, leading to low therapeutic efficacy. More specific and targeted therapies are still required.

The chronic nature of these diseases and thus the potential requirement for long-term delivery of therapeutic molecules makes gene therapy an

*Correspondence: Dept of Pharmacology & Clinical Pharmacology, University of Auckland, Auckland, New Zealand.
Email: ds.young@auckland.ac.nz

attractive treatment option. In particular, the ability of many viral vectors to mediate gene transfer to neurons leading to long-term production of a therapeutic protein helps circumvent the problems of repeated dosing regimens that are the hallmark of conventional drug treatments. Delivery of therapeutic molecules to the CNS has always been a significant challenge due to the physical barriers imposed by both the skull and the blood–brain barrier, which prevents the passage of large molecules from the bloodstream into brain tissue. Furthermore, the brain is composed of both neurons (that are post-mitotic) and non-neuronal cell types. Cell-to-cell communication occurs locally but also via interconnecting circuits throughout the brain. Moreover, there is global involvement of the brain in certain diseases (e.g. AD) but the involvement of very specific regions in other diseases (e.g. PD). These aspects have been taken into consideration when designing gene therapy approaches and in some cases these unique attributes have been exploited to develop more rational, targeted treatment approaches for these diseases. In the following sections, we provide an overview of the major gene therapy strategies developed for diseases affecting the brain.

2. Gene Therapy for Parkinson's Disease

PD is the second most common neurodegenerative disorder and affects 1% of the population over the age of 65. The majority of cases of PD are idiopathic, whereas rare familial forms are linked to gene mutations in α-synuclein and other proteins involved in the ubiquitin-proteasome pathway.[1] Both forms of PD share similar pathology, with the selective and progressive degeneration of dopamine neurons in the substantia nigra pars compacta (SNc) and the presence of abnormal intracellular protein aggregates called Lewy bodies in the remaining cells. Up to 50% of the dopamine-producing cells in the SNc that project axons to release the neurotransmitter dopamine in the caudate-putamen (striatum) can be lost before the manifestation of clinical symptoms. As striatal dopamine levels play a central role in the control of voluntary movement, dopamine depletion causes the characteristic debilitating motor symptoms associated with the disease.

The pharmacological mainstay for treatment of PD has been L-DOPA, the dopamine precursor that when taken orally readily crosses the blood–brain barrier to be converted to dopamine. While effective symptomatic relief can be provided in early stage PD, chronic L-DOPA use is associated with on–off effects and increasing doses of L-DOPA are required as the disease progresses.

Several features make PD an attractive gene therapy target: well-characterized and focal pathology that causes downstream effects on brain circuitry and clinical symptoms, well-established and characterized rodent and primate models of PD with predictive value regarding the dopamine deficiency, and the availability of non-invasive methods (e.g. motor function tests) for assessing phenotypic recovery following treatment. Thus in the early days of CNS gene therapy, PD was used as the prototype for evaluating the applicability of gene technology. Much of the data in these early studies provided insights into the types of gene delivery systems suitable for CNS gene therapy applications as well as the design of gene expression cassettes.

The first gene therapy strategies for PD involved boosting local dopamine production in the striatum. Conceptually, the aim was to circumvent the requirement of the surviving dopamine neurons to increase dopamine synthesis and instead use vectors to mediate transfer of genes involved in dopamine synthesis to neurons in the striatum, the region where dopamine levels are depleted. Thus the goal was to change the phenotype of striatal neurons, which normally do not produce dopamine, to cells capable of producing and secreting dopamine locally where required. The primary focus for early studies involved delivery of a cDNA for tyrosine hydroxylase (TH), the rate-limiting enzyme in dopamine synthesis using both defective Herpes simplex virus (HSV-1) and adeno-associated viral (AAV) vectors.[2,3] This strategy was tested in a rat model of PD involving unilateral lesioning of the dopamine neurons in the SNc with the neurotoxin 6-hydroxydopamine. One advantage of this rat model is that significant numbers of dopamine neurons can be killed in one brain hemisphere only causing an asymmetry in dopamine levels relative to the unlesioned side. The ability of treatment strategies to normalize the asymmetrical bias in striatal dopamine levels can be assessed non-invasively using spontaneous motor function tests and drug challenge with dopamine agonists such as apomorphine and amphetamine. Striatal dopamine levels were increased by 120% in the striatum injected with vectors expressing TH and this was associated with a reduction in motor dysfunction. Striatal neurons producing TH were observed in the striatum, however HSV-mediated expression levels decreased over time and cytopathological effects were found, whereas no such decrease was found using the AAV vector. Subsequent studies attempted to optimize dopamine production by evaluating other enzymes involved in dopamine synthesis, either individually or when co-expressed (by using multiple vectors or by use of bicistronic cassettes). Thus combinations of TH and aromatic acid decarboxylase (AADC) were evaluated in a Parkinsonian non-human primate model, with the addition of AADC postulated to improve dopamine production, as AADC can be rate-limiting in the presence of increased TH activity.[4] Partial phenotypic correction of the dopamine deficit was observed in Parkinsonian monkeys as well as in rodent models. Many of these studies highlighted the deficiencies in vector gene transfer systems at the time, with only small numbers of transduced neurons found surrounding each injection site. One of the main limitations of this type of approach is that dopamine levels are largely unregulated. An alternative strategy relies on a pro-drug strategy to achieve regulated dopamine delivery by peripheral administration of L-DOPA following AADC gene transfer.[5] In this manner, L-DOPA dosage could potentially be titrated to a level to achieve therapeutic benefit and thus customized for individualized patient use.

The dopamine restoration strategies aim to provide symptomatic relief for PD, but they do not address the continual temporal progression of the disease. The decline in striatal dopamine function is estimated at a rate of 5-10% per year. This factor, combined with ongoing requirements to incorporate regulatory systems in order to maintain dopamine at therapeutic levels led to assessment of alternative strategies. The second line of strategies stemmed from the observation that a significant dopamine cell population is still present during early stages of the disease, and thus genetic manipulation and protection of an existing cellular population may be more effective. Gene therapy approaches have centered on the overexpression of growth factors known to be critical for dopamine neuron survival as a strategy to slow down disease progression and/or promoting functional recovery by modulating regeneration of the surviving dopamine axon terminals. Two neurotrophic factors that have featured prominently are glial-derived neurotrophic factor (GDNF) and brain-derived neurotrophic factor (BDNF). In particular, GDNF overexpression has shown great promise. Both intranigral and intra-striatal infusion strategies using AAV, lentiviral and adenoviral vectors have been evaluated with the intra-striatal infusion paradigm showing the greatest therapeutic benefit.[6] When injected into a partially lesioned striatum of rats or non-human primates, GDNF counteracted the further loss of dopamine cell bodies and promoted sprouting from surviving axonal terminals leading to a gradual reversal of motor deficits. These studies highlighted the importance of careful consideration of the choice of infusion sites for studies in the CNS as protection of dopamine cell bodies in the SNc in the absence of functional striatal reinnervation was not sufficient to preserve intact motor performance. Functional recovery was only achieved when vectors were used to overexpress GDNF in the striatum.[7] Clinical experience using a growth factor approach has been currently limited to a related growth family member. Neurturin was delivered into multiple sites into the caudate putamen via an AAV2 vector and although initial data supported the safety and tolerability of this approach, no appreciable therapeutic efficacy was found in neurturin-treated patients compared to controls in phase II trials.[8,9]

Although most therapeutic efforts have centered on modulating the nigro-striatal pathway, the recognition that dopamine neuron loss causes downstream consequences to overall basal ganglia circuitry has led to development of a strategy that has shown very promising clinical results. The subthalamic nucleus, a component of the basal ganglia circuit, is a small nucleus that becomes disinhibited as a consequence of dopamine cell loss, thus leading to pathological excitation of downstream target brain regions involved in movement control. Dampening the excitatory activity of the subthalamic nucleus appears to be an effective strategy for controlling the motor symptoms associated with the disease, as shown by deep brain stimulation of this nucleus. AAV2 vector expressing glutamic acid decarboxylase (GAD), the major enzyme involved in synthesis of the inhibitory transmitter GABA, was delivered into the subthalamic nucleus as a strategy to phenotypically change these excitatory neurons into inhibitory neurons. Phase I clinical trials showed no adverse effects related to gene therapy and interestingly, significant improvements in motor function rating scores were observed.[10,11] Expansion to phase II clinical trials is currently underway.

Other approaches targeting mechanisms potentially involved in the disease process have also been evaluated including anti-apoptotic factors. Within the last decade, a growing body of evidence has highlighted the importance of defective protein handling and quality control as a major pathogenic mechanism contributing to dopamine neuron demise. Toxic forms of α-synuclein, a major component of Lewy bodies, are involved in the cell death process and while the precise mechanisms are still being elucidated, it is evident that this avenue of research will yield additional gene therapy targets for this disease.

3. Gene Therapy for Temporal Lobe Epilepsy

The epilepsies are a collection of distinct syndromes characterized by the occurrence of spontaneous recurrent seizures. Temporal lobe epilepsy (TLE), the most common form of seizure disorder is an ideal disease target for gene therapy. Like PD, the clinical manifestations of the disease originate from neuropathological changes in a very specific region of the temporal lobe, the hippocampus. Hippocampal cell loss and the subsequent reorganization of neuronal circuitry in this region leads to development of an "epileptic focus", a hyperexcitable region that is a site for seizure initiation.[12] Pharmacological approaches are currently used to control this chronic debilitating condition but despite the development of new antiepileptic medication, approximately 30% of patients suffer from pharmaco-resistant epilepsy. Furthermore, antiepileptic drugs do not prevent disease progression and long-term drug-based management is often associated with debilitating side effects. Surgical resection of the epileptic focus is often the only effective therapeutic option available.

There are several well-characterized rodent models of TLE. Typically, excitotoxins like kainate or application of focal repetitive electrical stimulation to a specific brain region is used to elicit short bouts of seizure activity in naïve animals. Another widely used model involves using excitotoxins or electrical stimulation to induce status epilepticus, a period of prolonged seizure activity sufficient to cause a selective pattern of hippocampal cell loss and subsequent reorganization of neuronal circuitry of surviving neurons that contributes to the development of chronic spontaneous recurrent seizures. These post-status epilepticus models of TLE, although considerably more laborious, are considered to be the best parallel with human TLE, because the neuropathology and seizure phenotype is similar to that in humans.

Gene therapy strategies for epilepsy could provide opportunities for achieving not only a sustained anticonvulsant effect, but also an antiepileptogenic effect that will block disease progression.[13] Approaches have focused on manipulating neurotransmission in this region and have ranged from increasing local GABA levels in the epileptogenic zone as a strategy to increase the threshold for neuronal excitability and hence prevent seizure initiation and propagation, to modulating NMDA glutamate receptor function.[14] Enhancing GABA receptor function by direct gene transfer approaches or *ex vivo* gene transfer and transplantation of GABA-secreting cells in various brain regions and circuits involved in seizure propagation caused a decrease in seizure severity or anticonvulsant effects. However many of the effects observed following cell transplantation were short-lived due to problems associated with graft survival and rejection.

The other main gene therapy targets have been the delivery of the peptides Neuropeptide Y and galanin, and modulation of adenosine levels, components that are believed to constitute an endogenous system that controls epileptic activity.[13] AAV-mediated overexpression of galanin in the hippocampus prevented hippocampal cell loss and decreased the number and duration of kainate-induced seizures. Similarly, NPY overexpression in the hippocampus delayed seizure onset and reduced the number and duration of seizures induced by kainate and retarded the rate of kindling epileptogenesis. Antiepileptic efficacy in a chronic post-status epilepticus rat model of TLE was also reported, with NPY overexpression decreasing spontaneous seizure frequency and progression.[15,16] These studies clearly establish the antiepileptic potential of these two peptides. Modulation of the hippocampal adenosinergic system to increase adenosine levels also appears to be a good therapeutic target that has also been supported by data from *ex vivo* gene therapy approaches.[17–19] Further work is still required before evaluation of these types of strategies in humans.

4. Huntington's Disease Gene Therapy

HD is an inherited adult-onset neurodegenerative disease caused by a mutation in the IT15 gene encoding huntingtin (htt). The mutation causes the abnormal expansion of a polyglutamine tract that confers a toxic gain of function to an otherwise benign protein. Mutant htt is prone to aggregation and forms intracellular protein aggregates that sequester other important cellular proteins leading to neuronal dysfunction and the demise of specific neuronal populations in the striatum and cerebral cortex.[20] HD typically strikes afflicted individuals in midlife and with no effective treatments currently available, individuals usually die within 10-20 years following disease onset. Classical clinical symptoms of HD include chorea, progressive cognitive decline and psychiatric disturbances. Several animal models that reproduce many of the neuropathological and behavioral characteristics of the human disease are available for screening new therapies. These include transgenic mouse lines that express N-terminal fragments of mutant htt (e.g. R6 mouse lines), knock-in mice or mouse lines that express full-length mutant htt (e.g. yeast artificial chromosome-transgenic lines), as well as chemical lesion or viral vector-mediated overexpression of mutant htt N-terminal fragments in rodents and non-human primates.

Early gene therapy strategies for HD were similar to those for other neurodegenerative diseases, with the main focus on the delivery neurotrophic factor genes as a direct means for protecting vulnerable striatal neurons against mutant htt-mediated toxicity. Alternative approaches have focused on the delivery of molecules aimed at directing neurogenesis, the production of new adult neurons to replace neurons lost in the disease.[21]

More recently, the development of effective gene silencing approaches using RNA interference technology has led to evaluation of strategies aimed at selectively reducing mutant htt expression. Genetic screening to identify individuals that have inherited the HD mutation provides the opportunity to intervene in the pathogenic process prior to onset of disease symptoms. Thus repressing expression of the disease-causing protein is expected to delay disease onset and mitigate severity of the disease.

A few recent studies have highlighted the feasibility and promise of such a gene silencing approach. AAV vector-mediated delivery of short-interfering RNAs or microRNA sequences that target huntingtin in a non-allele specific manner led to a concomitant reduction in mutant as well as normal htt gene product. This resulted in the attenuation of neuropathological and behavior deficits in HD transgenic and viral overexpression models.[9,22–25] Importantly, these studies suggest that complete elimination of the mutant allele may not be required. Although allele-specific targeting of the mutant htt allele would be ideal, thus preserving the biological functions of wild-type huntingtin, no single nucleotide polymorphisms resident in the mutant transcript have been identified to distinguish mutant versus normal allele. No toxicity was found in a knock-in mouse model of HD in which levels of htt expression were decreased by more than 50% for up to 4 months.[24] Further investigation is required to determine to what extent gene expression can be reduced without incurring toxicity due to loss of normal htt function.

5. Amyotrophic Lateral Sclerosis (ALS)

ALS is the most devastating and common adult-onset motor neuron disease in humans. The clinicopathological features of ALS are the progressive and selective loss of cortical, and spinal/bulbar motor neurons that causes skeletal muscle wasting, paralysis and ultimately death. Ninety-five percent of ALS cases are sporadic and of unidentifiable etiology, with the remaining familial cases dominantly inherited. Approximately 15% of familial ALS is associated with missense mutations in the SOD1 gene encoding the cytosolic enzyme Cu/Zn-superoxide dismutase. Indeed, the major animal models used for screening novel therapies are transgenic mouse lines expressing mutant SOD1 (mSOD1) that share many pathological features observed in human ALS. Defects in SOD1 function however, are not found in sporadic ALS suggesting involvement of other pathogenic mechanisms including oxidative stress, autoimmunity and glutamate-mediated excitotoxicity due to a decrease in excitatory amino acid transporter function, in the demise of motor neurons.[26] This needs to be borne in mind when extrapolating the potential efficacy of a promising treatment in mSOD1 mouse models to sporadic forms of ALS in humans. Currently there is no effective pharmacological treatment for ALS, with the glutamate antagonist riluzole the only approved drug treatment. Most drugs tested so far have only produced modest benefits suggesting alternative treatments are needed.

Apoptotic pathways are a target for disease intervention in ALS and the neuroprotective efficacy of a range of growth factors that support motor neuron survival such as vascular endothelial growth factor (VEGF), insulin growth factor (IGF1), GDNF and ciliary neurotrophic factor (CNTF) have been evaluated in ALS animal models, with effects on delaying disease onset and increasing lifespan of these mice one of the primary outcome measures. As motor neuron degeneration in genetic mouse models develops over several months, long-term chronic delivery of therapeutic molecules might be necessary to confer significant protection. Long-term systemic infusion of growth factors is fraught with problems. For example, although CNTF is a potent survival factor for motor neurons, poor bioavailability and toxicity following systemic administration has complicated its therapeutic use for ALS. Targeted delivery methods such as that mediated by gene therapy approaches may overcome some of these obstacles.

Gene therapy approaches for ALS have included overexpression of neu-rotrophic factors, anti-apoptotic genes and antioxidants via direct injection into the affected areas or in remote intramuscular sites that rely on retro-grade transport of the protective molecule to motor neuron cell bodies. Most of these have been evaluated in the mutant SOD1 mouse model, with AAV and lentiviral-mediated delivery of IGF1, GDNF and VEGF, Bcl-2 effec-tive in delaying disease onset and/or prolonging mouse lifespan (reviewed in Ref. 27). To date, most therapeutic approaches are targeted towards the genetic manipulation of neurons but the discovery that disease progres-sion occurs in a non-cell autonomous manner suggests that vector-mediated manipulation of astrocytes that contribute to disease pathogenesis may yield greater therapeutic efficacy.[28] The development of astrocyte-targeting vec-tors will help address these issues.

Similarly, gene silencing approaches using RNA interference sequences or antisense oligonucleotides that may have a lower risk profile for human usage due to the short-lived nature of these molecules could also be investigated.[29] Indeed approval for clinical trials using an anti-sense approach to knockdown mutant SOD1 expression is currently being sought.[30]

6. Gene Therapy for Canavan Disease

The world's first gene therapy trial for a CNS disorder was undertaken for Canavan disease, an inherited and incurable leukodystrophy.[31] Symptoms of the disease appear early in infancy leading to loss of early motor skills and mental retardation. Most Canavan disease patients live only into childhood. The underlying cause of the disease is a defective aspartoacylase gene that normally functions to break down N-acetyl aspartate. As a result of the loss of gene function, N-acetyl aspartate breakdown is impaired and accumulates to levels that are toxic to myelin, the fatty insulation around nerve fibers in the brain. Degeneration of the white matter of the brain occurs leading to a spongy appearance.[32] Thus a logical treatment approach for this disease would entail expressing a functional copy of the aspartoacylase gene. A liposomal-based gene transfer system and subsequently an AAV2 vector was used to deliver an aspartoacylase gene to the brains of infants with the genetic defect in two phase I clinical trials. Both treatments appeared to be well tolerated.[31,33,34] Long-term monitoring and extension to phase II/III trials will be required before definitive statements can be made on efficacy.

7. Gene Therapy for Alzheimer's Disease

AD, the leading cause of dementia in the elderly, is a progressive neurodegenerative disorder identified clinically by a characteristic decline in cognitive function. These cognitive changes are accompanied by characteristic anatomical changes — both familial and sporadic forms of AD display the same neuropathological hallmarks including accumulation of extracellular Aβ plaques and intracellular neurofibrillary tangles (NFTs), synaptic and neuronal loss in cortical brain regions, basal forebrain cholinergic deficits and inflammation.[35]

Given its pivotal role in AD pathogenesis, cerebral Aβ is a target for gene therapy of AD. Aβ, the primary component of the extracellular plaques found in AD brain, is the product of proteolytic processing of amyloid precursor protein (APP) by beta- and gamma-secretases. Numerous gene mutations linked to AD alter APP metabolism to increase Aβ production, resulting in its accumulation.[36] However, the overall concentration of Aβ in the brain is modulated not only by Aβ production, but by its clearance, with several endogenous proteases known to degrade cerebral Aβ. Both production and clearance processes could be manipulated by gene therapy to prevent Aβ accumulation, and several strategies have been tested in AD transgenic mice to reduce Aβ load. For example, reducing secretase activity could decrease Aβ production. Indeed, a lentiviral vector expressing siRNA targeting BACE1 (beta-secretase) led to decreased BACE1 expression and activity, reduced amyloid production and fewer neurodegenerative and behavioural deficits following injection into the brain of APP transgenic mice.[37]

Likewise, overexpression of $A\beta$ proteases could be expected to reduce $A\beta$ load by increasing clearance of extracellular $A\beta$. This has been achieved by gene transfer of neprilysin, a transmembrane protein that functions to degrade both intra- and extra-cellular $A\beta$. In APP transgenic mice, short-term overexpression of neprilysin (1 week) reduced amyloid deposits by 50%.[38] Long-term neprilysin gene transfer (6 months) to APP transgenic mice not only lowered the amyloid plaque load but was associated with reduced intracellular $A\beta$ levels, amelioration of dendritic and synaptic pathology and improved performance in cognitive tests.[39] Others have used an *ex vivo* gene delivery approach — following implantation of genetically modified fibroblasts expressing a secreted form of neprilysin, robust plaque clearance was observed in the hippocampus of aged APP transgenic mice with advanced plaque deposition.[40] The protease endothelin-converting enzyme (ECE-1) has also shown promise as an $A\beta$ clearing agent in a murine AD model.[41]

An additional and quite different gene therapy strategy for reduction of $A\beta$ deposits involves using AAV to immunize against the $A\beta$ peptide, resulting in clearance of $A\beta$ and attenuation of memory deficits in APP transgenic mice.[42]

In addition to $A\beta$, the basal forebrain cholinergic system is a major target for gene therapy treatment of AD. Reduced activity and loss of these neurons is associated with the cognitive decline seen in AD — treating this cholinergic hypofunction and degeneration is the mechanism of action for most currently available pharmacological treatments for AD. This can also be achieved by gene therapy — gene transfer to increase levels of neurotrophins like nerve growth factor (NGF) could be used to prevent cholinergic neurons from dying and to improve the functioning of these neurons by preserving their synapses. Early clinical studies using infusion of recombinant NGF protein directly into brain resulted in adverse side-effects such as pain and weight loss, and highlighted the need for delivery of NGF in pharmacologically relevant concentrations in target brain areas, while preventing access to non-target areas. This localized NGF delivery could be achieved by use of gene therapy, and there is much preclinical evidence from studies on both AD transgenic mice and aged primates which confirm a role for NGF in both cognitive enhancement and protection of basal forebrain cholinergic neurons.[43]

NGF gene therapy has proceeded to at least three phase I clinical trials, each utilizing a different method of NGF gene transfer.[43] The first trial used *ex vivo* NGF gene transfer — the patients' own fibroblasts, obtained by skin biopsy, were cultured and genetically modified using a retrovirus to produce and secrete human NGF. The cells were transplanted back into the cholinergic basal forebrain by stereotaxic surgery and patients followed up for 22 months. This phase I trial was completed, with encouraging clinical outcomes and prompted a second clinical trial (currently ongoing) in which *in vivo* NGF gene therapy is being used — subjects receive bilateral stereotactic injections of AAV2 NGF ("CERE-110", CereGene) into the basal forebrain and are being assessed for 24 months. In a third ongoing NGF gene therapy trial, genetically modified encapsulated cells that secrete NGF (NsG0202, NSGene) are being trialled. A catheter-like device containing genetically modified human fibroblasts enclosed behind a semi-permeable membrane that allows for the influx of nutrients and outflow of NGF is implanted into the brain. This form of delivery allows controlled, site-specific, local delivery of NGF but does not allow for direct contact between the foreign therapeutic cells and host tissue.

To date development of gene therapy treatments for AD has largely been based on gene transfer for enhancement of $A\beta$ clearance or on gene transfer to increase trophic support of vulnerable cholinergic neurons. However the coupling of gene knockdown technology (RNA interference) to viral vector expression systems means we now have the ability to not only over-express, but to knockdown gene expression and alternative strategies such as BACE1, APP and amyloid β-binding death inducing protein (AB-DIP) knockdown can now be evaluated.

8. Conclusions and Outlook

This chapter focuses on current developments and the application of gene therapy in the central nervous system. The chronic nature of the neurodegenerative diseases, and the paucity of good effective treatments for many of these diseases makes gene therapy an attractive option for the future management of these diseases. There are still several hurdles to overcome such as the refinement and incorporation of regulatory systems into gene cassettes for safety purposes and development of vector systems capable of widespread global transduction of large brain regions for certain neurological diseases. This field is still in its early days, but data from early clinical trials have shown promising outlooks for this technology, paving the way for more widespread evaluation of the technology in humans.

References

1. Skovronsky DM, Lee VM, Trojanowski JQ (2006). Neurodegenerative diseases: new concepts of pathogenesis and their therapeutic implications. *Annu Rev Pathol* **1**: 151–170.
2. Kaplitt MG, *et al.* (1994). Long-term gene expression and phenotypic correction using adeno-associated virus vectors in the mammalian brain. *Nat Genet* **8**: 148–154.
3. During MJ, Naegele JR, O'Malley KL, Geller AI (1994). Long-term behavioral recovery in parkinsonian rats by an HSV vector expressing tyrosine hydroxylase. *Science* **266**: 1399–1403.
4. During MJ, *et al.* (1998). *In vivo* expression of therapeutic human genes for dopamine production in the caudates of MPTP-treated monkeys using an AAV vector. *Gene Ther* **5**: 820–827.
5. Bankiewicz KS, *et al.* (2000). Convection-enhanced delivery of AAV vector in parkinsonian monkeys; *in vivo* detection of gene expression and restoration of dopaminergic function using pro-drug approach. *Exp Neurol* **164**: 2–14.
6. Bjorklund A, Kirik D, Rosenblad C, Georgievska B, Lundberg C, Mandel RJ (2000). Towards a neuroprotective gene therapy for Parkinson's disease: use of adenovirus, AAV and lentivirus vectors for gene transfer of GDNF to the nigrostriatal system in the rat Parkinson model. *Brain Res* **886**: 82–98.
7. Kirik D, Rosenblad C, Bjorklund A, Mandel RJ (2000). Long-term rAAV-mediated gene transfer of GDNF in the rat Parkinson's model: intrastriatal but not intranigral transduction promotes functional regeneration in the lesioned nigrostriatal system. *J Neurosci* **20**: 4686–4700.
8. Marks WJ, Jr., *et al.* (2008). Safety and tolerability of intraputaminal delivery of CERE-120 (adeno-associated virus serotype 2-neurturin) to patients with idiopathic Parkinson's disease: an open-label, phase I trial. *Lancet Neurol* **7**: 400–408.
9. Franich NR, Fitzsimons HL, Fong DM, Klugmann M, During MJ, Young D (2008). AAV vector-mediated RNAi of mutant huntingtin expression is neuroprotective in a novel genetic rat model of Huntington's disease. *Mol Ther* **16**: 947–956.
10. Kaplitt MG, *et al.* (2007). Safety and tolerability of gene therapy with an adeno-associated virus (AAV) borne GAD gene for Parkinson's disease: an open label, phase I trial. *Lancet* **369**: 2097–2105.
11. Luo J, *et al.* (2002). Subthalamic GAD gene therapy in a Parkinson's disease rat model. *Science* **298**: 425–429.

12. Dudek FE, Sutula TP (2007). Epileptogenesis in the dentate gyrus: a critical perspective. *Prog Brain Res* **163**: 755–773.

13. Riban V, Fitzsimons HL, During MJ (2009). Gene therapy in epilepsy. *Epilepsia* **50**: 24–32.

14. Haberman R, *et al.* (2002). Therapeutic liabilities of *in vivo* viral vector tropism: adeno-associated virus vectors, NMDAR1 antisense, and focal seizure sensitivity. *Mol Ther* **6**: 495–500.

15. Richichi C, *et al.* (2004). Anticonvulsant and antiepileptogenic effects mediated by adeno-associated virus vector neuropeptide Y expression in the rat hippocampus. *J Neurosci* **24**: 3051–3059.

16. Noe F, *et al.* (2008). Neuropeptide Y gene therapy decreases chronic spontaneous seizures in a rat model of temporal lobe epilepsy. *Brain* **131**: 1506–1515.

17. Young D, Dragunow M (1994). Status epilepticus may be caused by loss of adenosine anticonvulsant mechanisms. *Neuroscience* **58**: 245–261.

18. Boison D, Huber A, Padrun V, Deglon N, Aebischer P, Mohler H (2002). Seizure suppression by adenosine-releasing cells is independent of seizure frequency. *Epilepsia* **43**: 788–796.

19. Huber A, Padrun V, Deglon N, Aebischer P, Mohler H, Boison D (2001). Grafts of adenosine-releasing cells suppress seizures in kindling epilepsy. *Proc Natl Acad Sci USA* **98**: 7611–7616.

20. Ross CA (2002). Polyglutamine pathogenesis: emergence of unifying mechanisms for Huntington's disease and related disorders. *Neuron* **35**: 819–822.

21. Cho SR, Benraiss A, Chmielnicki E, Samdani A, Economides A, Goldman SA (2007). Induction of neostriatal neurogenesis slows disease progression in a transgenic murine model of Huntington disease. *J Clin Invest* **117**: 2889–2902.

22. DiFiglia M, *et al.* (2007). Therapeutic silencing of mutant huntingtin with siRNA attenuates striatal cortical neuropathology and behavioral deficits. *Proc Natl Acad Sci USA* **104**: 17204–17209.

23. Harper SQ, *et al.* (2005). RNA interference improves motor and neuropathological abnormalities in a Huntington's disease mouse model. *Proc Natl Acad Sci USA* **102**: 5820–5825.

24. McBride JL, *et al.* (2008). Artificial miRNAs mitigate shRNA-mediated toxicity in the brain: implications for the therapeutic development of RNAi. *Proc Natl Acad Sci USA* **105**: 5868–5873.

25. Machida Y, Okada T, Kurosawa M, Oyama F, Ozawa K, Nukina N (2006). rAAV-mediated shRNA ameliorated neuropathology in Huntington disease model mouse. *Biochem Biophys Res Commun* **343**: 190–197.

26. Cleveland DW, Rothstein JD (2001). From Charcot to Lou Gehrig: deciphering selective motor neuron death in ALS. *Nat Rev Neurosci* **2**: 806–819.

27. Goodall EF, Morrison KE (2006). Amyotrophic lateral sclerosis (motor neuron disease): proposed mechanisms and pathways to treatment. *Expert Rev Mol Med* **8**: 1–22.

28. Clement AM, *et al.* (2003). Wild-type nonneuronal cells extend survival of SOD1 mutant motor neurons in ALS mice. *Science* **302**: 113–117.

29. Ralph GS, *et al.* (2005). Silencing mutant SOD1 using RNAi protects against neurodegeneration and extends survival in an ALS model. *Nat Med* **11**: 429–433.

30. La Spada AR (2009). Getting a handle on Huntington's disease: silencing neurodegeneration. *Nat Med* **15**: 252–253.

31. Leone P, *et al.* (2000). Aspartoacylase gene transfer to the mammalian central nervous system with therapeutic implications for Canavan disease. *Ann Neurol* **48**: 27–38.

32. Matalon R, Michals-Matalon K (1999). Biochemistry and molecular biology of Canavan disease. *Neurochem Res* **24**: 507–513.

33. McPhee SW, *et al.* (2006). Immune responses to AAV in a phase I study for Canavan disease. *J Gene Med* **8**: 577–588.

34. Janson C, *et al.* (2002). Clinical protocol. Gene therapy of Canavan disease: AAV-2 vector for neurosurgical delivery of aspartoacylase gene (ASPA) to the human brain. *Hum Gene Ther* **13**: 1391–1412.

35. Van Dam D, De Deyn PP (2006). Drug discovery in dementia: the role of rodent models. *Nat Rev Drug Discov* **5**: 956–970.

36. Selkoe, DJ (2008). Soluble oligomers of the amyloid beta-protein impair synaptic plasticity and behavior. *Behav Brain Res* **192**: 106–113.

37. Singer O, *et al.* (2005). Targeting BACE1 with siRNAs ameliorates Alzheimer disease neuropathology in a transgenic model. *Nat Neurosci* **8**: 1343–1349.

38. Marr RA, *et al.* (2003). Neprilysin gene transfer reduces human amyloid pathology in transgenic mice. *J Neurosci* **23**: 1992–1996.

39. Spencer B, *et al.* (2008). Long-term neprilysin gene transfer is associated with reduced levels of intracellular Abeta and behavioral improvement in APP transgenic mice. *BMC Neurosci* **9**: 109.

40. Hemming ML, Patterson M, Reske-Nielsen C, Lin L, Isacson O, Selkoe DJ (2007). Reducing amyloid plaque burden via *ex vivo* gene delivery of an Abeta-degrading protease: a novel therapeutic approach to Alzheimer disease. *PLoS Med* **4**: e262.

41. Carty NC, *et al.* (2008). Adeno-associated viral (AAV) serotype 5 vector mediated gene delivery of endothelin-converting enzyme reduces Abeta deposits in APP + PS1 transgenic mice. *Mol Ther* **16**: 1580–1586.

42. Mouri A, Noda Y, Hara H, Mizoguchi H, Tabira T, Nabeshima T (2007). Oral vaccination with a viral vector containing Abeta cDNA attenuates age-related Abeta accumulation and memory deficits without causing inflammation in a mouse Alzheimer model. *FASEB J* **21**: 2135–2148.

43. Cattaneo A, Capsoni S, Paoletti F (2008). Towards non invasive nerve growth factor therapies for Alzheimer's disease. *J Alzheimers Dis* **15**: 255–283.

Chapter 12
Gene Therapy of Hemoglobinopathies

Angela E. Rivers and Arun Srivastava*

Human hemoglobinopathies, such as β-thalassemia and sickle cell disease (SCD), are the most common human genetic diseases worldwide, and are an attractive target for potential gene therapy. This potential could be realized if a functional β-globin gene could be safely and efficiently introduced into hematopoietic stem cells (HSCs), and lineage-restricted expression of the β-globin protein exceeding 15% could be achieved in erythroid progenitor cells. The use of first generation retroviral vectors remained limited due to the inability to achieve therapeutic levels of β-globin expression. Although therapeutic levels of β-globin expression could be achieved by second-generation retroviral vectors, their use in gene therapy trials for immuno-deficiency resulted in insertional mutagenesis leading to leukemia in several children, which raised serious concerns. The use of lentiviral vectors expressing β-globin has led to phenotypic correction of β-thalassemia and SCD in mouse models, but long-term safety issues with lentiviral vectors remain to be evaluated. This chapter will provide a brief review of obstacles that were encountered, and achievements that have occurred over the past two decades, including the development of an alternative vector system based on a non-pathogenic human parvovirus, the adeno-associated virus (AAV), for the potential gene therapy of β-thalassemia and sickle cell disease.

---

*Correspondence: Divisions of Hematology and Oncology, and Cellular and Molecular Therapy, Departments of Pediatrics and Molecular Genetics and Microbiology, University of Florida College of Medicine, Gainesville FL 32610.
E-mail: aruns@peds.ufl.edu

1. Introduction

As a group, hemoglobinopathies are the most common genetic diseases that afflict humans worldwide.[1] Severe forms cause a congenital anemia, which is only curable by allogeneic hematopoietic stem cell transplantation (HSCT).[2,3] Hemoglobinopathies can be divided into two groups. The first group, thalassemia, is a quantitative disorder in which there is a decrease in the α-globin chain (α-thalassemia) or in the β-globin chain (β-thalassemia).[4] The second is a qualitative disorder in which there are structural variants in the hemoglobin molecule. In qualitative hemoglobinopathies, individuals inherit hemoglobins with structural defects that lead to abnormal pathophysiology, such as in sickle cell disease (SCD).[5,6] The majority of hemoglobinopathy research focuses on β-thalassemia and SCD. In β-thalassemia, patients have severe anemia associated with cachexia, fatigue, congestive heart failure, growth retardation, and bone abnormalities. SCD is associated with painful and potentially life-threatening crises, which result from the inability of erythrocytes with sickle hemoglobin to flow effectively in blood vessels and an inflammatory response from chronic hemolysis. This results in end organ ischemia and crises such as acute chest syndrome, stroke, retinal hemorrhage, splenic sequestration, and myocardial infarction.[7] Although HSCT from bone marrow or cord blood could potentially be therapeutic, a vast majority of patients with hemoglobinopathies do not have suitable family donors. Furthermore, the risks of complications associated with unrelated bone marrow donors or cord blood transplant do not make this treatment a practical option for most patients. A safe and effective strategy involving gene therapy would be of significant benefit to a large number of patients who could be cured of their disease.

2. β-Thalassemia

In β-thalassemia, the excessive α-globin chain precipitates and leads to damage to the red blood cell membrane. Ineffective erythropoiesis leads to anemia, bone marrow expansion, extramedullary hematopoiesis and increased intestinal iron absorption. Clinically, β-thalassemia are divided into three categories: β-thalassemia^{minor}, β-thalassemia^{intermedia}, and

β-thalassemia^{major}. Individuals with β-thalassemia^{minor} have a mild anemia and are usually asymptomatic. Individuals with β-thalassemia^{intermedia} have moderate anemia, which requires intermittent transfusion. Individuals with β-thalassemia^{major} are severely anemic and require monthly transfusions. In β-thalassemia^{major}, the goal of transfusion therapy is to achieve total hemoglobin of 13–14 g/dl. At this hemoglobin level, there is improved growth and development and bone marrow complications are prevented. Patients can receive as much as 3–8 g of iron per year depending on their age, but since iron-loss is fixed, patients become iron-overloaded. Unbound iron accumulates in organs and may also be converted to a free radical that damages the lipid membrane of cells. This leads to cardiac (cardiomyopathies and conduction disturbance), endocrine (hypothyroidism, growth failure, and diabetes mellitus), hepatic (hepatic fibrosis and cirrhosis), and joint toxicities. Patients with β-thalassemia^{major} are protected *in utero* by the expression of fetal hemoglobin and thus are usually born healthy. This is very unlike α-thalassemia in which the fetus usually develops *hydrops fetalis* and dies *in utero*. However, recent advancements in umbilical cord sampling allow *in utero* diagnosis and treatment with intrauterine transfusions. Surviving infants are treated with chronic transfusion therapy or hematopoietic stem cell transplant (HSCT).[8] With these new advances, patients would benefit from the development of gene therapy approaches involving the globin genes.

3. Sickle Cell Disease

Sickle cell disease (SCD) is an autosomal recessive disorder affecting millions of people worldwide. The causative mutation is an A–T transversion in the sixth codon of the β-globin gene which leads to the substitution of valine for glutamic acid, resulting in the formation of abnormal hemoglobin, known as hemoglobin S (Hb S).[6] Following deoxygenation in red blood cells (RBCs), Hb S forms polymers causing the RBCs to become deformed (sickled) and adherent, leading to vaso-occlusive events, which may result in splenic infarct, kidney failure, stroke, painful crises, and chronic anemia.[5,7,9–13] At physiologically relevant conditions, deoxygenated Hb S has a lower solubility than oxygenated Hb S, deoxygenated Hb A, or oxygenated Hb F.[14] The reason that patients with heterozygous genotype (β^S/β^A) experience no symptoms is two-fold: (1) The cellular concentration of Hb S in β^S/β^A is $\sim$15 g/dL. This is below the 17 g/dL solubility under physiological conditions. (2) The predominant hemoglobin is the hybrid heterotetramer $[(\alpha\beta^A)(\alpha\beta^S)]$. Thus, the probability that the mixed hybrid $(\alpha\beta^A)(\alpha\beta^S)$, which forms within the cell in proportion to the binomial distribution, will enter the polymer is about half that of Hb S $[(\alpha\beta^S)(\alpha\beta^S)]$. Treatment modalities for SCD involves allogeneic bone marrow transplantation, induction of fetal hemoglobin, control of infections, and pain management.[11,15−19] Allogeneic transplantation usually requires myeloablative conditioning with cyclophosphamide, busulfan, antithymocyte globulin or total lymphoid radiation. In addition, cyclosporine and methotrexate are used post-transplant to induce immunosuppression.[20] Hydroxyurea (HU) is the only drug approved by the US Food and Drug Administration (FDA) to treat SCD. HU increases Hb F in SCD by inducing erythroid regeneration and augmenting γ-globin gene expression in a nitric oxide-dependent pathway.[21] Other drugs that increase Hb F and that are currently in phase I/II clinical trials for SCD are as follows. Decitibine, an analog of 5-azacytidine, induces Hb F by causing hypomethylation of the γ-globin genes.[9] Histone deacetylase (HDAC) inhibitors have also been shown to enhance γ-globin gene expression through histone hyperacetylation and alteration of chromatin structures.[22] Again, a gene therapy approach involving hematopoietic stem cells transduced with anti-sickling globin genes followed by HSCT would potentially be therapeutic.

4. Gene Therapy

As illustrated in Table 12.1, human hemoglobinopathies are the most common monogenic disease worldwide, and are among the likely candidates for gene-based treatment provided that the pluripotent hematopoietic stem cell can be stably transduced, and that long-term, regulated expression of a functional β-globin gene in the erythroid progenitor cell can be achieved.

Both the qualitative and quantitative hemoglobinopathies are only curable by HSCT. However, the majority of patients do not have an HLA-matched bone marrow donor, and due to concerns regarding conditioning regimens and graft versus host disease (GVHD), not all patients are eligible for bone marrow transplant. There is 10-20% mortality and morbidity rate of

Table 12.1.   Relative incidence of monogenic human genetic diseases.

Disease	Incidence	Defective gene	Target
Severe Combined Immunodeficiency	Rare	ADA	Bone Marrow
Hemophilia B	1:30,000	Factor IX	Liver
Hemophilia A	1:10,000	Factor VIII	Liver
Duchenne's Muscular Dystrophy	1:7,000	Dystrophin	Muscle
Inherited Emphysema	1:3,500	α1-antitrypsin	Lung/Liver
Cystic Fibrosis	1:2,000	CFTR	Lung
Lysosomal Storage Disease	1:1,500	Enzymes	Disorder-dependent
Sickle cell disease and β-thalassemia	1:600	Globin	Bone Marrow

patients who receive transplantation.[23] GVHD is a common complication of allogeneic bone marrow transplantation in which functional immune cells in the transplanted marrow recognize the recipient as "foreign" and mount an immunologic attack. In theory, cells that have undergone gene transfer would not elicit an immune response since they would be from the same patient. Therefore, effective and safe gene therapy would potentially increase the number of patients who could be cured of hemoglobinopathies and decrease their mortality and morbidity during transplant. Once gene therapy is developed, patients with a family history could have HSCs collected from one of three sources: (i) umbilical cord cells collected at birth, (ii) HSCs collected from peripheral blood following mobilization with Granulocyte Colony-Stimulating Factor (G-CSF), and (iii) reprogramming of human fibroblasts into HSCs.[24] The goals for effective gene therapy in hemoglobinopathies are similar to goals for all gene therapies in that it must be safe and effective. Additionally, the globin gene must be expressed in high amounts at a specific time in the erythroid lineage. In the case of β-thalassemia, the goal would be to achieve a normal hemoglobin level of 13–14 g/dl. In the case of SCD, it would be to achieve an anti-sickling hemoglobin level of at least 20% without unbalancing the α/β

globin ratio in red blood cells.[25,26] There has been much work in the past 30 years, and it appears that these goals are on the verge of being achieved.

4.1 *Oncoretroviral Vector-Mediated Globin Gene Transfer*

Mouse leukemia virus (MLV)-derived vectors, also referred to as oncoretroviral vectors, were the initial vectors that were used for β-globin gene transfer in the 1980's. These initial vectors included the β-globin gene, β-globin gene promoter, and β-globin gene enhancer. These first vectors were able to express β-globin tissue-specifically; however, they were unable to achieve therapeutic levels of β-globin expression in the HSC progeny.[27-29] During this same period, the locus control region (LCR) was discovered.[30] LCR is a 20 kb enhancer, which is located 60 kb upstream of the β-globin gene. The β-globin LCR consists of 5 DNAse-hypersensitive sites (HS) upstream of the ε globin gene. There is a sixth HS site located downstream of the β-globin cluster at the 3' end. The LCR serves four known functions: it activates the entire β-globin domain, functions as a transcriptional enhancer, insulates globin genes from the effect of surrounding inactive chromatin, and confers copy number-dependant expression of linked genes.[31] Initially, subfragments of the LCR ($\sim$1.0 kb) were incorporated into oncoretroviral vectors (Fig. 12.1A); which resulted in low expression, and vectors that were prone to sequence rearrangements.[32-34] When HS2, HS3 and HS4 sites were incorporated into oncoretroviral vectors, increased expression levels were seen in murine erythroleukemia cells.[35,36] However, it failed to abolish positional variability of expression, which limited the number of genes

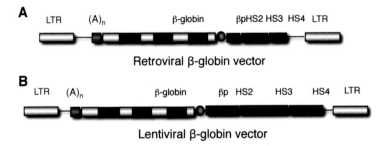

Fig. 12.1 Schematic representation of single-stranded recombinant AAV2-globin vectors: ssAAV2-HS2-βp-$^A\gamma$-globin (**A**) and ssAAV2-HS432-βp-β87$^+$-globin (**B**).

that could be expressed per construct.[36] Incorporation of larger LCR regions showed increased vector instability and genomic rearrangements. Some success was also seen using other erythroid-specific promoters: α locus HS-40, ankyrin, and hereditary persistence of fetal hemoglobin (HPFH) γ-globin. The ankyrin promoter driving a γ-globin gene was able to produce γ-globin gene to an average expression of 8% compared with α-globin gene.[37–39] The promoter of a patient with HPFH was also successfully used to reverse the transcriptional silencing of the γ-globin gene.[40,41] Although a recent report documents the successful use of retroviral vectors for gene therapy of severe combined immuno-deficiency (SCID) in patients with adenosine deaminase (ADA) deficiency,[42] their use in gene therapy trials for X-linked SCID has led to the development of leukemia in several children, and is thus unlikely that retroviral vectors will be utilized in potential gene therapy of hemoglobinopathies for the foreseeable future.

4.2 *Lentiviral Vector-Mediated Globin Gene Transfer*

To overcome the limitations of oncoretroviral-mediated globin gene transfer, 'self-inactivating' (SIN) lentiviruses have emerged as an alternative to oncoretroviruses. Lentiviruses can infect non-dividing cells and have improved RNA stability. A deletion of the promoters in the U3 region of both the 3′ and 5′ long terminal repeat (LTR), results in transcriptional inactivation of the LTR in integrated provirus. May *et al.*, were the first to show stable transduction of a lentiviral vector, TNS9, shown schematically in Fig. 12.1B, carrying the β-globin gene could achieve therapeutic correction of β-thalassemia^{intermedia}.[43] The TNS9 vector also contained a large LCR fragment (3.2 kb), deletion of a cryptic polyadenylation site within intron 2, β-globin 3′ proximal enhancer, and an extended promoter sequence at the flank. This study was highly significant because it was able to show improvement in hemoglobin levels to 11–13 g/dL, decreased reticulocyte count, improvement in extramedullary hematopoiesis, and markedly reduced hepatic iron accumulation. The studies that followed showed therapeutic correction of several β-thalassemia phenotypes, using SIN lentiviral vectors, which contain LCR made of HS2-3-4, expressing γ- or β/γ-globin, in various mouse models;[44–49] Rivella *et al.*, were able to achieve hemoglobin levels averaging 6.5 ± 2.9 g/dL. However, Pearson *et al.*, demonstrated a high variability, 7–90%, among their experimental mice in Hb F-containing RBCs. Imren *et al.*, showed pancellular correction of the thalassemic phenotype in a β-thalassemia[major] mouse model. Lisoski *et al.*, demonstrated that addition of the HS1 site could increase human β-globin expression in mice by 50%.[50] They were able to achieve hemoglobin levels of 9.5 g/dL. Puthenveetil *et al.*, and Malik *et al.*, using a NOD/SCID transplant mouse model, were able to correct CD34[+] cells from patients with β-thalassemia[major] to normal Hb A levels.[48,51] In sickle cell mouse models, therapeutic levels of anti-sickling β-globin expression has also been achieved by lentiviral vectors and phenotypic correction of SCD has been reported in mouse models.[52–55]

In contrast to the sustained expression of β-globin using lentiviral vector constructs, the few experiments done using similar lentiviral vector constructs containing the α-globin gene, the α-globin gene was only expressed at low levels for several months. The above body of work demonstrated that lentiviral vector appears to be effective; however its clinical safety is still under investigation. Several studies have reported the propensity of lentiviral vectors to integrate into active genes.[54,56] Genomic sequencing of vector containing fragments from CD34[+] cells transduced with a lentiviral vector expressing anti-sickling β-globin showed that 86% of proviral integration occurred in genes.[54] Although clinical trials with third generation of lentiviral vectors for β-thalassemia and SCD have been initiated,[57] and preliminary results from a phase I clinical trial with a lentiviral vector for HIV gene therapy has so far shown no adverse effects, the potential for insertional mutagenesis and long-term safety issues with lentiviral vectors remains to be addressed. Thus, we believe that further development of alternative vector systems, such as AAV, must continue to be pursued for the potential gene therapy of hemoglobinopathies, given the proven safety and efficacy of AAV vectors in several clinical trials in general, and efficacy in the gene therapy of Leber's congenital amaurosis (LCA) in particular.[58–61]

4.3 *Adeno-Associated Viral Vector-Mediated Globin Gene Transfer*

Several groups have demonstrated erythroid-restricted expression of globin genes from AAV vectors *in vitro*[62–64] as well as *in vivo*[65] using first generation single-stranded (ss) AAV2 vectors depicted schematically in

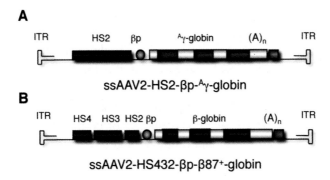

Fig. 12.2 Schematic representation of single-stranded recombinant AAV2-globin vectors: ssAAV2-HS2-βp-$^A\gamma$-globin (**A**) and ssAAV2-HS432-βp-$\beta87^+$-globin (**B**).

Fig. 12.2. However, when a ssAAV2 vector containing a human γ-globin gene under the control of the human β-globin gene promoter and the HS2 enhancer element (Fig. 12.2A) was used, transduction efficiencies of only ~8%, and transgene expression of ~4% at an MOI of 1, and 6% at an MOI of 10 in transplant recipient normal C57BL/6 mice could be achieved.[65] The levels of expression seen in these experiments were secondary to the use of low-density bone marrow cells (LDBM) and not enriched hematopoietic stem/progenitor cells. In addition, there was no selective pressure for expression of the transduced human globin gene since the recipient cells expressed high levels of the endogenous mouse globin genes. In subsequent studies, using a ssAAV vector containing the human β-globin gene under the control of its own promoter with an upstream mini-LCR cassette consisting of HS2, HS3 and HS4 enhancer elements (Fig. 12.2B), and enriched HSCs from homozygous β-thalassemic mice, long-term, erythroid lineage-restricted expression of a human β-globin gene could be achieved.[66] Although the expression level of the transduced human β-globin gene reached up to 35% of the endogenous murine β-globin gene, expression analyses were based on RT-PCR assays at the RNA level, that are not indicative of globin protein production.

In view of the low transduction efficiency of ssAAV2 vectors in murine HSCs,[67] additional ssAAV serotype vectors, AAV1 through AAV5, have been evaluated, and ssAAV1 serotype vector have been shown to be the most efficient in transducing primary murine HSCs, both *in vitro* and *in vivo*.[68] The efficacy of double-stranded, self-complimentary (sc) AAV serotype vectors, scAAV6 through scAAV10, containing the enhanced green-fluorescence protein (EGFP) reporter gene has also been evaluated. The use of self-complimentary AAV vectors circumvents the rate-limiting step of second-strand DNA synthesis of the transcriptionally-inactive single-stranded viral genomes.[69] These studies revealed that the transduction efficiency of scAAV7 serotype vectors was similar to that attained by scAAV1 serotype vectors,[70] and the use of erythroid lineage-specific promoters, such as the β-globin promoter and the human parvovirus B19 promoter at map unit 6 (B19p6), led to transgene expression levels ranging between 16–32% in erythroid progenitor cells.[71] Interestingly, higher level, erythroid-lineage restricted transgene expression was detected from the B19p6 promoter, even in the absence of the HS2 enhancer element, compared with that from the HS2-β-globin enhancer/promoter. No significant expression was detected in any other lineage in both primary and secondary transplant experiments in all serotypes tested. EGFP expression was readily detected in the erythroid lineage 9 months after primary transplantation, indicating that a stem cell with long-term repopulation ability had been transduced, and evidence of stable integration of the proviral genomes in mouse bone marrow cells 9 months post primary transplant suggested a random pattern in bone marrow progenitor cells.[70,71]

Recombinant AAV vectors containing a human anti-sickling β-globin gene expression cassette have been optimized.[72] In a comparative analysis of conventional ssAAV2 vectors (Fig. 12.3A, i) with scAAV2 vectors (Fig. 12.3A, ii), it became apparent that scAAV2 vectors were more efficient in mediating high-level expression of the anti-sickling β-globin gene in human K562 cells *in vitro*, even though they contained only one enhancer element (HS2). Since the transduction efficiency of scAAV2 vectors was better than that of ssAAV2 vectors, a vector containing β-globin under the control of the entire locus control cassette (HS2 + HS3 + HS4) in a double stranded genome would be better, but given the size-limitation of scAAV2 vectors ($\sim$3.3 kb),[73] a dual-vector approach has been utilized.[74] In this strategy, a single gene is split into two separate scAAV vectors, which then undergo intermolecular recombination following cellular entry and viral uncoating. Following co-infection of scAAV2-βp-globin vector (Fig. 12.3B, i) with either scAAV2-HS234 (Fig. 12.3B, ii), or scAAV2-HS432 (Fig. 12.3B, iii) vectors, transgene expression was observed from

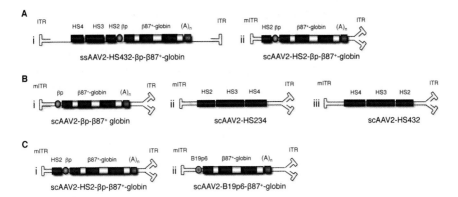

Fig. 12.3 Schematic structures of the single-stranded (ssAAV2-HS432-βp-β87$^+$-globin (i), and self-complementary (scAAV2-HS2-βp-β87$^+$-globin (ii) vectors (A); scAAV2-βp-β87$^+$-globin (i), scAAV2-HS234 (ii), and scAAV2-HS432 (iii) vectors (B); and scAAV-HS2-βp-β87$^+$-globin (i), and scAAV-B19p6-β87$^+$-globin (ii) vectors (C).

both groups. Although the dual vector approach worked, the extent of the transgene expression was not significantly enhanced compared with that from a single scAAV2-HS2-βp-globin vector. Since the human parvovirus B19 promoter at map unit 6 (B19p6) is a strong viral promoter, expression from which is restricted to differentiating hematopoietic erythroid progenitors,[75-77] coupled with its small size (~200 bp), prompted further studies to determine whether expression of the β-globin gene could be obtained from the B19p6 promoter alone, and whether expression from the scAAV-HS2-βp-globin vector (Fig. 12.3C, i) could be superseded by that from the scAAV-B19p6-β-globin vector (Fig. 12.3C, ii). These studies revealed that expression from the scAAV-B19p6-globin vector was ~4-fold higher than expression from the scAAV-HS2-βp-globin vector in both AAV2 and AAV1 serotypes. These results corroborate the ability of the B19p6 promoter to mediate high-efficiency β-globin gene expression in the complete absence of the β-globin gene enhancer elements. Additional studies have documented that ~16% of primary murine c-kit^+, lin^- cells expressed vector-derived β-globin protein in the absence of erythropoietin (EPO)-treatment, and ~24% of these cells expressed vector-derived β-globin protein following EPO-differentiation.[72]

Additional studies involving primary human HSC transduction by the optimal scAAV serotype vector containing the B19p6 promoter, followed by transplantation in NOD/SCID mice are warranted to substantiate whether expression from the B19p6 promoter is indeed restricted to human erythroid progenitor cells in these mice *in vivo*, and whether therapeutic levels of the β-globin protein can be expressed in erythroid progenitor cells. These questions notwithstanding, the recent development of the next generation of scAAV serotype vectors,[78,79] bode well for the utility of these vectors for safe, high-efficiency, and stable transduction of HSCs from patients with hemoglobinopathies, and lineage-restricted, long-term expression of therapeutic levels of β-globin protein in erythroid progenitor cells, which in turn, should lead to the potential gene therapy of human hemoglobinopathies in general, and β-thalassemia and sickle cell disease in particular.

References

1. Kan YW (1991). Molecular biology of hemoglobin: Its application to sickle cell anemia and thalassemia. *Schweiz Med Wochenschr Suppl* **43**: 51–54.
2. Walters MC, *et al.* (2005). Sibling donor cord blood transplantation for thalassemia major: Experience of the Sibling Donor Cord Blood Program. *Ann N Y Acad Sci* **1054**: 206–213.
3. Walters MC, *et al.* (1996). Bone marrow transplantation for sickle cell disease. *N Engl J Med* **335**: 369–376.
4. Atweh GF, *et al.* (2003). Hemoglobinopathies. *Hematology Am Soc Hematol Educ Program* 14–39.
5. Pauling L, Itano HA, Singer SJ, Wells IC (1949). Sickle Cell Anemia, a Molecular Disease. *Science* **110**: 543–547.
6. Ingram VM (1956). A specific chemical difference between the globins of normal and sickle-cell anemia haemoglobin. *Nature* **178**: 792–794.
7. Steinberg MH (2008). Sickle cell anemia, the first molecular disease: Overview of molecular etiology, pathophysiology, and therapeutic approaches. *ScientificWorldJournal* **8**: 1295–1324.
8. Remacha AF, *et al.* (2002). Hydrops fetalis-associated congenital dyserythropoietic anemia treated with intrauterine transfusions and bone marrow transplantation. *Blood* **100**: 356–358.
9. Saunthararajah Y, *et al.* (2003). Effects of 5-aza-2'-deoxycytidine on fetal hemoglobin levels, red cell adhesion, and hematopoietic differentiation in patients with sickle cell disease. *Blood* **102**: 3865–3870.
10. Bunn F (1987). Subunit Assembly of hemoglobin: An important determinant of Hematological Phenotype. *Blood* **69**: 1–6.
11. Steinberg M, *et al.* (1997). Fetal Hemoglobin in Sickle Cell Anemia: Determinants of Response to Hydroxyurea. *Blood* **89**: 1078–1088.

12. Steinberg MH, Dreiling BJ (1974). Glucose-6-phosphate dehydrogenase deficiency in sickle-cell anemia. *Ann Intern Med* **80**: 217.

13. Steinberg MH, West M, Gallager D, Mentzer W (1988). The effects of glucose-6-phosphate dehydrogenase deficiency upon sickle cell anemia. *Blood* **71**: 748–752.

14. Noguchi C (1996). Sickle Hemoglobin Polymer Structure — Function Correlates. In: Stephen H. Embury RPH, Narla Mohandas, Martin H Steinberg (ed). *Sickle Cell Disease: Basic Principles and Clinical Practice*. Lippincott-Raven: Philadelphia, Pennsylvania 19106, pp. 33–51.

15. Ferester A, *et al.* (1996). Hydroxyurea Treatment of severe sickle cell anemia in a pediatric clinical trial. *Blood* **88**: 1960–1964.

16. Raftopoulos HWM, Leboulch P, Bank A (1997). Long term transfer and expression of the human beta-globin gene in a mouse tranplant model. *Blood* **90**.

17. Walters M, Patience M, Leisenring W, *et al.* (1996). Bone Marrow transplantation for sickle cell disease. *N Engl J Med* **335**: 369.

18. Adams JI, Coleman M, Hayes J (1985). Modulation of fetal hemoglobin synthesis by iron deficiency. *N Engl J Med* **313**: 1402.

19. De Franceschi L, Corrocher R (2004). Established and experimental treatments for sickle cell disease. *Haematologica* **89**: 348–356.

20. Walter MC (2005). Stem cell therapy for sickle cell disease: Transplantation and gene therapy. *Am Soc of Hematol Educ Program* 66–73.

21. Buchanan GR, DeBaun MR, Quinn CT, Steinberg MH (2004). Sickle cell disease. *Hematology Am Soc Hematol Educ Program* 35–47.

22. Cao H (2004). Pharmacological induction of fetal hemoglobin synthesis using histone deacetylase inhibitors. *Hematology* **9**: 223–233.

23. Oringanje C, Nemecek E, Oniyangi O (2009). Hematopoietic stem cell transplantation for children with sickle cell disease. *Cochrane Database Syst Rev* CD007001.

24. Hanna J, *et al.* (2007). Treatment of sickle cell anemia mouse model with iPS cells generated from autologous skin. *Science* **318**: 1920–1923.

25. Leikin SL, *et al.* (1989). Mortality in children and adolescents with sickle cell disease. Cooperative Study of Sickle Cell Disease. *Pediatrics* **84**: 500–508.

26. Platt OS, *et al.* (1994). Mortality in sickle cell disease. Life expectancy and risk factors for early death. *N Engl J Med* **330**: 1639–1644.

27. Cone RD, Weber-Benarous A, Baorto D, Mulligan RC (1987). Regulated expression of a complete human beta-globin gene encoded by a transmissible retrovirus vector. *Mol Cell Biol* **7**: 887–897.

28. Dzierzak EA, Mulligan RC (1988). Lineage specific expression of a human beta-globin gene in murine bone marrow transplant recipients. *Adv Exp Med Biol* **241**: 41–43.

29. Bender MA, Gelinas RE, Miller AD (1989). A majority of mice show long-term expression of a human beta-globin gene after retrovirus transfer into hematopoietic stem cells. *Mol Cell Biol* **9**: 1426–1434.

30. Tuan D, London IM (1984). Mapping of DNase I-hypersensitive sites in the upstream DNA of human embryonic epsilon-globin gene in K562 leukemia cells. *Proc Natl Acad Sci USA* **81**: 2718–2722.

31. Tuan D, Abeliovich A, Lee-Oldham M, Lee D (1987). Identification of regulatory elements of human beta-like globin genes. *Prog Clin Biol Res* **251**: 211–220.

32. Plavec I, Papayannopoulou T, Maury C, Meyer F (1993). A human beta-globin gene fused to the human beta-globin locus control region is expressed at high levels in erythroid cells of mice engrafted with retrovirus-transduced hematopoietic stem cells. *Blood* **81**: 1384–1392.

33. Chang JC, Liu D, Kan YW (1992). A 36-base-pair core sequence of locus control region enhances retrovirally transferred human beta-globin gene expression. *Proc Natl Acad Sci USA* **89**: 3107–3110.

34. Novak U, *et al.* (1990). High-level beta-globin expression after retroviral transfer of locus activation region-containing human beta-globin gene derivatives into murine erythroleukemia cells. *Proc Natl Acad Sci USA* **87**: 3386–3390.

35. Leboulch P, *et al.* (1994). Mutagenesis of retroviral vectors transducing human beta-globin gene and beta-globin locus control region derivatives results in stable transmission of an active transcriptional structure. *Embo J* **13**: 3065–3076.

36. Sadelain M, *et al.* (1995). Generation of a high-titer retroviral vector capable of expressing high levels of the human beta-globin gene. *Proc Natl Acad Sci USA* **92**: 6728–6732.

37. Ren S, *et al.* (1996). Production of genetically stable high-titer retroviral vectors that carry a human gamma-globin gene under the control of the alpha-globin locus control region. *Blood* **87**: 2518–2524.

38. Sabatino DE, *et al.* (2000). Long-term expression of gamma-globin mRNA in mouse erythrocytes from retrovirus vectors containing the human gamma-globin gene fused to the ankyrin-1 promoter. *Proc Natl Acad Sci USA* **97**: 13294–13299.

39. Sabatino DE, *et al.* (2000). A minimal ankyrin promoter linked to a human gamma-globin gene demonstrates erythroid specific copy number dependent expression with minimal position or enhancer dependence in transgenic mice. *J Biol Chem* **275**: 28549–28554.

40. Katsantoni EZ, *et al.* (2003). Persistent gamma-globin expression in adult transgenic mice is mediated by HPFH-2, HPFH-3, and HPFH-6 breakpoint sequences. *Blood* **102**: 3412–3419.

41. Fragkos M, Anagnou NP, Tubb J, Emery DW (2005). Use of the hereditary persistence of fetal hemoglobin 2 enhancer to increase the expression of oncoretrovirus vectors for human gamma-globin. *Gene Ther* **12**: 1591–1600.

42. Aiuti A, *et al.* (2009). Gene therapy for immunodeficiency due to adenosine deaminase deficiency. *N Engl J Med* **360**: 447–458.

43. May C, *et al.* (2000). Therapeutic haemoglobin synthesis in beta-thalassaemic mice expressing lentivirus-encoded human beta-globin. *Nature* **406**: 82–86.

44. Rivella S, *et al.* (2003). A novel murine model of Cooley anemia and its rescue by lentiviral-mediated human beta-globin gene transfer. *Blood* **101**: 2932–2939.

45. Hanawa H, *et al.* (2004). Extended beta-globin locus control region elements promote consistent therapeutic expression of a gamma-globin lentiviral vector in murine beta-thalassemia. *Blood* **104**: 2281–2290.

46. Imren S, *et al.* (2002). Permanent and panerythroid correction of murine beta thalassemia by multiple lentiviral integration in hematopoietic stem cells. *Proc Natl Acad Sci USA* **99**: 14380–14385.

47. Persons DA, *et al.* (2001). Functional requirements for phenotypic correction of murine beta-thalassemia: implications for human gene therapy. *Blood* **97**: 3275–3282.

48. Puthenveetil G, *et al.* (2004). Successful correction of the human beta-thalassemia major phenotype using a lentiviral vector. *Blood* **104**: 3445–3453.

49. Pestina TI, *et al.* (2009). Correction of murine sickle cell disease using gamma-globin lentiviral vectors to mediate high-level expression of fetal hemoglobin. *Mol Ther* **17**: 245–252.

50. Lisowski L, Sadelain M (2007). Locus control region elements HS1 and HS4 enhance the therapeutic efficacy of globin gene transfer in beta-thalassemic mice. *Blood* **110**: 4175–4178.

51. Malik P, *et al.* (1998). An *in vitro* model of human red blood cell production from hematopoietic progenitor cells. *Blood* **91**: 2664–2671.

52. Pawliuk R, *et al.* (2001). Correction of sickle cell disease in transgenic mouse models by gene therapy. *Science* **294**: 2368–2371.

53. Levasseur DN, Ryan TM, Pawlik KM, Townes TM (2003). Correction of a mouse model of sickle cell disease: Lentiviral/antisickling beta-globin gene transduction of unmobilized, purified hematopoietic stem cells. *Blood* **102**: 4312–4319.

54. Imren S, *et al.* (2004). High-level beta-globin expression and preferred intragenic integration after lentiviral transduction of human cord blood stem cells. *J Clin Invest* **114**: 953–962.

55. Oh IH, *et al.* (2004). Expression of an anti-sickling beta-globin in human erythroblasts derived from retrovirally transduced primitive normal and sickle cell disease hematopoietic cells. *Exp Hematol* **32**: 461–469.

56. Schroder AR, *et al.* (2002). HIV-1 integration in the human genome favors active genes and local hotspots. *Cell* **110**: 521–529.

57. Bank A, Dorazio R, Leboulch P (2005). A phase I/II clinical trial of beta-globin gene therapy for beta-thalassemia. *Ann N Y Acad Sci* **1054**: 308–316.

58. Bainbridge JW, *et al.* (2008). Effect of gene therapy on visual function in Leber's congenital amaurosis. *N Engl J Med* **358**: 2231–2239.

59. Maguire AM, *et al.* (2008). Safety and efficacy of gene transfer for Leber's congenital amaurosis. *N Engl J Med* **358**: 2240–2248.

60. Cideciyan AV, *et al.* (2008). Human gene therapy for RPE65 isomerase deficiency activates the retinoid cycle of vision but with slow rod kinetics. *Proc Natl Acad Sci USA* **105**: 15112–15117.

61. Hauswirth W, *et al.* (2008). Phase I Trial of Leber Congenital Amaurosis due to RPE65 mutations by ocular subretinal injection of adeno-associated virus gene vector: Short-term results. *Hum Gene Ther.*

62. Walsh CE, *et al.* (1992). Regulated high level expression of a human gamma-globin gene introduced into erythroid cells by an adeno-associated virus vector. *Proc Natl Acad Sci USA* **89**: 7257–7261.

63. Einerhand MP, *et al.* (1995). Regulated high-level human beta-globin gene expression in erythroid cells following recombinant adeno-associated virus-mediated gene transfer. *Gene Ther* **2**: 336–343.

64. Miller JL, *et al.* (1994). Recombinant adeno-associated virus (rAAV)-mediated expression of a human gamma-globin gene in human progenitor-derived erythroid cells. *Proc Natl Acad Sci USA* **91**: 10183–10187.

65. Ponnazhagan S, Yoder MC, Srivastava A (1997). Adeno-associated virus type 2-mediated transduction of murine hematopoietic cells with long-term repopulating

ability and sustained expression of a human globin gene *in vivo*. *J Virol* **71**: 3098–3104.

66. Tan M, *et al.* (2001). Adeno-associated virus 2-mediated transduction and erythroid lineage-restricted long-term expression of the human beta-globin gene in hematopoietic cells from homozygous beta-thalassemic mice. *Mol Ther* **3**: 940–946.

67. Zhong L, *et al.* (2004). Impaired nuclear transport and uncoating limit recombinant adeno-associated virus 2 vector-mediated transduction of primary murine hematopoietic cells. *Hum Gene Ther* **15**: 1207–1218.

68. Zhong L, *et al.* (2006). Evaluation of primitive murine hematopoietic stem and progenitor cell transduction *in vitro* and *in vivo* by recombinant adeno-associated virus vector serotypes 1 through 5. *Hum Gene Ther* **17**: 321–333.

69. McCarty DM (2008). Self-complementary AAV vectors; advances and applications. *Mol Ther* **16**: 1648–1656.

70. Han Z, *et al.* (2008). Stable integration of recombinant adeno-associated virus vector genomes after transduction of murine hematopoietic stem cells. *Hum Gene Ther* **19**: 267–278.

71. Maina N, *et al.* (2008). Recombinant self-complementary adeno-associated virus serotype vector-mediated hematopoietic stem cell transduction and lineage-restricted, long-term transgene expression in a murine serial bone marrow transplantation model. *Hum Gene Ther* **19**: 376–383.

72. Maina N, *et al.* (2008). Optimization of recombinant adeno-associated viral vectors for human beta-globin gene transfer and transgene expression. *Hum Gene Ther* **19**: 365–375.

73. Wu J, *et al.* (2007). Self-complementary recombinant adeno-associated viral vectors: Packaging capacity and the role of rep proteins in vector purity. *Hum Gene Ther* **18**: 171–182.

74. Duan D, Yue Y, Engelhardt JF (2001). Expanding AAV packaging capacity with trans-splicing or overlapping vectors: a quantitative comparison. *Mol Ther* **4**: 383–391.

75. Ozawa K, Kurtzman G, Young N (1987). Productive infection by B19 parvovirus of human erythroid bone marrow cells *in vitro*. *Blood* **70**: 384–391.

76. Srivastava A, Lu L (1988). Replication of B19 parvovirus in highly enriched hematopoietic progenitor cells from normal human bone marrow. *J Virol* **62**: 3059–3063.

77. Ponnazhagan S, *et al.* (1996). Differential expression in human cells from the p6 promoter of human parvovirus B19 following plasmid transfection and recombinant adeno-associated virus 2 (AAV) infection: human megakaryocytic leukaemia cells are non-permissive for AAV infection. *J Gen Virol* **77**: 1111–1122.

78. Zhong L, *et al.* (2008). Next generation of adeno-associated virus 2 vectors: point mutations in tyrosines lead to high-efficiency transduction at lower doses. *Proc Natl Acad Sci USA* **105**: 7827–7832.

79. Petrs-Silva H, *et al.* (2009). High-efficiency transduction of the mouse retina by tyrosine-mutant AAV serotype vectors. *Mol Ther* **17**: 463–471.

Chapter 13
Gene Therapy for Primary Immunodeficiencies

Aisha Sauer, Barbara Cassani and Alessandro Aiuti[*,†,‡]

Primary immunodeficiencies are a heterogeneous group of inherited disorders that affect distinct components of the innate and adaptive immune system. Gene therapy with hematopoietic stem cells (HSC) represents an attractive therapeutic strategy for immunodeficient patients who lack a compatible allogeneic donor. In the last decade, significant improvements have been made in transferring genes by means of retroviruses and lentiviruses into HSC. Results from clinical studies with retroviral vectors showed restoration of immune competence and clinical benefit in patients affected by adenosine deaminase-deficient SCID, SCID due IL2RG deficiency (SCID-X1), and chronic granulomatous disease. The inclusion of a reduced-intensity conditioning regimen before HSC reinfusion resulted in long-term engraftment of multipotent gene corrected HSC. However, vector-related adverse events have occurred in the form of leukemic proliferation in the SCID-X1 and clonal expansion in CGD trials. This has underlined the need for accurate monitoring of the associated risks and for improvements in vector design. Gene therapy has shown substantial progress in preclinical studies for other immunodeficiency disorders, such as Wiskott Aldrich Syndrome, and lentiviral vectors are currently being translated into new clinical approaches.

*Correspondence: [†]San Raffaele Telethon Institute for Gene Therapy and [‡]University of Rome "Tor Vergata", Via Olgettina 58 20132 Milano.
E-mail: a.aiuti@hsr.it

1. Introduction

Primary immunodeficiencies (PID) represent a paradigm for gene therapy approaches to inherited disorders. PID are a genetically heterogeneous group of inherited disorders that affect distinct components of the innate and adaptive immune system, with impairment of their differentiation and/or functions.[1] Here we will discuss gene therapy approaches for Severe Combined Immunodeficiencies (SCID), the most severe forms of PID, and for two other forms of PID, which have been the subject of intensive investigations, Wiskott-Aldrich Syndrome (WAS) and Chronic Granulomatous Disease (CGD).

SCID comprise about 15% of PID and have frequency of 1:75,000 to 1:100,000 live births. SCIDs are excellent candidates for gene therapy due to the severity of the disease and the unmet medical need. Transplant of hematopoietic stem cells (HSC) from an HLA-identical sibling donors is available only for a minority of patients. Despite substantial improvements in protocol and procedures for HSC transplants, some patients continue to experience long-term complications after transplant,[2] and the use of alternative donors is still associated with high morbidity and mortality.[3] The current approach is based on *ex vivo* gene transfer in HSC with integrating vectors to reconstitute the patients' immune system. This approach is facilitated by the selective advantage for the survival and/or proliferation of lymphoid cells that carry the therapeutic gene, as observed in preclinical studies and HSC transplantion.[4]

Gene therapy has several potential advantages over existing treatment methods suitable for patients who do not have access to an HLA-identical sibling donor.[4] Transplantation of gene corrected HSC is potentially applicable to all PID patients, independent from the availability of a donor, since it is an autologous procedure and there is no delay for donor search. Moreover, the use of autologous gene corrected stem cells avoids rejection and graft-versus-host disease due to HLA-mismatches or minor antigen incompatibility. Finally, gene therapy does not require the use of immunosuppressive prophylaxes or high dose conditioning regimens associated with organ toxicity (liver, lung, kidney, CNS), prolonged period of myelosuppression, and increased risks of infections. On the other hand, the risk of insertional mutagenesis due to viral vector should be carefully weighted in the risk-benefit evaluation in consideration of the type of vector, the nature of the transgene and the specific disease background.

2. Adenosine Deaminase (ADA)-deficient SCID

ADA-deficiency is a form of SCID characterized by impaired T, B, NK cell development and function, recurrent infections, and failure to thrive. In addition, non-immunological abnormalities occur in several organs as the consequence of the systemic metabolic defect due to the accumulation of purine toxic metabolites.[5,6] ADA-SCID children who do not have access to a compatible donor are often being treated with enzyme replacement therapy (PEG-ADA).[6] PEG-ADA results in clinical improvement with about 70–80% survival, but the immunological reconstitution is often incomplete.[6] In addition, the high costs of lifelong treatment represent a major drawback for the national health system.

ADA-deficiency was the first genetic disease treated with gene therapy in the early 90s.[7,8] Several clinical studies have investigated the safety and efficacy of ADA gene transfer into autologous hematopoietic cells using retroviral vectors. In the initial trials, 19 patients received infusions of transduced lymphocytes or hematopoietic progenitor cells.[4] No toxicity was observed, and in most patients transduced T cells persisted in the circulation several years after infusion. However, the low gene transfer efficiency and engraftment levels observed in these patients did not allow to achieve a significant correction of the immunological and metabolic defects, and all patients were maintained on enzyme replacement therapy.

A major improvement was obtained after the introduction of a non-myeloablative dose of busulfan chemotherapy prior cell reinfusion, to make space in the bone marrow for gene corrected HSC (Fig. 13.1).[9] Ten patients without an HLA-identical sibling donor were treated with transduced autologous bone marrow CD34+ cells according to the protocol developed at HSR-TIGET (Milano).[10] The majority of patients had displayed an inadequate response to PEG-ADA or had failed an haploidentical transplant. To exploit the selective advantage for ADA$^+$ cells in a toxic environment, enzyme replacement therapy was not administered after gene therapy and eight patients remained off PEG-ADA. Busulfan induced a transient myelosuppression without organ toxicity, which allowed stable and efficient

215

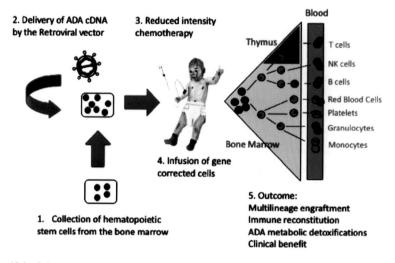

Fig. 13.1 Schematic representation of gene therapy for ADA-deficiency. Source: Ref. 11, with permission.

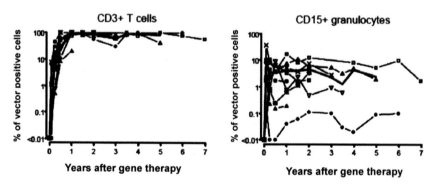

Fig. 13.2 Long-term engraftment of gene corrected T cells and granulocytes in ADA-SCID patients after gene therapy. Source: Ref. 10, with permission.

engraftment of transduced HSC (Fig. 13.2).[9] The dose of infused CD34[+] cells and the efficiency of gene transfer were shown to be critical factors in allowing a higher fraction of gene corrected HSC to engraft in patients.

The large majority of lymphocytes were ADA-transduced, confirming that PEG-ADA withdrawal favors the selective survival of gene corrected cells (Fig. 13.2). Clonal analysis of long-term repopulating cells demonstrated the presence of vector integrations shared among multiple

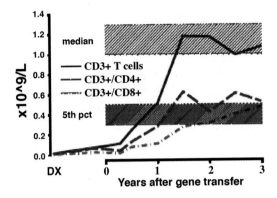

Fig. 13.3 T cell reconstitution after gene therapy for ADA-SCID. The graph indicates median CD3[+], CD4[+], and CD8[+] T cell counts after gene therapy. Median values and 5th percentiles of healthy controls in age class 2–5 years (upper boundary of shaded area) and age class 5–10 years (lower boundaries of shaded area) are shown for CD4[+] T cells. Source: Ref. 10, with permission.

hematopoietic lineages, thus proving the engraftment of multipotent HSC.[12] Vector-derived ADA was expressed in lymphocytes and red blood cells, leading to an efficient systemic detoxification up to 8 years after treatment.

Nine patients displayed recovery of polyclonal thymopoiesis, significant increase in T cell counts (Fig. 13.3) and normalization of their functions, including susceptibility to apoptosis and proliferative responses to mitogens and antigens.[13] Evidence of antigen-specific antibodies to vaccination antigens and pathogens was obtained in five children who discontinued IVIg treatment. The progressive restoration of immune and metabolic functions led to significant improvement of patients' development and protection from severe infections, without adverse events related to gene therapy.

In the clinical trials conducted recently in the US, the inclusion of busulfan conditioning and withdrawal of PEG-ADA also lead to improved immunological and metabolic outcome with respect to earlier protocols. However, one patient experienced a prolonged cytopenia following busulfan conditioning, as the consequence of a pre-existing cytogenetic abnormality, pointing out a potential limitation for patients subject to autologous gene transfer.[14] Efficient metabolic and immunological correction was reported in an ADA-SCID patient who received gene therapy combined with a single dose of melphalan as conditioning regimen.[15] The use of melphalan resulted in a lower proportion of transduced granulocytes and B cells, suggesting that type of chemotherapy may influence HSC engraftment.

Results of the above clinical trials indicate that gene therapy with non-myeloablative conditioning is associated with clinical benefit and is now an option to be considered for all ADA-SCID patients lacking an HLA-identical sibling donor. None of the patients enrolled in the different clinical trials worldwide showed clonal proliferation or adverse events related to gene transfer, indicating that, as opposed to SCID-X1 (see below), gene therapy for ADA-SCID has a favorable risk-benefit profile.[16,17] The development of self-inactivating vectors, such as lentiviral vectors might further improve the safety of this approach.[18,19]

3. X-linked Severe Combined Immunodeficiency (SCID-X1)

X-linked SCID (SCID-X1) is the most common form of severe combined immunodeficiency, accounting for around 60% of all cases. SCID-X1 is caused by mutations in the common cytokine receptor gamma chain (γc) gene, which forms part of the receptors for IL-4, IL-7, IL-9, IL-15 and IL-21. In the absence of γc-mediated signaling, the differentiation and activation of immune cells is compromised, resulting in lack of T and NK cells as well as functionally impaired B cells.[1]

The successful correction of γc-deficient mice by gene therapy lead A. Fischer and colleagues at Hôpital Necker (Paris) to assess the potential of gene therapy for the treatment of this disease. The protocol was based on *ex vivo* retrovirally-mediated γc gene transfer into autologous CD34+ bone marrow cells.[20] Progenitor cells were then re-infused into patients, who lacked a matched sibling or unrelated donors, in the absence of any preparative conditioning. A similar study was initiated by A. Thrasher and colleagues at Great Hormond Street Hospital (London).[21] This protocol varied from the French study essentially for the use of GALV-pseudotyped instead of a conventional amphotropic MLV-based vector. Overall, 18 out of the 20 SCID-X1 patients enrolled in both clinical trials benefited from gene therapy. Within 3 months after gene therapy, T cell counts reached normal level in nearly all the patients. Around 10 years of follow-up available for the first treated patients has provided evidence for still active thymopoiesis, with broadly diversified TCR repertoire.[16] The evidence that virtually all T and NK cells but fewer B cells and myeloid cells carried the transgene (1–2 trangene copies per cell) clearly demonstrates that γc expression confers on T and NK progenitors a strong selective growth advantage. Nevertheless, several patients discontinued immunoglobulin infusions and showed antigen specific responses following vaccination.

When gene therapy was attempted in older SCID-X1 patients, this resulted in partial efficacy or inability to recover T cell immunity due to failure of re-activating thymopoiesis.[22,23] The decline of thymic output could be likely related to the clinical history of chronic infection, GVHD or to the age of the patients at the time of intervention. This most probably reflects a crucial time-dependency of the capacity to reinitiate thymopoiesis, since the prolonged absence of thymocyte and thymic epithelial cell interaction can cause irreversible thymic hypoplasia and disaggregation. Thus, gene therapy should be considered as early as possible after diagnosis.

From a safety point of view, the trials for SCID-X1 represented a major setback for the safety of gene therapy. Four patients in the French trial and 1 in the English trial developed leukemia 3–6 years after treatment. This leukemia-like disease was a result of vector-mediated up-regulation of host cellular oncogenes by the MLV LTR.[16,17,24] Gamma-retroviral vectors naturally insert into active genes and detailed molecular analysis demonstrates a preference to insert into and around the transcriptional start site. By inserting into such regions, the strong enhancer elements found in the viral LTR can transactivate nearby and neighboring promoters leading to aberrant gene expression. In the SCID-X1 clinical trials the development of leukemia or genomic instability are directly related to such mechanisms.[16,17,24] In these cases initial aberrant expression of an oncogene (e.g. mainly LMO2) led to proliferation of specific clones, whereas additional genetic events eventually resulted in leukemic transformation. Chemotherapy led to sustained remission in 4 of the 5 cases of T cell leukemia, but one patient died likely due to refractory leukemia.[16,17]

Several possible explanations may account for the differences between the ADA-SCID and the SCID-X1 GT trials. First, ADA is a constitutively expressed enzyme of purine metabolism while IL2RG is a signal transducing receptor chain which is upregulated upon T cell activation and induces cell proliferation. Transgenic IL2RG may be expressed at inappropriate levels in different stages of differentiation. It should be noted that the kinetics of T cell reconstitution was substantially different, being the one of ADA-SCID more slow and progressive than the one observed for the SCID-X1 trials. Furthermore, the SCID-X1 background may favor the accumulation of mutations in progenitor cells blocked in their development or in gene corrected thymocytes following their rapid expansion occurring after restoration of IL2RG expression. Finally, combinatorial mechanisms leading to leukemogenesis may result from the interaction of IL2RG with cellular proto-oncogenes activated by vector insertions, such as LMO2.

The occurrence of these severe adverse events lead to the development of a new area of research focused on virus-mediated oncogenesis and significantly contributed to the current safety standards for gene therapy vectors. Most efforts to improve safety and efficiency for SCID GT have focused on the development of new self-inactivated (SIN) gammaretroviral and lentiviral vectors, which carry a reduced intrinsic risk. Potential candidates to future clinical trials include SIN-gammaretroviral vectors driven by the human EF1-alpha promoter,[25] and SIN-Lentiviral Vectors (SIN-LV) based on HIV-1 incorporating the ubiquitously acting chromatin opening element (UCOE).[26] These vectors induced stable γc gene expression and fully restored lymphoid differentiation and functions. In summary, SCID-X1 clinical trials were of major importance to verify the potential of gene therapy as a real alternative for PID therapy and to define the requirement of more efficient and safer vectors.

4. Gene Therapy for Other SCIDs

Here we will discuss the most recent advances in the application of somatic cell gene therapy to the correction of various molecular defects responsible for other types of SCIDs.

4.1 V(D)J Recombination Defects

To obtain the necessary level of diversity, B and T cell receptors are created by recombining preexisting gene segments, a process called V(D)J recombination. In the first steps of the process, the two lymphocyte-specific proteins

RAG1 and RAG2, collaborate to join the various variable (V), diversity (D), and joining (J) segments. Additionally, a large number of proteins participate in DNA repair during the late phases of V(D)J recombination, such as Ku70, Ku80, DNA Ligase IV, Artemis, XRCC4, TdT or the newly identified Cernunnos/XLF.[27] V(D)J recombination defects comprise 30% of human T⁻B⁻NK⁺ SCIDs. They are caused by mutations in the RAG1 or RAG2 genes resulting in an early block of V(D)J recombination or mutations in the Artemis gene leading to a general defect of nonhomologous end-joining repair.[28] The latter defect, referred to as radiosensitive-SCID (RS-SCID), is characterized, in addition to the absence of mature T and B lymphocytes, by an increased cellular sensitivity to ionizing radiation. Moreover, hypomorphic mutations in the RAG1/2 or Artemis genes may generate a different phenotype, named Omenn syndrome (OS), which is characterized by residual T and B cell differentiation and function.[29] Due to overwhelming infections and organ damage, if untreated, patients have a dismal prognosis. Current treatment requires both immune suppression and allogeneic HSC transplantation. However this is only partially successful, particularly for RS-SCID, in which the radiosensitive characteristic of the disease may preclude the use of standard conditioning regimens.[2] Consequently, gene therapy is considered a promising alternative strategy for these diseases.

Gene transfer approaches to correct V(D)J recombination defects have been initially explored in RAG1 or RAG2-deficient murine models. MLV-based gammaretroviral vectors have been shown to reconstitute RAG2-deficient mice effectively in the absence of detectable toxicity.[30] Similar efforts to correct RAG1 deficiency were successful but required high vector copy numbers to be efficient.[31]

More recently, studies have employed SIN-LVs in order to avoid potential side effects associated with the use of LTR-driven RVs. A single copy of an Artemis-expressing LV enabled long-term correction of both T- and B-cell deficiencies in recipient mice.[32] As a further step towards clinical application, it has been demonstrated that LV-corrected BM CD34⁺ cells isolated from Artemis-deficient patients are able to promote functional B-cell development in SCID mice. Using the same transduction procedure but a slightly different vector, correction of the B-cell defect in RAG1-deficient patients was much less efficient, suggesting that further improvements in transgene expression via modified vector design must be considered for this pathology.[33]

4.2 *Purine Nucleoside Phosphorylase (PNP) Deficiency*

Like ADA, Purine nucleoside phosphorylase (PNP) is an important enzyme for the degradation of purine nucleosides into uric acid or their salvage into nucleic acids.[5] PNP-deficient patients suffer from severe T cell immune dysfunction, increased susceptibility to infections, failure to thrive and death in the first years of life, which may be delayed with supportive treatments.[5] B cells are less sensitive than T cells to the metabolic abnormalities because of the different metabolic pathway utilized.[5] HSC transplants from closely matched donors have corrected the immune abnormalities in some PNP deficient patients. However, HLA-matched donors are not readily available, and transplants using other donors are frequently complicated by procedure-related morbidity, development of graft-versus-host disease or graft loss.[34]

Hopes for successful gene therapy for PNP deficiency have been supported by the success of gene therapy for ADA-SCID. Promising results have recently been achieved using a SIN HIV-1-based LV vector encoding for human PNP under the control of the human EF1-alpha promoter, which provided stable transgene expression in both human and murine cells. Importantly, after HSC gene therapy in the murine model, PNP activity increased over time suggesting a selective advantage for gene corrected cells. Although this study provided proof-of-principle that LVs could be used for PNP gene therapy, the observed effect was only transient, so that better BM engraftment and improved gene expression will be required to fulfill the potential of gene therapy for PNP deficiency in the future.[35]

4.3 *Janus Kinase 3 (Jak3) Deficiency*

Jak3 belongs to a family of signal transducing tyrosine kinases primarily expressed in hematopoietic cells, it associates with γc and is required for signal transduction by γc-containing receptors.[36] Therefore, the phenotype arising from rare mutations in the Jak3 gene is nearly identical to that of SCID-X1,[37] which is characterized by absence of circulating T and natural killer (NK) cells with normal numbers of poorly functioning B cells

(T-B+NK-). Different *in vitro* and *in vivo* preclinical studies have demonstrated that retroviral Jak3 gene transfer can correct developmental and functional defects in hematopoietic cells.[38] Jak3 gene correction of bone marrow CD34$^+$ cells was attempted in a single Jak3-SCID patient, who failed HSC transplantation, without achieving immune reconstitution.[39] In Jak3 deficient mice, direct intrafemoral injection of retroviral vectors has been postulated as an alternative to avoid deleterious effects of *ex-vivo* manipulation.[40] Although not all treated mice were fully reconstituted, this study demonstrated the feasibility of direct vector injection.

4.4 *IL-7R Deficiency*

The finding that mutations of γc and Jak3 led to SCID suggested that other disruptions in this pathway might result in similar disease phenotypes. Cytokines that utilize γc were logical candidates, such as mutations of IL-7 or the IL-7 receptor α chain (IL-7Rα). Mutations in IL-7Rα were subsequently identified in patients with autosomal recessive SCID, underlying about 10% of SCIDs.[1] IL-7Rα deficiency in humans lead to SCID manifested by isolated absence of T cells (T-B+NK+), underlining the importance of IL-7Rα-dependent signaling for T cell, but not B or NK cell, development.[1] IL-7 plays a critical role in the development and differentiation of thymocytes, and survival of both naïve and mature T cells.[41] Unlike its human counter-part, mouse IL-7 is also required for the development of B lymphocytes.[41] In IL-7Rα deficient mice gene transfer using a retroviral vector encoding for the murine IL-Rα gene allowed the reconstitution of T cell lineage in the thymus and periphery, while B cell recovery remained partial.[42] Remarkably, upon IL-7 stimulation, the ectopic expression of IL-7Rα on multipotent hematopoietic progenitors has been shown to induce a striking *in vitro* expansion of myeloid cells.[42] This evidence would suggest the requirement for more lineage-specific or autologous promoters for future clinical applications.

4.5 *Zeta Associated 70 kDa Phosphoprotein (ZAP-70) Deficiency*

Zap-70 deficiency is a rare autosomal recessive form of SCID, characterized by a selective inability to produce CD8$^+$ T cells and a signal transduction defect in peripheral CD4$^+$ cells. Peripheral T cells from affected patients demonstrate defective T cell signaling and abnormal thymic ontogeny caused by inherited mutations in the TCR-associated protein tyrosine kinase (PTK) ZAP-70. SCID due to ZAP-70 deficiency contributed the first evidence, that PTKs, particularly ZAP-70, are required for normal human T cell development and function.[1]

The feasibility of a HSC-based gene therapy strategy has been assessed only in preclinical mouse models, in which the engraftment of retrovirally transduced bone marrow progenitor cells allowed the correction of T cell-defect without affecting the differentiation and function of B cells.[43] More recently alternative gene delivery approaches were tested in order to lower the risk of insertional mutagenesis in HSCs and other hematopoietic precursors that physiologically do not express and thus require ZAP-70. New delivery approaches were based on direct intrathymic injection of a T cell-specific LV or thymic electro-gene transfer of plasmid DNA both encoding for ZAP-70 into deficient mice.[44,45] Preliminary results indicated long-term reconstitution of T cell differentiation and function with a diversified T cell receptor repertoire.

5. Wiskott-Aldrich-Syndrome (WAS)

WAS is a severe X-linked immunodeficiency caused by mutations in the gene encoding for WASP, a key regulator of signaling and cytoskeletal reorganization in hematopoietic cells. Mutations in WASP result in a wide spectrum of clinical manifestations ranging from the relatively mild X-linked thrombocytopenia (XLT) to the classic full-blown WAS phenotype characterized by thrombocytopenia, immunodeficiency, eczema, and high susceptibility to developing tumors and autoimmune manifestations. The life expectancy of patients affected by severe WAS is reduced, unless they are successfully cured by HSC transplantation. Since many patients lack a compatible bone marrow donor, the administration of WAS gene corrected autologous HSCs was considered an alternative therapeutic approach.[46] Increasing evidence suggests that WASP-expressing cells have a proliferative or survival advantage over their WASP-deficient counterparts. A selective accumulation of revertant T cells has been observed in several WAS patients who underwent spontaneous somatic reversion in a lymphoid progenitor. Ideally, this selective advantage will facilitate the efficacy of gene therapy.[46]

Preclinical studies for WAS gene therapy were initially based on gammaretroviruses and then moved to LV vector approaches in order to increase efficacy and safety. Protocols for the application of gene therapy for WAS are currently under development at several centers in Europe. The first gene therapy study has been initiated in 2007 in Germany to assess the feasibility, toxicity, and potential benefit of HSC gene therapy with MLV-derived retroviral vector encoding WAS*p* under the control of a viral promoter.[47] Preliminary results from the German trial seem to be encouraging, but the safety profile of the retroviral vector still has to be carefully evaluated.

The LV-based approach included native regulatory sequences to target the expression of the therapeutic WAS transgene physiologically to the hematopoietic system. This vector was shown to be effective for correction of multiple cellular defects *in vitro* and *in vivo*.[48,49] In the WAS mouse model, engraftment of gene corrected bone marrow HSC yielded long-term WASP expression and the correction of immune, inflammatory and cytoskeletal defects,[49,50] without adverse events linked to gene transfer. This vector is in development for clinical studies that are planned in several European countries. Due to its design, the WAS-LV vector may have significant advantages for clinical application in humans in terms of natural gene regulation, and reduction in the potential for adverse mutagenic events.[51]

6. Chronic Granulomatous Disease

The encouraging results of the gene therapy trials for SCID demonstrated the curative potential of gene transfer and represented a strong rationale for the development of gene therapy protocols for other immunodeficiencies, such as CGD. This PID is an inherited disorder of innate immunity in which phagocytic leukocytes are unable to generate microbicidal oxidants. Consequently, affected patients are susceptible to recurrent opportunistic bacterial and fungal infections, leading to the formation of chronic granulomas. CGD is caused by mutation or deletions in any of four genes encoding for essential subunits of the phagocytic nicotinamide dinucleotide phosphate (NADPH) oxidase complex (gp91phox, p22phox, p47phox and p67phox). NADPH oxidase, which is localized in the phagosomal and plasma membrane of phagocytes, generates superoxide and related toxic oxygen metabolites to kill invading bacteria and fungi by respiratory burst.

Approximately 70% of CGD patients carry defects in the X-linked gene encoding for gp91phox (X-CGD).[52] Although lifelong prophylaxis reduced the incidence of infection in CGD patients, the overall annual mortality is still high (2–5%) and the success rate of HSC transplant is limited by graft-versus-host disease (GVHD) and inflammatory flare-ups at infectious sites.[52] Therefore CGD has been proposed as an ideal candidate for gene therapy, also considering that low levels of gene correction may be sufficient for curative effects, since healthy female CGD carriers can have < 10% of the normal levels of neutrophils.[53]

Although significant long-term correction has been obtained in murine models by retroviral-mediated gene transfer into HSC in combination with nonmyeloablative conditioning or bone marrow ablation,[54,55] similar results have not been achieved in unconditioned human CGD patients.[56] In the US clinical trial, a total of ten patients were enrolled, five with a p47phox deficiency and five with gp91phox-deficient CGD. The autologous CD34$^+$ cells mobilized from peripheral blood were transduced with a pseudotyped MLV retrovirus encoding for p47phox or gp91phox, respectively. Eight patients demonstrated very low levels of reconstituted neutrophils for up to 14 months after treatment.[56]

A clinical trial for X-CGD using a RV has recently been reported by M. Grez and colleagues. Two X-CGD patients were treated with *ex vivo* RV gene transfer of gp91phox into mobilized peripheral blood. Importantly non-myeloablative conditioning was administered before the infusion of genetically modified HSCs. After a period of myelosuppression, cell counts recovered gradually and patients recovered from severe infections.[57] However, an unexpected clonal expansion of transduced myeloid cells, occurred at approximately 5 months post-treatment. Vector transduced myelopoiesis expanded 3 to 5 fold in both patients due to activating integrations in the zinc finger transcription factor homologs MDS1/EVI1, PRDM16 or SETBP1. Both patients subsequently developed myelodysplasia with Monosomy 7. One patient died 2.5 years after gene therapy as a result of a severe bacterial infection.[57] Though the number of transduced cells remained high at the time of death, gp91phox expression was almost completely diminished due to the predominance of a new myeloid clone with very little oxidase function. Subsequent studies of the proviral SFFV LTR demonstrated the methylation of CpG dinucleotides within the promoter but not within the SFFV enhancer sequence. Despite these adverse events, it is clear that therapeutic benefit, although transient, can be achieved by gene therapy for CGD.[4]

Several factors could account for the clinical outcome observed in the CGD trial. The use of a vector with SFFV LTR sequences, which contains potent enhancer elements for gene expression, favored the activation of specific genes and led to the observed clonal expansion. The case for CGD is different from SCID due to the absence of a selective advantage for gene corrected cells, which facilitates the progressive loss of transgene expression due to promoter methylation.

Future clinical trials will require superior vector design to improve safety of gene transfer into HSC while ensuring adequate transgene expression in myeloid lineages. SIN lentiviral vectors encoding gp91[phox] under constitutive promoters have been generated, achieving significant restoration of the oxidase enzyme activity and neutrophil counts in a human/mouse xenograft model.[58]

7. Conclusions and Outlook

Allogeneic HSC transplantation has proved to be a successful treatment for PID but it is limited by complications in mismatched donor settings. In the last decade, gene therapy has been developed as a successful alternative strategy. Results from clinical trials employing gene corrected HSC showed immunological improvement and clinical benefit in patients affected by ADA-deficient SCID, SCID-X1, and CGD. The introduction of a low intensity conditioning regimen has been identified as a crucial factor in achieving adequate engraftment of HSC and therapeutic levels of transgenes. However, the occurrence of serious complications in SCID-X1 and CGD trials have highlighted the risks of insertional mutagenesis with the retroviral vector technology. These findings have stimulated a large number of studies into mechanisms and ways to prevent this activity by the development of novel vector technology. Several clinical trials are being developed to treat various forms of PID including SCID and WAS.

References

1. Geha RS, *et al.* (2007). Primary immunodeficiency diseases: An update from the International Union of Immunological Societies Primary Immunodeficiency Diseases Classification Committee. *J Allergy Clin Immunol* **120**: 776–794.
2. Neven B, *et al.* (2009). Long-term outcome after haematopoietic stem cell transplantation of a single-centre cohort of 90 patients with severe combined immunodeficiency: Long-term outcome of HSCT in SCID. *Blood* **113**: 4114–4124.
3. Antoine C, *et al.* (2003). Long-term survival and transplantation of haemopoietic stem cells for immunodeficiencies: Report of the European experience 1968–1999. *Lancet* **361**: 553–560.
4. Kohn, DB (2008). Gene therapy for childhood immunological diseases. *Bone Marrow Transplant* **41**: 199–205.
5. Hirschorn R, Candotti, F (2006). Immunodeficiency Due to Defects of Purine Metabolism In: Ochs H, C Smith, J Puck (eds). *Primary immunodeficiency diseases*. Oxford University Press: Oxford. pp. 169–196.
6. Booth C, *et al.* (2007). Management options for adenosine deaminase deficiency: Proceedings of the EBMT satellite workshop (Hamburg, March 2006). *Clin Immunol* **123**: 139–147.
7. Kohn DB, *et al.* (1995). Engraftment of gene-modified umbilical cord blood cells in neonates with adenosine deaminase deficiency. *Nat Med* **1**: 1017–1023.
8. Bordignon C, *et al.* (1995). Gene therapy in peripheral blood lymphocytes and bone marrow for ADA- immunodeficient patients. *Science* **270**: 470–475.
9. Aiuti A, *et al.* (2002). Correction of ADA-SCID by stem cell gene therapy combined with nonmyeloablative conditioning. *Science* **296**: 2410–2413.
10. Aiuti A, *et al.* (2009). Gene therapy for immunodeficiency due to adenosine deaminase deficiency. *N Engl J Med* **360**: 447–458.
11. Aiuti A, *et al.* (2009). Hematopoietic stem cell gene therapy for adenosine deaminase deficient-SCID. *Immunol Res.* **4**: 150–159.
12. Aiuti A, *et al.* (2007). Multilineage hematopoietic reconstitution without clonal selection in ADA-SCID patients treated with stem cell gene therapy. *J Clin Invest* **117**: 2233–2240.
13. Cassani B, *et al.* (2008). Altered intracellular and extracellular signaling leads to impaired T-cell functions in ADA-SCID patients. *Blood* **111**: 4209–4219.
14. Engel BC, *et al.* (2007). Prolonged pancytopenia in a gene therapy patient with ADA-deficient SCID and trisomy 8 mosaicism: A case report. *Blood* **109**: 503–506.
15. Gaspar HB, *et al.* (2006). Successful reconstitution of immunity in ADA-SCID by stem cell gene therapy following cessation of PEG-ADA and use of mild preconditioning. *Mol Ther* **14**: 505–513.
16. Hacein-Bey-Abina S, *et al.* (2008). Insertional oncogenesis in 4 patients after retrovirus-mediated gene therapy of SCID-X1. *J Clin Invest* **118**: 3132–3142.
17. Howe SJ, *et al.* (2008). Insertional mutagenesis combined with acquired somatic mutations causes leukemogenesis following gene therapy of SCID-X1 patients. *J Clin Invest* **118**: 3143–3150.

18. Mortellaro A, *et al.* (2006). *Ex vivo* gene therapy with lentiviral vectors rescues adenosine deaminase (ADA)-deficient mice and corrects their immune and metabolic defects. *Blood* **108**: 2979–2988.

19. Carbonaro DA, *et al.* (2008). Neonatal bone marrow transplantation of ADA-deficient SCID mice results in immunological reconstitution despite low levels of engraftment and an absence of selective donor T lymphoid expansion. *Blood* **111**: 5745–5754.

20. Hacein-Bey-Abina S, *et al.* (2002). Sustained correction of X-linked severe combined immunodeficiency by *ex vivo* gene therapy. *N Engl J Med* **346**: 1185–1193.

21. Gaspar HB, *et al.* (2004). Gene therapy of X-linked severe combined immunodeficiency by use of a pseudotyped gammaretroviral vector. *Lancet* **364**: 2181–2187.

22. Chinen J, *et al.* (2007). Gene therapy improves immune function in preadolescents with X-linked severe combined immunodeficiency. *Blood* **110**: 67–73.

23. Thrasher AJ, *et al.* (2005). Failure of SCID-X1 gene therapy in older patients. *Blood* **105**: 4255–4257.

24. Hacein-Bey-Abina S, *et al.* (2003). A serious adverse event after successful gene therapy for X-linked severe combined immunodeficiency. *N Engl J Med* **348**: 255–256.

25. Thornhill SI, *et al.* (2008). Self-inactivating gammaretroviral vectors for gene therapy of X-linked severe combined immunodeficiency. *Mol Ther* **16**: 590–598.

26. Zhang F, *et al.* (2007). Lentiviral vectors containing an enhancer-less ubiquitously acting chromatin opening element (UCOE) provide highly reproducible and stable transgene expression in hematopoietic cells. *Blood* **110**: 1448–1457.

27. Sobacchi C, Marrella V, Rucci F, Vezzoni P, Villa, A (2006). RAG-dependent primary immunodeficiencies. *Hum Mutat* **27**: 1174–1184.

28. Moshous D, *et al.* (2001). Artemis, a novel DNA double-strand break repair/V(D)J recombination protein, is mutated in human severe combined immune deficiency. *Cell* **105**: 177–186.

29. Villa A, Notarangelo LD, Roifman, CM (2008). Omenn syndrome: Inflammation in leaky severe combined immunodeficiency. *J Allergy Clin Immunol* **122**: 1082–1086.

30. Yates F, *et al.* (2002). Gene therapy of RAG-2-/- mice: sustained correction of the immunodeficiency. *Blood* **100**: 3942–3949.

31. Lagresle-Peyrou C, *et al.* (2006). Long-term immune reconstitution in RAG-1-deficient mice treated by retroviral gene therapy: A balance between efficiency and toxicity. *Blood* **107**: 63–72.

32. Mostoslavsky G, Fabian AJ, Rooney S, Alt FW, Mulligan, RC (2006). Complete correction of murine Artemis immunodeficiency by lentiviral vector-mediated gene transfer. *Proc Natl Acad Sci USA* **103**: 16406–16411.

33. Lagresle-Peyrou C, *et al.* (2008). Restoration of human B-cell differentiation into NOD-SCID mice engrafted with gene-corrected CD34$^+$ cells isolated from Artemis or RAG1-deficient patients. *Mol Ther* **16**: 396–403.

34. Grunebaum E, *et al.* (2006). Bone marrow transplantation for severe combined immune deficiency. *Jama* **295**: 508–518.

35. Liao P, Toro A, Min W, Lee S, Roifman CM, Grunebaum E (2008). Lentivirus gene therapy for purine nucleoside phosphorylase deficiency. *J Gene Med* **10**: 1282–1293.

36. Russell SM, *et al.* (1995). Mutation of Jak3 in a patient with SCID: essential role of Jak3 in lymphoid development. *Science* **270**: 797–800.

37. Macchi P, *et al.* (1995). Mutations of Jak-3 gene in patients with autosomal severe combined immune deficiency (SCID). *Nature* **377**: 65–68.

38. Bunting KD, Lu T, Kelly PF, Sorrentino BP (2000). Self-selection by genetically modified committed lymphocyte precursors reverses the phenotype of JAK3-deficient mice without myeloablation. *Hum Gene Ther* **11**: 2353–2364.

39. Gaspar HB, Thrasher AJ (2005). Gene therapy for severe combined immunodeficiencies. *Expert Opin Biol Ther* **5**: 1175–1182.

40. McCauslin CS, *et al.* (2003). *In vivo* retroviral gene transfer by direct intrafemoral injection results in correction of the SCID phenotype in Jak3 knock-out animals. *Blood* **102**: 843–848.

41. Hofmeister R, Khaled AR, Benbernou N, Rajnavolgyi E, Muegge K, Durum, SK (1999). Interleukin-7: physiological roles and mechanisms of action. *Cytokine Growth Factor Rev* **10**: 41–60.

42. Jiang Q, Li WQ, Aiello FB, Klarmann KD, Keller JR, Durum SK (2005). Retroviral transduction of IL-7Ralpha into IL-7Ralpha-/- bone marrow progenitors: Correction of lymphoid deficiency and induction of neutrophilia. *Gene Ther* **12**: 1761–1768.

43. Otsu M, *et al.* (2002). Reconstitution of lymphoid development and function in ZAP-70-deficient mice following gene transfer into bone marrow cells. *Blood* **100**: 1248–1256.

44. Adjali O, *et al.* (2005). *In vivo* correction of ZAP-70 immunodeficiency by intrathymic gene transfer. *J Clin Invest* **115**: 2287–2295.

45. Irla M, *et al.* (2008). ZAP-70 restoration in mice by *in vivo* thymic electroporation. *PLoS ONE* **3**: e2059.

46. Bosticardo M, Marangoni F, Aiuti A, Villa A, Roncarolo MG (2009). Recent advances in understanding the pathophysiology of Wiskott-Aldrich syndrome. *Blood* **113**: 6288–6295.

47. Boztug K, Dewey RA, Klein C (2006). Development of hematopoietic stem cell gene therapy for Wiskott-Aldrich syndrome. *Curr Opin Mol Ther* **8**: 390–395.

48. Charrier S, *et al.* (2007). Lentiviral vectors targeting WASp expression to hematopoietic cells, efficiently transduce and correct cells from WAS patients. *Gene Ther* **14**: 415–428.

49. Marangoni F, *et al.* (2009). Evidence for Long-term Efficacy and Safety of Gene Therapy for Wiskott-Aldrich Syndrome in Preclinical Models. *Mol Ther* **17**: 1073–1082.

50. Dupre L, *et al.* (2006). Efficacy of gene therapy for Wiskott-Aldrich syndrome using a WAS promoter/cDNA-containing lentiviral vector and nonlethal irradiation. *Hum Gene Ther* **17**: 303–313.

51. Galy A, Roncarolo MG, Thrasher AJ (2008). Development of lentiviral gene therapy for Wiskott Aldrich syndrome. *Expert Opin Biol Ther* **8**: 181–190.

52. Stasia MJ, Li XJ (2008). Genetics and immunopathology of chronic granulomatous disease. *Semin Immunopathol* **30**: 209–235.

53. Woodman RC, *et al.* (1995). A new X-linked variant of chronic granulomatous disease characterized by the existence of a normal clone of respiratory burst-competent phagocytic cells. *Blood* **85**: 231–241.

54. Dinauer MC, Li LL, Bjorgvinsdottir H, Ding C, Pech N (1999). Long-term correction of phagocyte NADPH oxidase activity by retroviral-mediated gene transfer in murine X-linked chronic granulomatous disease. *Blood* **94**: 914–922.

55. Sadat MA, *et al.* (2003). Long-term high-level reconstitution of NADPH oxidase activity in murine X-linked chronic granulomatous disease using a bicistronic vector expressing gp91phox and a Delta LNGFR cell surface marker. *Hum Gene Ther* **14**: 651–666.

56. Malech HL, *et al.* (1997). Prolonged production of NADPH oxidase-corrected granulocytes after gene therapy of chronic granulomatous disease. *Proc Natl Acad Sci USA* **94**: 12133–12138.

57. Ott MG, *et al.* (2006). Correction of X-linked chronic granulomatous disease by gene therapy, augmented by insertional activation of MDS1-EVI1, PRDM16 or SETBP1. *Nat Med* **12**: 401–409.

58. Naumann N, *et al.* (2007). Simian immunodeficiency virus lentivector corrects human X-linked chronic granulomatous disease in the NOD/SCID mouse xenograft. *Gene Ther* **14**: 1513–1524.

Chapter 14
Gene Therapy for Hemophilia

David Markusic, Babak Moghimi and Roland Herzog*

Hemophilia A and B are X-linked bleeding disorders resulting from a deficiency in coagulation factor VIII (F.VIII) or factor IX (F.IX), respectively. Patients experience spontaneous bleeding episodes into soft tissues and joints, which is managed by intravenous administration of recombinant coagulation factor protein. Hemophilia is well suited for correction by gene transfer, providing prophylactic treatment. As little as 1-2% of normal circulating levels of coagulation factor can provide partial correction of disease, no regulation of gene expression is required, and therapeutic efficacy is easily evaluated. Although F.VIII and F.IX are naturally synthesized in the liver, many cell types are capable of synthesizing biologically active factor. Numerous animal models have demonstrated long-term expression and disease correction using different target tissues and vectors. Viral gene transfer in hemophilia B dogs resulted in therapeutic expression for at least 8 years, while transient therapeutic expression has been demonstrated in clinical trials. Future investigations will attempt to achieve long-term expression in humans using modified protocols.

1. Introduction

Hemophilia A and B are X-linked bleeding disorders, which result from a deficiency of F.VIII or F.IX, respectively, in the coagulation cascade (Fig. 14.1). Severe disease is defined as <1% factor activity, whereas 1 to 5% and >5% of normal are defined as moderate and mild disease, respectively.[1] The incidence of hemophilia A is 1 in 5,000 live male births

---
*Correspondence: Department of Pediatrics, Div. Cellular and Molecular Therapy, University of Florida, Gainesville, Florida.
E-mail: rherzog@ufl.edu

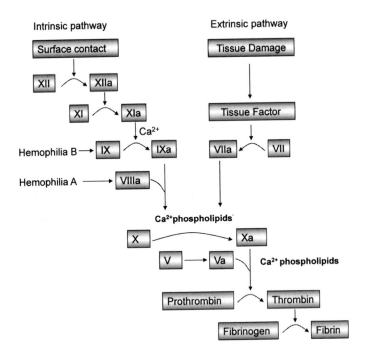

Fig. 14.1   Coagulation cascade.

with two-thirds of affected patients having the severe form of the disease. In contrast, almost one-half of individuals with hemophilia B (1 in 25,000–30,000 births) have F.IX levels above 1 percent (moderate or mild disease). The most common sites of bleeding are into joints, muscles, and the gastrointestinal tract. Hemarthrosis–bleeding into the joints — is the primary cause of morbidity in hemophilic patients resulting in chronic pain and physical debilitation.[2] Intracranial hemorrhage is a leading cause of mortality in hemophilic patients.

Standard treatment is periodical administration of recombinant coagulation factors to maintain therapeutic levels in moderate and severe patient's plasma. Patients have a risk of forming inhibitory antibodies (inhibitors), which occur in approximately 25 percent of patients with hemophilia A and 3 to 5 percent of those with hemophilia B. These inhibitors complicate bleeding episodes because of diminished responsiveness to factor concentrates.[3,4] The predilection for patients with severe disease is consistent with observations that inhibitors primarily occur in patients with large deletions and stop mutations, compared to small deletions or mis-sense mutations.[5] Hemophilia patients with high levels of inhibitor require treatment with agents such as prothrombin complex concentrates or recom-binant factor VIIa, which bypass the intrinsic pathway (Fig. 14.1). Long-term management of hemophilic patients with inhibitors requires immune tolerance induction (ITI) via the administration of repetitive high doses of F.VIII or F.IX with or without immunosuppressive therapy, which are highly expensive protocols.

2. Limitations of Hemophilia Treatment With Coagulation Factor Concentrates or Recombinant Coagulation Factors

Treatment regimens for hemophilic patients consist of either prophylaxis or treatment on demand, where the former is recommended to be the standard of care by the National Hemophilia Foundation, the World Federation of Hemophilia, and the World Health Organization.[6] Due to the high costs asso-ciated with prophylactic treatment, however, many patients are treated on demand, which can result in progressive joint damage over time. In the past, there was a significant risk for the transmission of blood borne pathogens (HBV, HCV, and HIV) in contaminated plasma derived factor concentrates. Careful screening and preparation of plasma derived factor concentrates and the introduction of recombinant protein coagulation factors have virtu-ally eliminated the risk of blood borne pathogen transmission. Both F.VIII and F.IX have a short biologically active half-life, which requires frequent infusions to maintain therapeutic levels.

Hemophilia is a monogenic recessive disorder, caused by the loss of function of a single gene and is well suited for correction by transfer of a functional copy of the genes encoding F.VIII or F.IX. Both F.VIII and F.IX are secreted into the plasma and circulate in an inactive form (zymogen), and therefore do not require strict regulation for expression. Based on clin-ical data from patients treated with plasma derived or recombinant protein, circulating levels as low as 1% of normal promote a transition from severe to moderate disease, and levels reaching 5% of normal can lead to complete elimination of spontaneous bleeding episodes. Although F.VIII and F.IX are naturally expressed from the liver, other tissues can be used to express and secrete functional protein. Clinical outcome from gene transfer is easily measured using standard clinical assays for circulating coagulation factor levels in plasma and coagulation assays.

3. Gene Transfer for Correction of Hemophilia

3.1 *Ex Vivo Gene Transfer of F.VIII and F.IX*

Ex vivo gene transfer utilizes autologous cells from the patient such as fibroblasts,[20,21] hematopoietic stem cells (HSC),[22] or myoblasts,[23] which are harvested from the affected individual and infected with a gene transfer vector expressing either F.VIII or F.IX. This may be followed by *in vitro* selection and expansion. Ultimately, the transduced cells are transplanted back into the patient, engraft, and provide therapeutic expression of the respective coagulation factor (Fig. 14.2). *Ex vivo* gene transfer allows for a significant reduction in the amount of vector required for gene transfer, since gene transfer is performed directly on the target cell under defined conditions. Analysis and selection of cells with high levels of F.VIII or F.IX

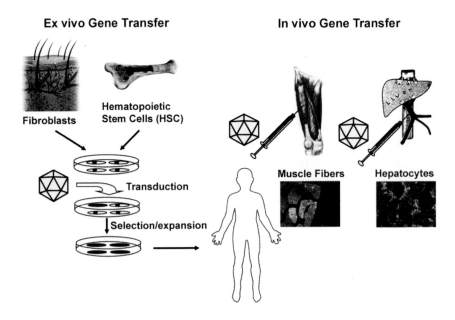

Fig. 14.2 *Ex vivo* and *in vivo* gene transfer schematic.

expression allows for optimal therapeutic benefit following transplantation. From an immunological standpoint, gene transfer to off-target cells, such as professional antigen presenting cells (APC), is prevented, thus greatly reducing the risk of inducing a humoral and/or cytotoxic T lymphocyte (CTL) responses against F.VIII or F.IX. Additionally, complications from activation of an innate immune response to the vector may be reduced, thereby reducing the risk of inflammation directed against the site of gene transfer. Lastly, *ex vivo* gene transfer allows for efficient gene transfer if the patient has neutralizing antibodies against the viral vector,[24] or if the viral vector is inactivated in human sera.[25]

The choice of target cells is limited by the ability to culture the cells *in vitro* while maintaining viability and functionality. Target cells should ideally be long-lived stem cells (e.g. hematopoietic stem cells) that can provide expression in differentiated progeny, or differentiated cells such as fibroblasts, which can be efficiently and stably transduced and expanded *in vitro*. And lastly, the protocol requires efficient engraftment of gene-modified target cells following transplantation in order to provide long-term therapeutic benefit. From a practical standpoint, *ex vivo* gene transfer can be more labor intensive, require specialized facilities for the culturing and manipulation of the patient's cells, and may subject patients to multiple minor surgical procedures, which poses a risk for patients with severe hemophilia.

3.2 In Vivo Gene Transfer of F.VIII and F.IX

The liver is a preferred target for gene transfer, because it represents the natural site of F.VIII and F.IX biosynthesis (Fig. 14.2). However, since both F.VIII and F.IX are secreted into the circulation, liver specific expression is not required. The main limitation is that the intended target tissue should be capable of performing the post-translational modifications required for expression of biologically active F.VIII and F.IX protein.[26] Adult hemophilia patients with liver disease (e.g. hepatitis B and C from prior exposure to contaminated coagulation factor concentrates) require alternative tissues for gene transfer. Skeletal muscle has been extensively investigated as an alternative target tissue for adeno-associated viral (AAV) gene transfer of F.IX, and has been shown to be capable of expressing biologically active protein (Fig. 14.2). The poor stability of F.VIII and potential difficulties with reaching the circulation has been prohibitive in obtaining therapeutic levels of F.VIII from gene transfer to skeletal muscle.[27]

Liver gene transfer is achieved by intravascular (IV) delivery via the peripheral circulation, or by infusion into the portal vein or hepatic artery. Ideally, the vector administration procedure should be minimally invasive, such as IV administration to a peripheral vein or an angiographic procedure for vector infusion into the hepatic artery. Hepatic gene transfer using an AAV serotype 2 vector has led to long-term therapeutic expression of F.VIII and F.IX in both murine and canine models of hemophilia. One draw back for administration of AAV2 vectors into a blood vessel is that humans are a natural host for AAV2, and there is a high incidence of neutralizing antibodies, which can block gene transfer.[28]

Intramuscular (IM) delivery of a gene transfer vector expressing F.IX has the advantage of being minimally invasive to the patient, and local administration limits biodistribution to other tissues. Although long-term F.IX expression has been obtained by IM injection of an AAV-F.IX vector in murine and canine hemophilia models, this was limited to animal models with missense mutations.[29,30] Additionally, a higher incidence of developing inhibitors was observed when using a high dose per injection site.[31] At higher doses, this may require too many injections to be practically applied in a clinical treatment protocol. Because of these practical restrictions on IM viral administration in humans, vascular delivery techniques have been developed to substantially enhance gene transfer to skeletal muscle, e.g. by performing isolated limb perfusion.[32] This results in wide spread gene transfer to skeletal muscle from a single bolus injection of viral vector. Additionally, alternative AAV serotypes (AAV-1, -6, -7, 8-, and -9)[33,34] with a higher tropism for muscle are currently being evaluated in animal models and human clinical trials for improved gene transfer efficiency.

4. AAV is a Preferred Gene Therapy Vector for *In Vivo* Gene Transfer to Correct of Hemophilia

For *in vivo* gene transfer, an ideal vector would provide stable long-term expression in terminally differentiated cells of the liver or skeletal muscle. Of the commonly used gene transfer vectors, adenoviral, lentiviral vectors, and AAV vectors are capable of efficient and stable gene transfer to non-dividing cells. Early versions of adenoviral vectors and later gutted adenoviral vectors, which do not express any adenoviral proteins, proved to be too immunogenic for *in vivo* gene transfer, excluding these vectors for use in humans. Murine retroviral vectors require cell cycling for efficient gene transfer and have only shown practical use for *ex vivo* gene transfer and administration to neonatal animals, where hepatocytes are still dividing. In addition, these vectors pose a higher risk for insertional mutagenesis. Lentiviral vectors — retroviral vectors that efficiently transduce nondividing cells — are currently being evaluated in small and large animal models with some promising results. Safety concerns regarding insertional mutagenesis and feasibility of scaling up vector production to therapeutic doses still have to be addressed.

In contrast, AAV based vectors have led to long-term expression in murine and canine hemophilia models and are considered to be less immunogenic with regard to innate immune responses, and these vectors are comparatively inefficient in transducing antigen presenting cells (APC) compared to adeno- or lentiviral vectors. Different serotypes of AAV are available with different tissue tropism. The AAV2 serotype has been the most extensively studied for safety and efficacy in pre-clinical and clinical trials. The AAV vector genome predominantly remains in an episomal form in hepatocytes and muscle fibers, reducing the risk of insertional mutagenesis. The main limitation of AAV vectors is a reduced packaging capacity as compared to adenoviral or retroviral vectors. This has significantly limited AAV based studies for F.VIII gene transfer due to the large cDNA coding for the full length or B domain deleted (BDD) F.VIII variant.[35] The coding sequence for BDD deleted F.VIII is ~4.3 kb compared to 1.4 kb for F.IX.

5. Immunological Considerations for Efficient F.IX Gene Transfer

Gene transfer to an immunocompetent recipient can cause a variety of immune responses including an innate and adaptive immune response to viral structural proteins and the transgene product. Factors affecting the immune response to gene transfer in hemophilia include host factors, gene transfer protocol/route of administration, and vector design and dosing.

Some examples of host factors effecting gene transfer are immuno-profile/ genetic modifiers, prior exposure to virus, nature of the underlying mutation, and coexisting inflammation.

The immune system detects viral gene transfer at different stages. For example, Toll-like receptors (TLRs) specifically activate downstream innate immune response against both single and double stranded viral DNA and RNA. The ability of viral vectors to transduce APCs increases the chance of being presented by MHC class I to cytotoxic T lymphocytes (CTLs), which can lead to elimination of transduced cells. Finally, the transgene product can be presented on MHC class II and activate further adaptive immune responses, including antibody formation. Although many of the immunogenic components of viral vectors have been removed to reduce these immune responses, fundamental innate responses to viral infection still exist and can prime the immune system for adaptive responses. Another important factor modulating immune responses to the transgene product is the transfer protocol. Extensive investigations performed in animal models have indicated that vehicle characteristics such as route of administration, vector dose, transgene promoter, and preparation contaminants can influence the type and potency of the immune reaction. Hepatic administration of an AAV2 F.IX vector results in sustained expression and absence of inhibitors to F.IX protein in hemophilic mice deficient in circulating F.IX protein suggesting that expression in the liver may lead to tolerance.[12] Studies in different strains of mice showed that hepatic gene transfer of AAV vectors induced antigen specific $CD4^+CD25^+$ regulatory T cells, which are both necessary and sufficient for tolerance induction to the transgene product.[36]

In addition to potential immune responses to transgene product (F.VIII or F.IX), two clinical trials with an AAV2 vector demonstrate that immune responses to the viral vector can affect long-term efficacy of gene therapy. AAV2 mediated gene transfer to skeletal muscle in mouse and dog hemophilia models demonstrated both safety and efficacy for F.IX expression, leading to the initiation of a human clinical trial for IM AAV2 F.IX gene transfer.[37] In the human trial, circulating levels of F.IX were below 1 percent due to restrictions on total administered viral dose. All patients receiving an IM injection experienced an increase of 2-3 logs in neutralizing antibody titer to viral capsid after vector administration, effectively preventing re-administration of virus to boost F.IX expression levels.[38]

Based on success in pre-clinical studies, a human trial with hepatic artery delivery of an AAV2 F.IX vector under the control of a liver specific promoter was initiated. One subject in the highest dose cohort reached therapeutic levels of F.IX (10–12%). However, 4 weeks after vector infusion, F.IX levels began to fall, gradually returning to baseline (<1%) at 10 weeks.[39] ELISPOT analysis of T cell responses in peripheral blood mononuclear cells (PBMC) isolated from a subsequently treated subject demonstrated IFN-γ production in response to AAV2 capsid peptides, but not to F.IX peptides.[39] The identified AAV2 capsid peptide epitopes were used to synthesize MHC class I pentamers for the direct detection of capsid-specific CD8$^+$ T cell populations in peripheral blood. Expansion of AAV capsid-specific CD8$^+$ T cells *in vitro* and *in vivo* was demonstrated.[40] These experimental results provided further evidence that a CTL response directed against AAV capsid (presented on transduced hepatocytes) may have resulted in a decline of F.IX levels. The time course of the observed spike in liver enzymes correlated with the time required for expansion and contraction of a capsid-specific CD8$^+$ T cell population. The inconsistency between animal models and humans in duration of expression may be explained by the fact that humans are natural hosts for AAV infection; and co-infection with immunogenic helper viruses, such as adenovirus, may generate memory T cells, thus leading to concomitant immune response after renewed exposure by gene transfer.

Several strategies are currently being evaluated to address immunological issues raised from the two AAV2 clinical trials. Exon-shuffled AAV variants or variants isolated from non human-primate species may avoid both neutralizing antibodies against viral capsid and potential CTL responses against viral capsid. Pre-treatment with transient immune suppression (IS) has been studied in rodents and primates. Regimens based on mycophenolate mofetil (MMF) and rapamycin resulted in long-term expression of transgene without antibody formation against F.IX, confirming the safety of AAV-2 gene transfer in the context of an IS regimen. There is increasing evidence that pre-treatment with rapamycin provides selective expansion of CD4$^+$CD25$^+$Foxp3$^+$ regulatory T cells, and finally blocking of co-stimulatory signals on T cells could lead tolerance in hemophilia gene transfer, pointing toward novel strategies for tolerance to the transgene and blockage of responses to the vector.[41]

6. Advancements from Small and Large Animal Models of Hemophilia

The availability of small murine and large canine animal models of hemophilia (Table 14.1) have greatly facilitated the development and testing of gene transfer vectors. These animal models mimic some of the disease phenotypes observed in human hemophilia patients and therefore are ideal for the evaluation of optimal gene transfer vectors, routes of administration, and target tissues for correction of disease phenotype. Additionally, the murine and canine hemophilia disease models have provided a means to assess potential immunological responses to the transgene (F.VIII or F.IX) and viral vector, and develop methodologies to allow for long-term therapeutic expression of F.VIII and F.IX.

6.1 *Murine Hemophilia Models*

Murine models for hemophilia A and B have allowed for the demonstration efficacy of gene transfer both *ex vivo* and *in vivo* from retroviral, adenoviral, and AAV vectors and naked DNA for the correction of hemophilia. Gene transfer studies in mice showed that F.IX could be efficiently expressed from skeletal muscle, providing evidence for the production of biologically active F.IX protein from a tissue that does not normally synthesize F.IX. Restricting expression of F.IX to hepatocytes using an AAV vector with a liver specific promoter results in induction of tolerance to F.IX in murine hemophilia models.[12,42] The generation of murine F.IX knockout mice transgenic for commonly observed human F.IX mutations has allowed for the investigation of the relative risk of inhibitor formation with CRIM — and CRIM + F.IX mutants following gene transfer.[11,13]

6.2 *Canine Hemophilia Models*

Canine models of hemophilia have been instrumental in the implementation of human gene therapy clinical trials. Therapeutic gene transfer that is effective in inbred mice often fails to be reproducible in humans. Dogs are

Table 14.1. Animal models of severe hemophilia.

Murine Models	Strain background	Genotype/Phenotype	Ref.
Hemophilia A	C57BL/6 +129sv	Partial F.VIII deletion, activity <1%	7
Hemophilia B	C57BL/6 +129sv	F.IX gene deletion, Spontaneous bleeds	8–10
Hemophilia B R333QhFIX	C57BL/6 +129sv	Expresses non-functional, crm+ protein (missense mutation)	11
Hemophilia B	C3H/HeJ	F.IX gene deletion, spontaneous bleeds, increased risk of inhibitor formation	12
Hemophilia B hF.IX[a]-transgenic	C57LB/6	Several transgenic lines expressing different non-functional forms of hF.IX[a] (missense and nonsense mutations)	13
Canine Models	CRIM[b] status	Phenotype	Ref.
Hemophilia A (Chapel Hill)	Low	Gene inversion, <1% F.VIII activity with spontaneous bleeds	14, 15
Hemophilia A (Queens)	Negative	Gene inversion, <1% F.VIII activity with spontaneous bleeds	16, 17
Hemophilia B (Alabama)	Negative	Complex "null" mutation with increased risk of inhibitor formation	18
Hemophilia B (Chapel Hill)	Negative	Missense mutation, F.IX deficient with spontaneous bleeds	19

[a]hF.IX; human factor IX; [b]CRIM (Cross-reacting Immunological Material).

much closer in terms of size to humans and provide a more realistic evaluation for the feasibility of scaling up vector production to provide therapeutic
expression. Therapeutic doses of AAV2 vectors obtained in canines closely
predicted doses that lead to transient therapeutic expression in humans.[39]
Since dogs have a longer lifespan then mice, long-term studies for efficacy
and safety from gene transfer can be performed. Indeed, a single gene transfer of an AAV2 vector expressing canine F.IX to the liver has resulted in
stable expression and correction of major complications from hemophilia
for at least eight years,[43] demonstrating the feasibility of long-term, perhaps even life-long correction from a single AAV vector administration in
humans.

7. Gene Therapy Trials for Hemophilia Past, Present, and Future

There have been six clinical gene therapy trials for the treatment of either
hemophilia A or hemophilia B, summarized in Table 14.2. *Ex vivo* gene
transfer to autologous fibroblasts with plasmid transfection of F.VIII[21] or
retroviral vector transduction of F.IX respectively[45] followed by transplantation lead to a transient increase in F.VIII and F.IX levels to approximately 4% of normal levels followed by a decline to pre-treatment levels.
In vivo gene transfer of a retroviral vector[44] or a gutless adenoviral vector expressing F.VIII by peripheral intravenous delivery was not further
pursued. Manno *et al.* investigated IM[46] and hepatic artery[39] delivery of
AAV2 vectors expressing F.IX. Long-term expression of F.IX was documented in muscle biopsy samples from treated patients, but circulating
levels of F.IX were sub-therapeutic. One patient in the high dose cohort of
the hepatic artery trial had a transient therapeutic increase in F.IX expression as discussed above[40] Based on the data accumulated from the liver
AAV2 clinical trial, new trials are designed using AAV-F.IX gene transfer
to liver. A regimen of immune suppression prior to and briefly following
vector injection may prevent activation of a memory CTL response during
the time that capsid protein antigen will be presented to the immune system. Self-complementary and alternate serotype vectors may be effective
at lower vector doses and may help circumvent immune responses to vector antigens. For example, the AAV8 serotype, derived from non-human

Table 14.2. Hemophilia A and B gene therapy clinical trials.

Trial	Endpoint	Status	Reference
Ex vivo transfection autologous fibroblasts with F.VIII plasmid	Transient increase in F.VIII activity up to 4% normal	Closed	Roth *et al.*[21]
Intravenous delivery of F.VIII retroviral vector	Some persistence of vector genome in PBMC. Intermittent detection of F.VIII levels >1%.	Closed	Powell *et al.*[44]
Intravenous delivery of a F.VIII gutless adenoviral vector	Transient elevation in F.VIII activity with elevation in liver enzymes	Closed	Not published
Ex vivo transduction of autologous fibroblasts with F.IX retroviral vector	Transient increase in F.IX levels to approximately 4% normal	Closed	Qiu *et al.*[45]
Intramuscular delivery of a F.IX AAV vector	Persistent F.IX expression in muscle. No long-term rise above 1% normal	Closed	Manno *et al.*[46]
Hepatic artery delivery of a F.IX AAV vector	F.IX levels peaked around 10% in high dose cohort followed by return to baseline	Closed	Manno *et al.*[39]

primates,[47] has shown superior gene transfer efficiency in murine liver and reduced activation of capsid-specific T cells.

8. Conclusions

Pre-clinical studies in animals have allowed for optimized protocols, including viral vector selection, route of administration, and dosing for initiating clinical trials. Of the different viral vectors being evaluated, AAV gene transfer has shown the greatest efficacy in both animal and human trials. Recent developments with a lentiviral vector, incorporating a miRNA target to suppress F.IX gene expression in APCs, have resulted in long-term correction in a murine model of hemophilia following liver gene transfer.[48] Human clinical trials using AAV gene transfer of F.IX for hemophilia B gene therapy have shown much promise, but also encountered hurdles. Refinements in gene transfer protocols, including the use of alternative viral serotypes, routes of delivery, and transient immune suppression may be required to obtain long-term correction of this bleeding disorder in humans. Improved understanding of vector biology will likely result in the production of superior vectors that can express F.VIII at therapeutic levels and lead to the development of human clinical trials for hemophilia A.

References

1. White GC 2nd, Rosendaal F, Aledort LM, Lusher JM, Rothschild C, Ingerslev J (2001). Definitions in hemophilia. Recommendation of the scientific subcommittee on factor VIII and factor IX of the scientific and standardization committee of the International Society on Thrombosis and Haemostasis. *Thromb Haemost* **85**: 560.
2. Avina-Zubieta JA, Galindo-Rodriguez G, Lavalle C (1998). Rheumatic manifestations of hematologic disorders. *Curr Opin Rheumatol* **10**: 86–90.
3. Hoyer LW, Scandella D (1994). Factor VIII inhibitors: structure and function in autoantibody and hemophilia A patients. *Semin Hematol* **31**: 1–5.
4. Hay CR (1998). Factor VIII inhibitors in mild and moderate-severity haemophilia A. *Haemophilia* **4**: 558–563.
5. Gouw SC, van der Bom JG, van den Berg Marijke H (2007). Treatment-related risk factors of inhibitor development in previously untreated patients with hemophilia A: the CANAL cohort study. *Blood* **109**: 4648–4654.
6. Berntorp E *et al.* (2003). Consensus perspectives on prophylactic therapy for haemophilia: summary statement. *Haemophilia* **9 Suppl 1**: 1–4.
7. Bi L, Lawler AM, Antonarakis SE, High KA, Gearhart JD, Kazazian HH Jr. (1995). Targeted disruption of the mouse factor VIII gene produces a model of haemophilia A. *Nat Genet* **10**: 119–121.
8. Lin HF, Maeda N, Smithies O, Straight DL, Stafford DW (1997). A coagulation factor IX-deficient mouse model for human hemophilia B. *Blood* **90**: 3962–3966.
9. Wang L, Zoppe M, Hackeng TM, Griffin JH, Lee KF, Verma IM (1997). A factor IX-deficient mouse model for hemophilia B gene therapy. *Proc Natl Acad Sci USA* **94**: 11563–11566.
10. Kundu RK, *et al.* (1998). Targeted inactivation of the coagulation factor IX gene causes hemophilia B in mice. *Blood* **92**: 168–174.

11. Jin DY, Zhang TP, Gui T, Stafford DW, Monahan PE (2004). Creation of a mouse expressing defective human factor IX. *Blood* **104**: 1733–1739.
12. Mingozzi F, *et al.* (2003). Induction of immune tolerance to coagulation factor IX antigen by *in vivo* hepatic gene transfer. *J Clin Invest* **111**: 1347–1356.
13. Sabatino DE, *et al.* (2004). Novel hemophilia B mouse models exhibiting a range of mutations in the Factor IX gene. *Blood* **104**: 2767–2774.
14. Graham JB, Buckwalter JA, *et al.* (1949). Canine hemophilia; observations on the course, the clotting anomaly, and the effect of blood transfusions. *J Exp Med* **90**: 97–111.
15. Lozier JN, *et al.* (2002). The Chapel Hill hemophilia A dog colony exhibits a factor VIII gene inversion. *Proc Natl Acad Sci USA* **99**: 12991–12996.
16. Hough C, *et al.* (2002). Aberrant splicing and premature termination of transcription of the FVIII gene as a cause of severe canine hemophilia A: similarities with the intron 22 inversion mutation in human hemophilia. *Thromb Haemost* **87**: 659–665.
17. Giles AR, Tinlin S, Greenwood R (1982). A canine model of hemophilic (factor VIII:C deficiency) bleeding. *Blood* **60**: 727–730.
18. Mauser AE, Whitlark J, Whitney KM, Lothrop CD Jr. (1996). A deletion mutation causes hemophilia B in Lhasa Apso dogs. *Blood* **88**: 3451–3455.
19. Evans JP, Brinkhous KM, Brayer GD, Reisner HM, High, KA (1989). Canine hemophilia B resulting from a point mutation with unusual consequences. *Proc Natl Acad Sci USA* **86**: 10095–10099.
20. Hansen J, Qing K, Srivastava A (2001). Adeno-associated virus type 2-mediated gene transfer: altered endocytic processing enhances transduction efficiency in murine fibroblasts. *J Virol* **75**: 4080–4090.
21. Roth DA, Tawa NE Jr., O'Brien JM, Treco DA, Selden RF (2001). Nonviral transfer of the gene encoding coagulation factor VIII in patients with severe hemophilia A. *N Engl J Med* **344**: 1735–1742.
22. Chang AH, Stephan MT, Lisowski L, Sadelain M (2008). Erythroid-specific Human Factor IX Delivery From *In Vivo* Selected Hematopoietic Stem Cells Following Non-myeloablative Conditioning in Hemophilia B Mice. *Mol Ther* **16**: 1745–1752.
23. Yao SN, Kurachi K (1992). Expression of human factor IX in mice after injection of genetically modified myoblasts. *Proc Natl Acad Sci USA* **89**: 3357–3361.
24. Halbert CL, *et al.* (2006). Prevalence of neutralizing antibodies against adeno-associated virus (AAV) types 2, 5, and 6 in cystic fibrosis and normal populations: Implications for gene therapy using AAV vectors. *Hum Gene Ther* **17**: 440–447.
25. DePolo NJ, *et al.* (2000). VSV-G pseudotyped lentiviral vector particles produced in human cells are inactivated by human serum. *Mol Ther* **2**: 218–222.
26. Kaufman RJ (1998). Post-translational modifications required for coagulation factor secretion and function. *Thromb Haemost* **79**: 1068–1079.
27. Rosenberg JB, *et al.* (1998). Intracellular trafficking of factor VIII to von Willebrand factor storage granules. *J Clin Invest* **101**: 613–624.
28. Scallan CD, *et al.* (2006). Human immunoglobulin inhibits liver transduction by AAV vectors at low AAV2 neutralizing titers in SCID mice. *Blood* **107**: 1810–1817.
29. Nathwani AC, Davidoff A, Hanawa H, Zhou JF, Vanin EF, Nienhuis AW (2001). Factors influencing *in vivo* transduction by recombinant adeno-associated viral vectors expressing the human factor IX cDNA. *Blood* **97**: 1258–1265.

30. Fields PA, *et al.* (2001). Risk and prevention of anti-factor IX formation in AAV-mediated gene transfer in the context of a large deletion of F9. *Mol Ther* **4**: 201–210.

31. Herzog RW, *et al.* (2002). Influence of vector dose on factor IX-specific T and B cell responses in muscle-directed gene therapy. *Hum Gene Ther* **13**: 1281–1291.

32. Arruda VR, *et al.* (2005). Regional intravascular delivery of AAV-2-F.IX to skeletal muscle achieves long-term correction of hemophilia B in a large animal model. *Blood* **105**: 3458–3464.

33. Arruda VR, *et al.* (2004). Safety and efficacy of factor IX gene transfer to skeletal muscle in murine and canine hemophilia B models by adeno-associated viral vector serotype 1. *Blood* **103**: 85–92.

34. Hauck B, Xiao W (2003). Characterization of tissue tropism determinants of adeno-associated virus type 1. *J Virol* **77**: 2768–2774.

35. Pittman DD, Alderman EM, Tomkinson KN, Wang JH, Giles AR, Kaufman RJ (1993). Biochemical, immunological, and *in vivo* functional characterization of B-domain-deleted factor VIII. *Blood* **81**: 2925–2935.

36. Cao O, *et al.* (2007). Induction and role of regulatory CD4+CD25+ T cells in tolerance to the transgene product following hepatic *in vivo* gene transfer. *Blood* **110**: 1132–1140.

37. Herzog RW, *et al.* (1997). Stable gene transfer and expression of human blood coagulation factor IX after intramuscular injection of recombinant adeno-associated virus. *Proc Natl Acad Sci USA* **94**: 5804–5809.

38. Jiang H, *et al.* (2006). Evidence of multiyear factor IX expression by AAV-mediated gene transfer to skeletal muscle in an individual with severe hemophilia B. *Mol Ther* **14**: 452–455.

39. Manno CS, *et al.* (2006). Successful transduction of liver in hemophilia by AAV-Factor IX and limitations imposed by the host immune response. *Nat Med* **12**: 342–347.

40. Mingozzi F, *et al.* (2007). CD8(+) T-cell responses to adeno-associated virus capsid in humans. *Nat Med* **13**: 419–422.

41. Peng B, *et al.* (2008). Transient blockade of the inducible costimulator pathway generates long-term tolerance to factor VIII after nonviral gene transfer into hemophilia A mice. *Blood* **112**: 1662–1672.

42. Dobrzynski E, Fitzgerald JC, Cao O, Mingozzi F, Wang L, Herzog, RW (2006). Prevention of cytotoxic T lymphocyte responses to factor IX-expressing hepatocytes by gene transfer-induced regulatory T cells. *Proc Natl Acad Sci USA* **103**: 4592–4597.

43. Niemeyer GP, *et al.* (2008). Long term correction of inhibitor prone hemophilia B dogs treated with liver-directed AAV2 mediated factor IX gene therapy. *Blood* **113**: 797–806.

44. Powell JS, *et al.* (2003). Phase 1 trial of FVIII gene transfer for severe hemophilia A using a retroviral construct administered by peripheral intravenous infusion. *Blood* **102**: 2038–2045.

45. Qiu X, *et al.* (1996). Implantation of autologous skin fibroblast genetically modified to secrete clotting factor IX partially corrects the hemorrhagic tendencies in two hemophilia B patients. *Chin Med J (Engl)* **109**: 832–839.

46. Manno CS, *et al.* (2003). AAV-mediated factor IX gene transfer to skeletal muscle in patients with severe hemophilia B. *Blood* **101**: 2963–2972.

47. Gao GP, Alvira MR, Wang L, Calcedo R, Johnston J, Wilson JM (2002). Novel adeno-associated viruses from rhesus monkeys as vectors for human gene therapy. *Proc Natl Acad Sci USA* **99**: 11854–11859.

48. Brown BD, *et al.* (2007). A microRNA-regulated lentiviral vector mediates stable correction of hemophilia B mice. *Blood* **110**: 4144–4152.

Chapter 15

Gene Therapy for Obesity and Diabetes

Sergei Zolotukhin* and Clive H. Wasserfall

In spite of a complex multitrait etiology of diet-induced obesity, the case is made for supporting genetic approach in treating this chronic condition. With the recent progress in characterizing hundreds of obesity-related genes, and due to the development of comprehensive computational algorithms, a number of primary candidate target genes have been identified. The results of several *in vivo* studies supporting the notion of successful gene therapy for obesity are briefly described.

1. Introduction

Obesity is an important health issue worldwide, particularly in the United States where the prevalence of obesity has increased substantially over the last 2 decades. In 2005, among the total U.S. adult population surveyed, 60.5% were overweight, 23.9% were obese, and 3.0% were extremely obese. Research has shown that obesity increases the risk of developing a number of conditions including type 2 diabetes (T2D), hypertension, coronary heart disease, ischemic stroke, colon cancer, post-menopausal breast cancer, endometrial cancer, gall bladder-disease, osteoarthritis, and obstructive sleep apnea.

In spite of a heavy burden imposed by the obesity epidemic, and in the face of a massive effort from various pharmaceutical companies, no reliable weight-reducing drug is yet available. While many studies have highlighted

*Correspondence: Division of Cellular and Molecular Therapy, Dept. of Pediatrics, Cancer & Genetics Research Complex, University of Florida, 1376 Mowry Rd, PO Box 103610.
E-mail: szlt@ufl.edu

the importance of environmental and behavioral modifications to treat obesity, little has been written about the possibility of genetic approaches.

2. Understanding Obesity: Why We Get Fat

At first glance, the rising rate of obesity seems to be simply a consequence of modern life's access to large amounts of palatable, high calorie food with limited physical activity. On the other hand, the less susceptible part of populace, exposed to the same environmental factors, remains lean and metabolically healthy therefore implicating genetic makeup as essential component of body weight (BW) maintenance. So what 'bears more weight': nature or nurture, genes or environment? Numerous studies have described complex environmental, behavioral, and genetic influences leading to a chronic imbalance favoring energy accumulation and excessive weight gain. But there is also growing evidence that suggests a significant contribution of epigenetic factors into the development of insulin resistance and obesity in childhood through adult life.

2.1 *Genetic Factors: Human Obesity Gene Map*

Progress in the human genome sequencing project and mapping of obesity genes have raised the possibility of selectively targeting this disease at the nucleic acid level. As of October 2005, 176 human obesity cases due to single-gene mutations in 11 different genes have been reported, 50 loci related to Mendelian syndromes relevant to human obesity have been mapped to a genomic region, and causal genes or strong candidates have been identified for most of these syndromes.[1] There are 244 genes that, when mutated or expressed as transgenes in the mouse, result in phenotypes that affect body weight and adiposity. The number of quantitative trait loci (QTLs) reported from animal models currently reaches 408. The number of human obesity QTLs derived from genome scans continues to grow, and there are 253 QTLs for obesity-related phenotypes from 61 genome-wide scans. A total of 52 genomic regions harbor QTLs supported by two or more studies. The obesity gene map shows putative loci on all chromosomes except Y. Recent progress in characterizing obesity-related genes, in and of itself, does not explain epidemic trends in recent history, for it is the environment, not the human genome that has undergone dramatic changes.

2.2 Environmental Factors: The Big Two and Other Causal Contributors

There is a consensus that recent environmental changes are almost certainly responsible for the obesity epidemic. The most frequently mentioned factors include (1) food marketing practices technology, and (2) institution-driven reductions in physical activity — the "Big Two".[2] A multitude of other environmental factors, coming into play during the last 3 decades, apparently contributed considerably into this trend. These factors include: (1) sleep deprivation; (2) endocrine disruptors (lipophilic, environmentally stable, industrially produced substances that can affect endocrine functions); (3) reduction in variability in ambient temperature; (4) decreased smoking; (5) pharmaceutical iatrogenesis (weight gain induced by many psychotropic medications, anticonvulsants, antidiabetics, antihypertensives, steroid hormones contraceptives, antihistamines, protease inhibitors, and selective antidepressants); (6) changes in distribution of ethnicity and age; (7) increasing gravida age (age of the first pregnancy); (8) intrauterine and intergenerational effects; (9) greater body-mass index (BMI) that is associated with greater reproductive fitness yielding selection for obesity-predisposing genotypes; (10) assortative, non-random mating for adiposity trait; (11) composition of gut microbiom. Although the effect of any one of the above factors may be small, the combined effects appear to be significant. Of course, considering any environmental factor, it is important to be aware that such factors act in concert with individual genetic susceptibilities described in the sections above.

3. General Strategies in Gene Therapy for Obesity

In the age of biotechnology we have experienced un-paralleled progress in gene delivery techniques and control of gene expression, but can we successfully apply these technologies to treat obesity? The general perception is that conventional gene therapy is not appropriate for multi-trait disorder such as obesity. The argument is that even if one can achieve technical excellence in delivering and regulating genetic information in particular cell types or tissues, there are too many deregulated genes to balance, and that the subset of these genes are going to be different in each particular case of obesity anyway. Yet, a counter-argument could be made that precisely because of the complex cumulative nature of this disorder manifesting in variable expressivity and variegation of many genes, one needs to modulate only a limited number of shared pathways, or maybe even a single gene, as long as this particular gene occupies a key position in a crucial metabolic pathway. One might picture a recognizable energy scale balancing two cups with hundreds of genes combining for energy intake in one cup while as many genes acting together for energy expenditure in the opposite cup. To shift a balance in the desired direction one needs to either 'reduce the weight' in one cup by inhibiting the action of any gene in this pool or, conversely, induce the expression of any other gene in the opposite pool. In other words, regardless of the underlying causes, the treatment (gene target) could be a general one. In fact, a short list of such genes has been recently described by Tiffin *et al.*[3] The investigators reviewed seven independent computational disease gene prioritization methods, and then applied them in concert to the analysis of 9556 positional candidate genes for T2DM and the related trait obesity. As a result, a list of nine primary candidate genes for T2DM and five genes for obesity had been generated. Two genes, lipoprotein lipase precursor (LPL) and branched-chain α-keto acid dehydrogenase (BCKDHA), are common to these two sets.

What are the best gene targets among various genes associated with this complex disorder? In general, genes regulating anabolic (inducing conservation and uptake of energy) and catabolic (promoting energy expenditure and decreasing FI) pathways are legitimate targets. For example, to reduce BW one would aim to downregulate the activity of the former (e.g. by utilizing siRNAs) and/or upregulate the latter (either by exogenous gene delivery or by control of endogenous gene). Furthermore, the mechanism of action of the targeted gene may require central (brain) or peripheral administration of a vector. The hypothalamus, a satiety center within the brain, is an attractive target because of its involvement in the integration of peripheral metabolic signals. In a clinical setting, however, the choice will most likely involve targeting peripheral organs involved in energy metabolism (gut, liver, muscle, or fat).

4. Gene Delivery Vehicles

Excess adipose tissue accumulates over long periods of time, making obesity a chronic condition that discourages physical activity and energy expenditure further promoting the vicious cycle of weight gain. If gene therapy is to be successful, the expression of a therapeutic transgene should be persistent for as long as the condition persists. Hence the treatment should provide a long-term mode of transgene expression either through the integration into host cell chromosome, or, preferably, via persistent episomal expression. Among current gene delivery systems, only viral vectors can fulfill this requirement.

Viruses are naturally evolved vehicles which efficiently transfer their genes into host cells. This ability makes them particularly useful for the delivery of therapeutic genes as engineered vectors. Viral vectors for gene transfer are the topic of several chapters in this book, so the experimental examples provided here will highlight only recombinant Adeno-associated virus (rAAV) vectors for the treatment of obesity and provide a rationale for the use of rAAV.

5. Gene Targets for Obesity

The following selected list of genes does not represent a comprehensive catalog of potential targets, but reflects only the contributing author's experimental data and bias. It is safe to assume that many more gene targets will be tested and, hopefully, applied in clinical practice in the future.

5.1 *Leptin*

Despite the problematic issues related to leptin resistance, two applications may be appropriate for peripheral administration of leptin gene therapy: (1) congenital leptin deficiency, a rare disorder of morbid obesity; (2) lipodystrophic syndrome, an adiposopathy associated with low leptin levels. In both cases, clinical trials of leptin protein replacement therapy have been very successful. Leptin treatment of these patients improved metabolic abnormalities such as insulin resistance, hyperglycemia, hyperinsulinemia, dyslipidemia and hepatic steatosis.[4] Intramuscular administration of rAAV encoding the leptin gene could provide long-term regulation of leptin expression[5] and replace daily injectable drug regimens. For most obese patients, however, leptin therapy does not represent a viable option requiring the development of alternative strategies.

5.2 *Neurocytokines*

One strategy is to by-pass the leptin signaling pathway altogether. This could be accomplished utilizing CNTF, a cytokine belonging to the Interleukin 6 (IL6) family. Acute treatment with CNTF reduced the obesity-related phenotype of ob/ob and db/db mice which lack functional leptin and leptin receptor, respectively.

Since CNTF is under investigation in clinical trials, it is critical to study the long term effects of its action on the brain where it exerts anorexigenic effect. This was accomplished in DIO rat model using centrally administered rAAV encoding CNTF, or leukemia inhibitory factor (LIF), another neurocytokine belonging to the IL6 family.[6] Using DNA microarray analysis of gene expression in the hypothalamus, the authors presented evidence that constitutive expression of cytokines in the brain evokes a state of perceived chronic inflammation leading to either temporal weight reduction (CNTF) or severe cachexia (LIF). Neither of these neurocytokines appears to fulfill the requirements of a sustained (CNTF) and safe (LIF) therapy. These results convey a cautionary note regarding long-term use of neurocytokines in therapeutic applications.

5.3 *AMP-Activated Protein Kinase (AMPK)*

One of the genes activated downstream of leptin signaling in skeletal muscle is AMPK, a "master switch" that mediates a majority of metabolic functions. AMPK is also the downstream effector of a protein kinase cascade that is switched on by increases in the AMP:ATP ratio. Once activated, AMPK switches on catabolic pathways that generate ATP while switching off ATP-consuming processes. AMPK appears to be a key regulator in controlling metabolism in response to diet and exercise. This enzyme also regulates food intake and energy expenditure at the whole body level by mediating the effects of hormones and cytokines such as leptin, adiponectin and ghrelin. Complex multi-domain protein structure precludes its upregulation via transgene vector delivery making AMPK unlikely as a direct gene therapy target. Modifying the upstream steps of the signaling cascade could provide an alternative approach.

5.4 *Adiponectin*

One example of upstream targeting of the AMPK signaling cascade was demonstrated in the case of gene adiponectin. This hormone, also known as Acrp30, is a polypeptide hormone secreted by adipose tissue that shows insulin-sensitizing, anti-inflammatory, and anti-atherogenic properties.[7,8] In contrast to leptin, the expression of adiponectin is reduced in obese and diabetic mice, and plasma levels of Acrp30 are lower in obese compared to lean humans. Similarly, Acrp30 replenishment decreases body adiposity and improves insulin sensitivity in various models of obesity. The mechanism by which Acrp30 ameliorates insulin sensitivity and improves glucose metabolism remains obscure, but evidence suggests that Acrp30 increases fatty acid oxidation in muscle and liver, and decreases hepatic glucose production. When rAAV vector harboring Acrp30 gene was administered intraportally, ectopically expressed adiponectin counteracted the development of diet-induced obesity (DIO) and ameliorated insulin sensitivity for 41 weeks.[9] Experiments elucidating the mechanism of this sustained weight loss suggested that transgene adiponectin regulated hepatic lipogenesis and gluconeogenesis, dichotomizing downstream at the point of AMPK.

5.5 *Wnt-10b*

Adipose tissue is derived from the embryonic mesenchyme. In adult organisms, a population of mesenchymal progenitor stem cells (MSCs) can be induced to multilineage differentiation including adipo-, osteo-, neuro-, and myogenesis. Surprisingly, even mature human adipocytes, when cultured *in vitro*, can de-differentiate into fibroblasts, which, upon expansion, can be turned into lipid-synthesizing adipocytes again. These facts demonstrate amazing flexibility and plasticity of adipose tissue providing hope for a therapeutic, rather than surgical, reduction in fat mass.

At first glance inhibition of adipogenesis is an inappropriate approach to anti-obesity therapy. Indeed, enhanced adipogenesis cannot cause obesity. Adiposity represents increased energy storage, the result of an imbalance between energy intake and energy output. Blocking differentiation of pre-adipocytes into adipocytes does not alter this fundamental equation. Enhanced adipogenesis in obesity is the result of energy imbalance, not the cause. Adipocytes provide a safe place to store lipids; when these cells are absent, as in lipodystrophy, lipids accumulate in muscle, liver and other locations. This is believed to cause significant metabolic derangement, including insulin resistance and hepatosteatosis that leads to cirrhosis.

Following this rationale, it appears that any therapy directed towards blocking adipocyte differentiation will be counterproductive. And yet, the *in vivo* experimental data obtained in this author's laboratory contradicts conventional wisdom. In one study, the long-term metabolic consequences of the upregulated Wnt/β-catenin signaling in skeletal muscles of adult, DIO rats has been investigated.[10] The long-term expression of rAAV1-Wnt10b was tested after intramuscular injection in female DIO rat. Animals fed high-fat diet and treated with rAAV1-Wnt10b showed a sustained reduction of 16% in BW accumulation compared to controls, and expression of Wnt10b was accompanied by a reduction in hyperinsulinemia and triglycerides plasma levels, as well as improved glucose homeostasis.

5.6 Obesity Gene Menu à la Carte

Considering the polygenic character of obesity, the task of picking the most promising and efficient candidate target gene appears to be overwhelming. Nevertheless, the contributing author's experimental data points at the following list of potential promising gene targets: (i) Stearoyl-CoA Desaturase (SCD1); (ii) Anorexigenic gut peptides Oxyntomodulin and peptide tyrosine–tyrosine (PYY); and (iii) Fibroblast Growth Factors (FGF-19 and FGF-21). In addition, computational methods employed data from a variety of sources helped to identify the most likely candidate disease genes from vast gene sets. For example, Tiffin *et al.* utilized seven independent computational disease gene prioritization methods, and then applied them in concert to the analysis of 9556 positional candidate genes for type 2 diabetes (T2D) and the related trait obesity.[3] The analysis exploited the premise that genes selected by the most independent methods are least likely to be false positives or artefacts of the type of approach used. This study generated a list of 9 genes selected as potential T2D genes: four of the nine genes located in the mitochondrion (BCKDHA, OAT, ACAA2, ECHS1);

some were involved in the metabolism of fatty acids (ACAA2, ECHS1, LPL), lipids (LPL and ACAA2), amino acids (BCKDHA and OAT), and glycogen and glucose (PRKCSH and PGM1). Using the same approach, a total of five genes were selected as most likely candidate genes for obesity. Two of these, LPL and BCKDHA, overlap with the set of most likely T2D candidates, and the additional three selected as candidates for obesity only are CAT, NEU1 and VLDLR.

The described list of potential obesity related gene targets is not complete and will be undoubtedly expand in the near future. Embracing gene-based therapeutics will require overcoming many obstacles. It will take a significant effort from the scientific community to educate public and physicians alike on the molecular mechanisms of obesity and diabetes while encouraging behavioral compliance with healthier food choices, and promoting routine physical exercise.

5.7 *Obesity and Diabetes*

This chapter has discussed obesity and described the disease diabetes as a related consequence. Chapter XVIII on autoimmune diseases also describes a form of diabetes. It is important to put these two forms of diabetes in context. The word diabetes comes from ancient Greeks and means to "pass through", referring to the symptom of excessive urination. This is then subcategorized into two forms of diabetes, firstly diabetes insipidus, which as the name implies refers to "insipid" urine, and this is caused by an imbalance in antidiuretic hormone and not part of this discussion. Secondly we have diabetes mellitus which has many underlying causes.[11] Mellitus comes from Latin meaning "honey" and hence refers to glucose in the urine. The two major forms of diabetes mellitus are now referred to as Type 1 Diabetes and Type 2 diabetes (referred to earlier in the Chapter XVIII and this chapter as T1D and T2D). The other forms are rarer and related to either complications of other diseases (e.g. cystic fibrosis-related diabetes) or to specific genetic defects (e.g. mitochondrial genes in maturity onset diabetes of youth). These are all covered in the American Diabetes Associations publications.[11]

So while the diagnosis of diabetes mellitus essentially means that blood glucose levels are elevated, for T1D insulin is an absolute requirement because the underlying autoimmune destruction has all but eliminated the body's ability to produce insulin and for T2D there is effectively a resistance to the action of insulin and insulin may only be required later in the disease when other agents are failing.

Some of the long term complications for both T1D and T2D as mentioned before may be amenable to similar therapeutic approaches including gene therapy, but either preventing or treating the underlying causes will be quite different. In summary T1D is an autoimmune disease and this aspect has to be an integral part of therapy as discussed in Chapter XVIII. T2D has obesity and genetics as the central cause and these aspects need to be addressed as outlined earlier in this chapter.

References

1. Rankinen T, *et al.* (2006). The human obesity gene map: The 2005 update. *Obesity (Silver Spring)* **14**: 529–644.
2. Keith SW, *et al.* (2006). Putative contributors to the secular increase in obesity: Exploring the roads less traveled. *Int J Obes (Lond)*.
3. Tiffin N, *et al.* (2006). Computational disease gene identification: A concert of methods prioritizes type 2 diabetes and obesity candidate genes. *Nucleic Acids Res* **34**: 3067–3081.
4. Gorden P, Gavrilova O (2003). The clinical uses of leptin. *Curr Opin Pharmacol* **3**: 655–659.
5. Murphy JE, Zhou S, Giese K, Williams LT, Escobedo JA, Dwarki VJ (1997). Long-term correction of obesity and diabetes in genetically obese mice by a single intramuscular injection of recombinant adeno-associated virus encoding mouse leptin. *Proc Natl Acad Sci USA* **94**: 13921–13926.
6. Prima V, Tennant M, Gorbatyuk OS, Muzyczka N, Scarpace PJ, Zolotukhin S (2004). Differential modulation of energy balance by leptin, ciliary neurotrophic factor, and leukemia inhibitory factor gene delivery: Microarray deoxyribonucleic acid-chip analysis of gene expression. *Endocrinology* **145**: 2035–2045.
7. Scherer PE, Williams S, Fogliano M, Baldini G, Lodish HF (1995). A novel serum protein similar to C1q, produced exclusively in adipocytes. *J Biol Chem* **270**: 26746–26749.
8. Tsao TS, Lodish HF, Fruebis J (2002). ACRP30, a new hormone controlling fat and glucose metabolism. *Eur J Pharmacol* **440**: 213-221.
9. Shklyaev S, *et al.* (2003). Sustained peripheral expression of transgene adiponectin offsets the development of diet-induced obesity in rats. *Proc Natl Acad Sci USA* **100**: 14217–14222.
10. Aslanidi G, *et al.* (2007). Ectopic expression of Wnt10b decreases adiposity and improves glucose homeostasis in obese rats. *Am J Physiol* (in press).
11. (2009). Diagnosis and classification of diabetes mellitus. *Diabetes Care* **32(Suppl 1)**: S62–S67.

Chapter 16
Gene Therapy for Duchenne Muscular Dystrophy

Takashi Okada* and Shin'ichi Takeda

Duchenne muscular dystrophy (DMD) is caused by various distinct mutations, ranging from point mutations to large deletions, in the *dystrophin* gene. These mutations have led to a variety of therapeutic modalities for muscular dystrophy, including gene replacement, gene correction, and modification of the gene product. Gene replacement therapy provides an impersonal approach for treating DMD. Adeno-associated virus (AAV) vector-mediated truncated *micro-dystrophin* gene delivery has been successful in some animal models of DMD. However, recent evidence of immune-mediated loss of vector persistence in dogs and humans suggests that immune modulation might be necessary to achieve successful long-term transgene expression in these species. In this chapter, we focus on the methods that have been developed for gene replacement therapy using vectors based on the AAV.

1. Introduction

1.1 *Background of Duchenne Muscular Dystrophy*

Duchenne muscular dystrophy (DMD) is the most common form of childhood muscular dystrophy. While the focus of this chapter is gene therapy for DMD, it should be pointed out that substantial progress has also been made toward gene therapy for other muscular dystrophies. For example,

*Correspondence: Department of Molecular Therapy, National Institute of Neuroscience, National Center of Neurology and Psychiatry, 4-1-1 Ogawa-Higashi, Kodaira, Tokyo 187-8502, Japan.
E-mail: t-okada@ncnp.go.jp

sustained partial restoration of α-sarcoglycan deficiency in skeletal muscle has been reported in patients with a severe form of this disorder.[1]

DMD is an X-linked recessive disorder with an incidence of 1 in 3500 live male births.[2] DMD causes progressive degeneration and regeneration of skeletal and cardiac muscles due to mutations in the *dystrophin* gene, which encodes a 427-kDa subsarcolemmal cytoskeletal protein.[3] DMD is associated with severe progressive muscle weakness and typically leads to death between the ages of 20 and 35 years. Due to recent advances in respiratory care, much attention is now focused on treating the cardiac conditions suffered by DMD patients.

The approximately 2.5-megabase *dystrophin* gene is the largest gene identified to date, and because of its size, it is susceptible to a high sporadic mutation rate. Absence of dystrophin and the dystrophin-glycoprotein complex (DGC) from the sarcolemma leads to severe muscle wasting (Fig. 16.1). DMD is characterized by the absence of functional protein (in contrast to Becker muscular dystrophy, which is commonly caused by in-frame deletions of the *dystrophin* gene resulting in the synthesis of a partially functional protein).

2. Gene-replacement Strategies using Virus Vectors

2.1 *Choice of Vector*

(i) AAV vectors: Successful therapy for DMD requires the restoration of dystrophin protein in skeletal and cardiac muscles. While various viral vectors have been considered for the delivery of genes to muscle fibers, the adeno-associated virus (AAV)-based vector is emerging as the gene transfer vehicle with the most potential for use in DMD gene therapies. The advantages of the AAV vector include the lack of disease associated with a wild-type virus, the ability to transduce non-dividing cells, and the long-term expression of the delivered transgenes.[5] Serotypes 1, 6, 8 and 9 of recombinant AAV (rAAV) exhibit a potent tropism for striated muscles.[6] Since a 5 kb genome is considered to be the upper limit for a single AAV virion, a series of rod-truncated micro-dystrophin genes is used in this treatment.[7]

(ii) Adenovirus vectors are efficient delivery systems of episomal DNA into eukaryotic cell nuclei.[8] The utility of adenovirus vectors has been increased by capsid modifications that alter tropism, and by the generation

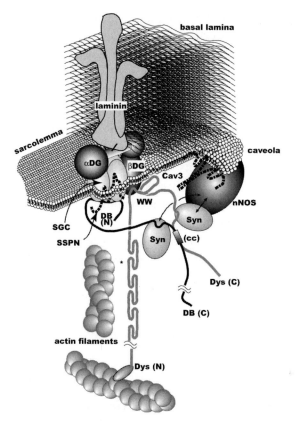

Fig. 16.1 Dystrophin-glycoprotein complex. Molecular structure of the dystrophin-glycoprotein complex and related proteins superimposed on the sarcolemma and subsarcolemmal actin network (redrawn from Yoshida *et al.*,[4] with modifications). cc, coiled-coil motif on dystrophin (Dys) and dystrobrevin (DB); SGC, sarcoglycan complex; SSPN, sarcospan; Syn, syntrophin; Cav3, caveolin-3; N and C, the N and C termini, respectively; G, G-domain of laminin; asterisk indicates the actin-binding site on the dystrophin rod domain; WW, WW domain.

of hybrid vectors that promote chromosomal insertion.[9] Also, gutted adenovirus vectors devoid of all adenoviral genes allow for the insertion of large transgenes, and trigger fewer cytotoxic and immunogenic effects than do those only deleted in the E1 regions (from bases 343 to 2270).[10]

(iii) Human artificial chromosomes (HACs) have the capacity to deliver a large gene (roughly 6-10 megabases) into host cells without integrating the gene into the host genome, thereby preventing the possibility of insertional mutagenesis and genomic instability.[11]

(iv) Vectors for targeted integration: A goal in clinical gene therapy is to develop gene transfer vehicles that can integrate exogenous therapeutic genes at specific chromosomal loci, so that insertional oncogenesis is prevented. AAV can insert its genome into a specific locus, designated AAVS1, on chromosome 19 of the human genome.[12] The AAV Rep78/68 proteins and the Rep78/68-binding sequences are the trans- and cis-acting elements needed for this reaction. A dual high-capacity adenovirus-AAV hybrid vector with full-length human dystrophin-coding sequences flanked by AAV integration-enhancing elements was tested for targeted integration.[13]

(v) Gene correction is a process whereby sequence alterations in genes can be corrected by homologous recombination-mediated gene conversion between the recipient target locus and a donor construct encoding the correct sequence.[14] The introduction of a corrective sequence together with a site-specific nuclease to induce a double-stranded break (DSB) at sites responsible for monogenic disorders would activate gene correction. Pairs of designated zinc-finger protein with tandem DNA binding sites fused to the cleavage domain of the Fok1 protein were introduced into model systems or cell lines and produced corrections in 10–30% of cases tested.[15]

2.2 *Modification of the Dystrophin Gene and Promoter*

Due to the large deletion in its genome, the gutted adenovirus vector can package 14 kb of full-length *dystrophin* cDNA. Multiple proximal muscles of seven-day-old utrophin/dystrophin double knockout mice (*dko* mice), which typically show symptoms similar to human DMD, were effectively transduced with the gutted adenovirus bearing full-length murine *dystrophin* cDNA.[16] However, further improvements are needed to regulate the virus-associated host immune response before clinical trials can be performed.

A series of truncated *dystrophin* cDNAs containing rod repeats with hinge 1, 2, and 4 were constructed (Fig. 16.2).[7] Although AAV vectors are too small to package the full-length *dystrophin* cDNA, AAV vector-mediated gene therapy using a rod-truncated *dystrophin* gene provides a promising approach.[17] The structure and, particularly, the length of the rod are crucial for the function of micro-dystrophin.[18] An AAV type 2 vector expressing

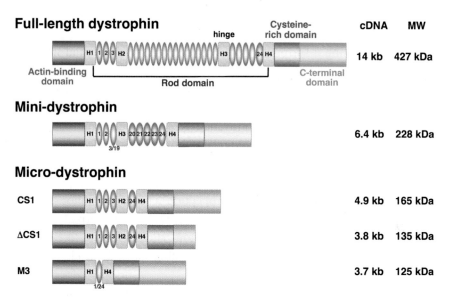

Fig. 16.2 Structures of full-length and truncated dystrophin. Helper-dependent adenovirus vector can package 14 kb of full-length dystrophin cDNA because of the large-sized deletion in its genome. A mini-dystrophin is cloned from a patient with Becker muscular dystrophy, which is caused by in-frame deletions resulting in the synthesis of partially functional protein. A series of truncated micro-dystrophin cDNAs harboring only four rod repeats with hinge 1, 2, and 4 (CS1); the same components, except that the C-terminal domain is deleted (delta CS1); or one rod repeat with hinge 1 and 4 (M3), are constructed to be packaged in the AAV vector.

micro-dystrophin (DeltaCS1) under the control of a muscle-specific MCK promoter was injected into the tibialis anterior (TA) muscles of dystrophin-deficient *mdx* mice,[19] and resulted in extensive and long-term expression of micro-dystrophin that exhibited improved force generation.

Codon usage may impact transgene expression. For example, expression and function of micro-dystrophin was investigated in the *mdx* mouse using two different configurations of the gene under the control of a muscle-restrictive promoter (Spc5-12) and optimized for codon usage.[20] Codon optimization of micro-dystrophin significantly increased micro-dystrophin mRNA and protein levels after intramuscular and systemic administration of plasmid DNA or rAAV8. By randomly assembling myogenic regulatory elements into synthetic promoter recombinant libraries, several artificial promoters were isolated whose transcriptional potencies greatly exceed those of natural myogenic and viral gene promoters.[21]

2.3 Use of Surrogate Genes

An approach using a surrogate gene would bypass the potential immune responses associated with the delivery of exogenous dystrophin. Methods to increase expression of utrophin, a dystrophin paralog, show promise as a treatment for DMD. AAV-6 vector harboring a murine codon-optimized micro-utrophin transgene was intravenously administered into adult *dko* mice to alleviate the pathophysiological abnormalities.[22] The paralogous gene efficiently acted as a surrogate for *dystrophin*. Myostatin has been extensively documented as being a negative regulator of muscle growth. Systemic gene delivery of myostatin propeptide, a natural inhibitor of myostatin, enhanced body-wide skeletal muscle growth in both normal and *mdx* mice.[23] The delivery of various growth factors, such as insulin-like growth factor-I (IGF-I), has been successful in promoting skeletal muscle regeneration after injury.[24]

Matrix metalloproteinases (MMPs) may also be useful in improving muscle regeneration. MMPs are key regulatory molecules in the formation, remodeling and degradation of all extracellular matrix (ECM) components in pathological processes. MMP-9 is involved predominantly in the inflammatory process during muscle degeneration.[25] In contrast, MMP-2 is associated with ECM remodeling during muscle regeneration and fiber growth.

3. AAV-Mediated Transduction of Animal Models

3.1 Vector Production

A potential hurdle for treatment of multiple muscle groups using viral gene transfer is the required large amount of vector and the related task of scaling up production. To gain acceptance as a medical treatment with a dose of over 1×10^{13} genome copies (g.c.)/kg body weight, AAV vectors require a scalable and economical production method. An example for a current production method using a helper virus-free transfection system is shown in Fig. 16.3 (see also Chapter 6). Alternatives include packaging cell lines expressing Rep/Cap with regulated Rep/Cap expression, which is complicated by toxic effects of early expression of Rep proteins.[26] A scalable method, using active gassing and large culture vessels, was developed to

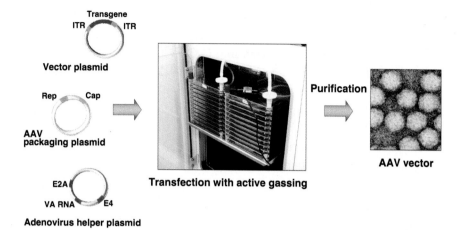

Fig. 16.3   A scalable triple plasmid transfection system using active gassing. When adenovirus helper plasmid is co-transfected into human embryonic kidney 293 cells along with a vector plasmid encoding the AAV vector and an AAV packaging plasmid harboring *rep-cap* genes, the AAV vector is produced as efficiently as when using adenovirus infection. A large-scale transduction method to produce AAV vectors with an active gassing system makes use of large culture vessels for labor- and cost-effective transfection in a closed system. Samples containing vector particles are further purified with a two-tier CsCl gradient or dual ion-exchange chromatography to obtain highly purified vector particles.

transfect rAAV in a closed system, in a labor- and cost-effective manner.[27] This vector production system achieved a yield of more than 5×10^{13} g.c./flask by improving gas exchange to maintain the physiological pH in the culture medium. Recent developments also suggest that AAV vector production in insect cells would be compatible with current good manufacturing practice production on an industrial scale.[28]

3.2 *Animal Models for the Gene Transduction Study*

Dystrophin-deficient canine X-linked muscular dystrophy was found in a golden retriever with a 3′ splice-site point mutation in intron 6.[29] The clinical and pathological characteristics of dystrophic dogs are more similar to those of DMD patients than are those of *mdx* mice. A beagle-based model of canine X-linked muscular dystrophy, which is smaller and easier to handle than the golden retriever-based muscular dystrophy dog (GRMD) model, has been established in Japan, and is referred to as CXMD_J.[30] The limb and temporal muscles of CXMD_J dogs are affected by two months of age, corresponding to the second peak of serum creatine kinase. Interestingly, extensive lymphocyte-mediated immune responses to rAAV2-*lacZ* occurred after direct intramuscular injection into CXMD_J dogs, despite successful delivery of the same viral construct into mouse skeletal muscle.[31] In contrast to rAAV2, rAAV8-mediated transduction of canine skeletal muscles produced significantly higher transgene expression with less lymphocyte proliferation than rAAV2.[32]

It is increasingly important to develop strategies to treat DMD that consider the effect on cardiac muscle. The pathology of the conduction system in CXMD_J was analyzed to establish the therapeutic target for DMD.[33] Although dystrophic changes of the ventricular myocardium were not evident at the age of 1 to 13 months, Purkinje fibers showed remarkable vacuolar degeneration when dogs were as young as four-months-old. Furthermore, degeneration of Purkinje fibers was coincident with overexpression of Dp71 at the sarcolemma. The degeneration of Purkinje fibers could be associated with the distinct deep Q waves present in ECGs and the fatal arrhythmias seen in cases of dystrophin deficiency.[33]

3.3 Immunological Issues of rAAV

Neo-antigens introduced by AAV vectors evoke significant immune reactions in DMD muscle, since increased permeability of sarcolemma allows leakage of the transgene products from the dystrophin-deficient muscle fibers.[34] AAV2-mediated transfer into skeletal muscles of normal dogs resulted in low and transient expression, together with intense cellular infiltration, and the marked activation of cellular and humoral immune responses.[31] Furthermore, an *in vitro* interferon-gamma release assay showed that canine splenocytes respond to immunogens or mitogens more strongly than do murine splenocytes. In fact, co-administration of immunosuppressants, cyclosporine (CSP) and mycophenolate mofetil (MMF) improved rAAV2 transduction. The AAV2 capsids can induce a cellular immune response via MHC class I antigen presentation with a cross-presentation pathway, and rAAV2 has also been proposed to have an effect on human dendritic cells (DCs). In contrast, other serotypes, such as rAAV8, induced T cell activation to a lesser degree.[32] Immunohistochemical analysis revealed that the rAAV2-injected muscles showed higher rates of infiltration of CD4$^+$ and CD8$^+$ T lymphocytes in the endomysium than the rAAV8-injected muscles.[32]

Resident antigen-presenting cells, such as DCs, myoblasts, myotubes and regenerating immature myofibers, might play a role in the immune response. A recent study also showed that mRNA levels of MyD88 and co-stimulating factors, such as CD80, CD86 and type I interferon, are elevated in both rAAV2- and rAAV8-transduced dog DCs *in vitro*.[32] A brief course of immunosuppression with a combination of anti-thymocyte globulin (ATG), CSP and MMF was effective in permitting AAV6-mediated, long-term and robust expression of a canine micro-dystrophin in the skeletal muscle of a dog DMD model.[35]

3.4 *Intravascular Vector Administration by Limb Perfusion*

Although recent studies suggest that vectors based on AAV are capable of body-wide transduction in rodents, translating this finding into large animals remains a challenge. Intravascular delivery can be performed as a form of limb perfusion, which might bypass the immune activation of DCs in the injected muscle.[32] Our laboratory performed limb perfusion-assisted intravenous administration of rAAV8-micro-dystrophin into the hind limb of CXMD$_J$ dogs resulting in extensive transgene expression in the distal limb muscles without obvious immune responses for as long as eight weeks after injection (Fig. 16.4).[32]

3.5 *Global Muscle Therapies*

In comparison with fully dystrophin-deficient animals, targeted transgenic repair of skeletal muscle, but not cardiac muscle, paradoxically elicits a five-fold increase in cardiac injury and dilated cardiomyopathy.[36] Because the dystrophin-deficient heart is highly sensitive to increased stress, increased activity by the repaired skeletal muscle provides the stimulus for heightened cardiac injury and heart remodeling. In contrast, a single intravenous injection of AAV9 vector expressing micro-dystrophin efficiently transduces the entire heart in neonatal *mdx* mice, thereby ameliorating cardiomyopathy.[37]

Since a number of muscular dystrophy patients can be identified through newborn screening, neonatal transduction may lead to an effective early intervention in DMD patients. After a single intravenous injection, robust

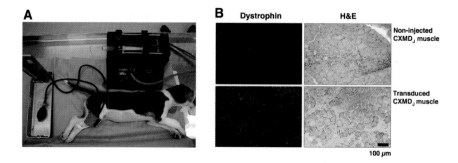

Fig. 16.4 Intravascular vector administration by limb perfusion. (A) A blood pressure cuff is applied just above the knee of an anesthetized CXMDɪ dog. A 24-gauge catheter is inserted into the lateral saphenous vein, connected to a three-way stopcock, and flushed with saline. With a blood pressure cuff inflated to over 300 mmHg, saline (2.6 ml/kg) containing papaverine (0.44 mg/kg) and heparin (16 U/kg) is injected by hand over a 10 second period. The three-way stopcock is connected to a syringe containing rAAV8 expressing micro-dystrophin (1×10^{14} vg/kg, 3.8 ml/kg). The syringe is placed in a PHD 2000 syringe pump. Five minutes after the papaverine/heparin injection, rAAV8 is injected at a rate of 0.6 ml/sec. (B) Administration of rAAV8-micro-dystrophin by limb perfusion produces extensive transgene expression in the distal limb muscles of CXMDɪ dogs without obvious immune responses at four weeks after injection. Modified from *Mol Ther* **17**: 73–80, 2008.

skeletal muscle transduction with AAV9 vector throughout the body was observed in neonatal dogs.[38] Systemic transduction was achieved in the absence of pharmacological intervention or immune suppression and lasted for at least six months, whereas cardiac muscle was barely transduced in the dogs.

4. Safety and Potential Impact of Clinical Trials

The initial clinical studies have laid the foundation for future studies, providing important information about vector dose, viral serotype selection, and immunogenicity in humans. The first virus-mediated gene transfer for muscle disease was carried out for limb-girdle muscular dystrophy type 2D using rAAV1. The study, consisting of intramuscular injection of virus into a single muscle, was discharged to establish the safety of this procedure in phase I clinical trials.[39] The first clinical gene therapy trial for DMD began in March 2006.[39] This was a phase I/IIa study in which an AAV vector was used to deliver micro-dystrophin to the biceps of boys with DMD. The study was conducted on six boys with DMD, each of whom was transduced with mini-dystrophin genes in a muscle of one arm in the absence of serious adverse events.

While low immunogenicity was considered a major strength supporting the use of rAAV in clinical trials, a number of observations have recently provided a more balanced view of this procedure.[40] An obvious barrier to AAV transduction is the presence of circulating neutralizing antibodies that prevent the virion from binding to its cellular receptor.[41] This potential threat can be reduced by prescreening patients for AAV serotype-specific neutralizing antibodies or by performing procedures such as plasmapheresis before gene transfer. Another challenge recently revealed is the development of a cytotoxic T cell (CTL) response to AAV capsid peptides. In the human factor IX gene therapy trial in which rAAV was delivered to the liver, only short-term transgene expression was achieved and levels of therapeutic protein declined to baseline levels 10 weeks after vector infusion.[40] This was accompanied by elevation of serum transaminase levels and a CTL response toward specific AAV capsid peptides. To overcome this response, transient immunosuppression may be required until AAV capsids are completely cleared. Additional findings suggest that T cell activation requires AAV2 capsid binding to the heparan sulfate proteoglycan (HSPG) receptor, which would permit virion shuttling into a DC pathway, as cross-presentation.[42] Exposure to vectors from other AAV clades, such as AAV8, did not activate capsid-specific T cells.

5. Development of Alternative Strategies

5.1 *Design of Read-through Drugs*

To suppress premature stop codon mutations, treatments involving aminoglycosides and other agents have been attempted. PTC124, a novel drug capable of suppressing premature termination and selectively inducing ribosomal read-through of premature, but not normal, termination codons, was recently identified using nonsense-containing reporters.[43] The selectivity of PTC124 for premature termination codons, its oral bioavailability and its pharmacological properties indicate that this drug may have broad clinical potential for the treatment of a large group of genetic disorders with limited or no therapeutic options.

5.2 *Modification of mRNA Splicing*

By inducing the skipping of specific exons during mRNA splicing, antisense compounds against exonic and intronic splicing regulatory sequences were shown to correct the open reading frame of the DMD gene and thus to restore truncated yet functional dystrophin expression *in vitro*.[44] Intravenous infusion of an antisense phosphorothioate oligonucleotide created an in-frame *dystrophin* mRNA via exon skipping in a 10-year-old DMD patient possessing an out-of-frame exon 20 deletion of the *dystrophin* gene.[45] Moreover, the adverse-event profile and local dystrophin-restoring effect of a single intramuscular injection of an antisense 2'-O-methyl phosphorothioate oligonucleotide, PRO051, in patients with DMD were explored.[46] Four patients received a dose of 0.8 mg of PRO051 in the TA muscle. Each patient showed specific skipping of exon 51 of dystrophin in 64 to 97% of myofibers, without clinically apparent adverse side effects.

The efficacy and toxicity of intravenous injection of stable morpholino phosphorodiamidate (morpholino)-induced exon skipping were tested using CXMD_J dogs, and widespread rescue of dystrophin expression to therapeutic levels was observed.[47] Furthermore, a morpholino oligomer with a designed cell-penetrating peptide can efficiently target a mutated *dystrophin* exon in cardiac muscles.[48]

Long-term benefits can be obtained through the use of viral vectors expressing antisense sequences against regions within the *dystrophin* gene. The sustained production of dystrophin at physiological levels in entire groups of muscles as well as the correction of muscular dystrophy were achieved by treatment with exon-skipping AAV1-U7.[49]

5.3 *Ex Vivo Gene Therapy*

Transplantation of genetically corrected autologous myogenic cells is a possible treatment for DMD. Freshly isolated satellite cells transduced with lentiviral vectors expressing micro-dystrophin were transplanted into the TA muscles of *mdx* mice, and these cells efficiently contributed to the regeneration of muscles with micro-dystrophin expression at the sarcolemma.[50] Mesoangioblasts are vessel-associated stem cells and might be candidates for future stem cell therapy for DMD.[51] Intra-arterial delivery of wild-type canine mesoangioblasts resulted in the extensive recovery of *dystrophin* expression, normal muscle morphology and function in the GRMD. Multipotent mesenchymal stromal cells (MSCs) are less immunogenic and have the potential to differentiate and display a myogenic phenotype.[52]

6. Future Perspectives

6.1 *Pharmacological Intervention*

The use of a histone deacetylase (HDAC) inhibitor would likely enhance the utility of rAAV-mediated transduction strategies in the clinic.[53] In contrast to adenovirus-mediated transduction, the improved transduction with rAAV induced by the HDAC inhibitor is due to enhanced transgene expression rather than to increased viral entry. Enhanced transduction may be related to the histone-associated chromatin form of the rAAV concatemer in transduced cells. Since various HDAC inhibitors are currently being tested in clinical trials for many diseases, the use of such agents in rAAV-mediated DMD gene therapy is theoretically and practically reasonable.

6.2 *Capsid Modification*

A DNA shuffling-based approach for developing cell type-specific vectors is an intriguing possibility to achieve altered tropism. Capsid genomes of AAV serotypes 1-9 were randomly reassembled using PCR to generate a chimeric capsid library.[54] A single infectious clone (chimeric-1829) containing genome fragments from AAV1, 2, 8, and 9 was isolated from an integrin minus hamster melanoma cell line previously shown to have low permissiveness to AAV.

7. Conclusions and Outlook

DMD remains an untreatable genetic disease that severely limits motility and life expectancy in affected children. The systemic delivery of rAAV to transduce truncated dystrophin is predicted to ameliorate the symptoms of DMD patients in the future. To translate gene transduction technologies into clinical practice, development of an effective delivery system with improved vector constructs as well as efficient immunological modulation must be established. A novel protocol that considers all of these issues would help improve the therapeutic benefits of DMD gene therapy.

References

1. Mendell JR, *et al.* (2009). LGMD 2D gene therapy restores alpha-sarcoglycan and associated proteins. *Annals Neurol*, in press.
2. Emery AE (1991). Population frequencies of inherited neuromuscular diseases–a world survey. *Neuromuscul Disord* **1**: 19–29.
3. Hoffman EP, Brown RH Jr., Kunkel LM (1987). Dystrophin: the protein product of the Duchenne muscular dystrophy locus. *Cell* **51**: 919–928.
4. Yoshida M, *et al.* (2000). Biochemical evidence for association of dystrobrevin with the sarcoglycan-sarcospan complex as a basis for understanding sarcoglycanopathy. *Hum Mol Genet* **9**: 1033–1040.
5. Okada T, *et al.* (2002). Adeno-associated viral vector-mediated gene therapy of ischemia-induced neuronal death. *Methods Enzymol* **346**: 378–393.
6. Inagaki K, *et al.* (2006). Robust systemic transduction with AAV9 vectors in mice: efficient global cardiac gene transfer superior to that of AAV8. *Mol Ther* **14**: 45–53.
7. Yuasa K, Miyagoe Y, Yamamoto K, Nabeshima Y, Dickson G, Takeda S (1998). Effective restoration of dystrophin-associated proteins *in vivo* by adenovirus-mediated transfer of truncated dystrophin cDNAs. *FEBS Lett* **425**: 329–336.
8. Okada T, Ramsey J, Munir J, Wildner O, Blaese M (1998). Efficient directional cloning of recombinant adenovirus vectors using DNA-protein complex. *Nucleic Acids Res* **26**: 1947–1950.
9. Okada T, *et al.* (2004). *In situ* generation of pseudotyped retroviral progeny by adenovirus-mediated transduction of tumor cells enhances the killing effect of HSV-tk suicide gene therapy *in vitro* and *in vivo*. *J Gene Med* **6**: 288–299.
10. Hammerschmidt DE (1999). Development of a gutless vector. *J Lab Clin Med* **134**: C3.
11. Hoshiya, H, *et al.* (2008). A highly Stable and Nonintegrated Human Artificial Chromosome (HAC) Containing the 2.4 Mb Entire Human Dystrophin Gene. *Mol Ther*.
12. Kotin RM, Linden RM, Berns KI (1992). Characterization of a preferred site on human chromosome 19q for integration of adeno-associated virus DNA by non-homologous recombination. *Embo J* **11**: 5071–5078.
13. Goncalves MA, *et al.* (2005). Transfer of the full-length dystrophin-coding sequence into muscle cells by a dual high-capacity hybrid viral vector with site-specific integration ability. *J Virol* **79**: 3146–3162.
14. Klug, A (2005). Towards therapeutic applications of engineered zinc finger proteins. *FEBS Lett* **579**: 892–894.
15. Porteus MH, Baltimore, D (2003). Chimeric nucleases stimulate gene targeting in human cells. *Science* **300**: 763.
16. Kawano R, Ishizaki M, Maeda Y, Uchida Y, Kimura E, Uchino M (2008). Transduction of full-length dystrophin to multiple skeletal muscles improves motor performance and life span in utrophin/dystrophin double knockout mice. *Mol Ther* **16**: 825–831.

17. Wang B, Li J, Xiao X (2000). Adeno-associated virus vector carrying human minidystrophin genes effectively ameliorates muscular dystrophy in mdx mouse model. *Proc Natl Acad Sci USA* **97**: 13714–13719.
18. Sakamoto M, *et al.* (2002). Micro-dystrophin cDNA ameliorates dystrophic phenotypes when introduced into mdx mice as a transgene. *Biochem Biophys Res Commun* **293**: 1265–1272.
19. Yoshimura M, *et al.* (2004). AAV vector-mediated microdystrophin expression in a relatively small percentage of mdx myofibers improved the mdx phenotype. *Mol Ther* **10**: 821–828.
20. Foster H, *et al.* (2008). Codon and mRNA sequence optimization of microdystrophin transgenes improves expression and physiological outcome in dystrophic mdx mice following AAV2/8 gene transfer. *Mol Ther* **16**: 1825–1832.
21. Li X, Eastman EM, Schwartz RJ, Draghia-Akli, R (1999). Synthetic muscle promoters: activities exceeding naturally occurring regulatory sequences. *Nat Biotechnol* **17**: 241–245.
22. Odom GL, Gregorevic P, Allen JM, Finn E, Chamberlain JS (2008). Microutrophin delivery through rAAV6 increases lifespan and improves muscle function in dystrophic dystrophin/utrophin-deficient mice. *Mol Ther* **16**: 1539–1545.
23. Qiao C, *et al.* (2008). Myostatin propeptide gene delivery by adeno-associated virus serotype 8 vectors enhances muscle growth and ameliorates dystrophic phenotypes in mdx mice. *Hum Gene Ther* **19**: 241–254.
24. Schertzer JD, Lynch GS (2006). Comparative evaluation of IGF-I gene transfer and IGF-I protein administration for enhancing skeletal muscle regeneration after injury. *Gene Ther* **13**: 1657–1664.
25. Fukushima K, *et al.* (2007). Activation and localization of matrix metalloproteinase-2 and -9 in the skeletal muscle of the muscular dystrophy dog (CXMDJ). *BMC Musculoskelet Disord* **8**: 54.
26. Okada T, *et al.* (2001). Development and characterization of an antisense-mediated prepackaging cell line for adeno-associated virus vector production. *Biochem Biophys Res Commun* **288**: 62–68.
27. Okada T, *et al.* (2005). Large-scale production of recombinant viruses by use of a large culture vessel with active gassing. *Hum Gene Ther* **16**: 1212–1218.
28. Cecchini S, Negrete A, Kotin RM (2008). Toward exascale production of recombinant adeno-associated virus for gene transfer applications. *Gene Ther* **15**: 823–830.
29. Valentine BA, Cooper BJ, de Lahunta A, O'Quinn R, Blue JT (1988). Canine X-linked muscular dystrophy. An animal model of Duchenne muscular dystrophy: clinical studies. *J Neurol Sci* **88**: 69–81.
30. Shimatsu Y, *et al.* (2005). Major clinical and histopathological characteristics of canine X-linked muscular dystrophy in Japan, CXMDJ. *Acta Myol* **24**: 145–154.
31. Yuasa K, *et al.* (2007). Injection of a recombinant AAV serotype 2 into canine skeletal muscles evokes strong immune responses against transgene products. *Gene Ther.*
32. Ohshima S, *et al.* (2008). Transduction Efficiency and Immune Response Associated With the Administration of AAV8 Vector Into Dog Skeletal Muscle. *Mol Ther.*
33. Urasawa N, *et al.* (2008). Selective vacuolar degeneration in dystrophin-deficient canine Purkinje fibers despite preservation of dystrophin-associated proteins with overexpression of Dp71. *Circulation* **117**: 2437–2448.

34. Yuasa K, *et al.* (2002). Adeno-associated virus vector-mediated gene transfer into dystrophin-deficient skeletal muscles evokes enhanced immune response against the transgene product. *Gene Ther* **9**: 1576–1588.

35. Wang Z, *et al.* (2007). Sustained AAV-mediated Dystrophin Expression in a Canine Model of Duchenne Muscular Dystrophy with a Brief Course of Immunosuppression. *Mol Ther* **15**: 1160–1166.

36. Townsend D, Yasuda S, Li S, Chamberlain JS, Metzger JM (2008). Emergent dilated cardiomyopathy caused by targeted repair of dystrophic skeletal muscle. *Mol Ther* **16**: 832–835.

37. Bostick B, Yue Y, Lai Y, Long C, Li D, Duan D (2008). Adeno-associated virus serotype-9 microdystrophin gene therapy ameliorates electrocardiographic abnormalities in mdx mice. *Hum Gene Ther* **19**: 851–856.

38. Yue Y, *et al.* (2008). A single intravenous injection of adeno-associated virus serotype-9 leads to whole body skeletal muscle transduction in dogs. *Mol Ther* **16**: 1944–1952.

39. Rodino-Klapac LR, Chicoine LG, Kaspar BK, Mendell JR (2007). Gene therapy for duchenne muscular dystrophy: expectations and challenges. *Arch Neurol* **64**: 1236–1241.

40. Manno CS, *et al.* (2006). Successful transduction of liver in hemophilia by AAV-Factor IX and limitations imposed by the host immune response. *Nat Med* **12**: 342–347.

41. Scallan CD, *et al.* (2006). Human immunoglobulin inhibits liver transduction by AAV vectors at low AAV2 neutralizing titers in SCID mice. *Blood* **107**: 1810–1817.

42. Vandenberghe LH, *et al.* (2006). Heparin binding directs activation of T cells against adeno-associated virus serotype 2 capsid. *Nat Med* **12**: 967–971.

43. Welch EM, *et al.* (2007). PTC124 targets genetic disorders caused by nonsense mutations. *Nature* **447**: 87–91.

44. Takeshima Y, Nishio H, Sakamoto H, Nakamura H, Matsuo M (1995). Modulation of *in vitro* splicing of the upstream intron by modifying an intra-exon sequence which is deleted from the dystrophin gene in dystrophin Kobe. *J Clin Invest* **95**: 515–520.

45. Takeshima Y, *et al.* (2006). Intravenous infusion of an antisense oligonucleotide results in exon skipping in muscle dystrophin mRNA of Duchenne muscular dystrophy. *Pediatr Res* **59**: 690–694.

46. van Deutekom JC, *et al.* (2007). Local dystrophin restoration with antisense oligonucleotide PRO051. *N Engl J Med* **357**: 2677–2686.

47. Yokota T, *et al.* (2009). Efficacy of systemic morpholino exon-skipping in Duchenne dystrophy dogs. *Annals Neurol.*

48. Wu B, *et al.* (2008). Effective rescue of dystrophin improves cardiac function in dystrophin-deficient mice by a modified morpholino oligomer. *Proc Natl Acad Sci USA* **105**: 14814–14819.

49. Goyenvalle A, *et al.* (2004). Rescue of dystrophic muscle through U7 snRNA-mediated exon skipping. *Science* **306**: 1796–1799.

50. Ikemoto M, *et al.* (2007). Autologous transplantation of SM/C-2.6(+) satellite cells transduced with micro-dystrophin CS1 cDNA by lentiviral vector into mdx mice. *Mol Ther* **15**: 2178–2185.

51. Sampaolesi M, *et al.* (2006). Mesoangioblast stem cells ameliorate muscle function in dystrophic dogs. *Nature* **444**: 574–579.

52. Dezawa M, *et al.* (2005). Bone marrow stromal cells generate muscle cells and repair muscle degeneration. *Science* **309**: 314–317.
53. Okada T, *et al.* (2006). A histone deacetylase inhibitor enhances recombinant adeno-associated virus-mediated gene expression in tumor cells. *Mol Ther* **13**: 738–746.
54. Li W, *et al.* (2008). Engineering and selection of shuffled AAV genomes: a new strategy for producing targeted biological nanoparticles. *Mol Ther* **16**: 1252–1260.

Chapter 17
Cancer Gene Therapy

Kirsten A.K. Weigel-Van Aken*

Cancer gene therapy, initially designed as the transfer of a single gene into a tumor, has long been viewed with skepticism, and ultimately has failed clinically. Gene transfer in the context of additional treatment modalities, such as radiation/chemotherapy, oncolytic activity of a replicating virus, anti-angiogenesis, or immune stimulation, however, has the potential to have a significant impact on current cancer treatment. Primary tumor destruction has been one of the most successful endeavors of modern oncology. The re-occurrence of disease due to dormant tumor cells that evade the primary therapy, however, still constitutes a death sentence for many patients. Due to their inherent resistance to conventional chemo- and radiation therapies, these relapsed tumor cells warrant extensive exploration of alternative therapeutic approaches. Localized treatments of metastases such as direct injection or electroporation of plasmid DNA or injection of oncolytic or gene transfer viruses have only resulted in localized successes, because the occult and multifocal nature of the metastatic process still leads to evasion from therapeutic intervention. Novel strategies of cancer gene therapy therefore focus on means to provoke an antitumor immune response in a tumor-tolerant host using a variety of methods for tumor antigen and costimulatory signal delivery and means to improve tumor-targeting and prevent immune clearance of therapeutic viruses.

*Correspondence: Department of Pediatrics, Division of Cellular and Molecular Therapy, University of Florida College of Medicine, Cancer & Genetics Research Complex, 1376 Mowry Road, Gainesville, FL 32610-3610.
E-mail: weigel@ufl.edu

1. Introduction

Gene therapy involves the transfer of genetic information into a patient's cells to replace, repair, or turn off an abnormal gene, a strategy originally developed for single-gene diseases and subsequently extended to complex, multigenic disease such as cancer. Some common strategies used for cancer gene therapy are:

1. Suppression of oncogene expression or transfer of a tumor suppressor gene. Examples include transfer of a normal p53 gene, suppression of the ErbB2/HER2 gene, suppression of c-myc and c-fos genes, and transfer of melanoma differentiation associated protein 7.
2. Augmentation of immunological responses to cancer cells through transfer of cytokine genes (IL-2, IL-12, GM-CSF, TNF-α), co-stimulatory molecules (CD80), and antigen genes (CA15-3, HER2)
3. Suicide gene therapy by transfer of prodrug-activating enzymes into tumor cells followed by treatment with the prodrug, resulting in a high concentration of the activated drug in the tumor tissues: for example, herpes simplex virus thymidine kinase (HSV-TK) gene and gancyclovir; cytosine deaminase gene and fluorocytosine; and cytochrome P450 2B6 gene and cyclophosphamide.

2. Targeting the Tumor Cell

2.1 *DNA Electroporation*

Electroporation, or electric-pulse-provoked permeabilization of the plasma membrane, has been defined as the modification of the membrane impermeability to ions and hydrophilic and/or charged molecules as a consequence of exposure of cells to appropriate electric pulses. Transfer of DNA across the cell membrane requires electrophoretic acceleration of the DNA molecules within the electric field. Tumor electroporation differs from other tissues such as skeletal muscle, in the duration of the physiological local, histamine-dependent hypoperfusion (vascular lock) that can last up to several hours in the former and only a few minutes in the latter. In 2008, the first human trial of gene transfer utilizing *in vivo* DNA electroporation was reported. Of 24 melanoma patients treated with electroporation after IL-12

DNA injections, 10% showed complete regression of treated lesions and 42% experienced either disease stabilization or partial responses.[1]

2.2 Non-Oncolytic Viral Vectors

2.2.1 Retrovirus

Retroviral vectors derived from the murine retrovirus Moloney Murine Leukemia Virus (MMLV) were the first and most extensively used vectors in gene therapy. They have a relatively limited size capacity for foreign DNA (up to 8 kb) and require a cell to go through mitosis with the consequent disintegration of the nuclear membrane to allow transduction to occur. The viral DNA randomly integrates into the cellular genome, a process that led to severe side effects in a human sever combined immune deficiency (SCID) trial in France.[2] Since the tumor selectivity of retroviral vectors solely relies on the dependence of transduction on the active proliferation of target cells, retroviral vectors have been preferentially used in tumors that reside within a tissue that itself has a low proliferative index, such as the brain. Clinical trial results from patients with malignant glioma that received stereotactic injections of either retroviral vectors encoding a prodrug-converting enzyme (HSV-*TK*) or vector-producing cells generating the vectors *in situ* reported extremely low levels of transduced tumor cells, less than 0.03%, while the threshold transduction level for clinical benefit was estimated to be around 10%.[3] The first clinical trial in the US using systemic, intravenous (i.v.) administration of a retroviral vector was done with a non-replicating retroviral vector that displayed a collagen-binding domain of von Willebrandt factor on its envelope and encoded a mutant human cyclin G1 gene (Rexin-G vector). Although this vector had demonstrated inhibition of pancreatic cancer cell proliferation *in vitro*, the results in pancreatic cancer patients were disappointing with no clinical tumor response detected.[4] Retroviral vectors have also been used to transduce hematopoietic stem cells with drug resistance genes to prevent bone marrow suppression following high dose chemotherapy. Clinical trials, however, demonstrated only a relatively modest protection, and alternative vector systems are currently being evaluated.

Replication-competent retrovirus vectors based on amphotropic MLV were shown to be capable of achieving highly efficient replicative spread and gene delivery through solid tumors in mice. However, the use of such a vector in humans needs to be critically evaluated.

2.2.2 *Lentivirus*

Lentiviruses, such as human and simian immunodeficiency viruses (HIV, SIV), were developed as gene therapy vectors to overcome some of the shortcomings of retroviruses, including viral instability and the inability to transduce nondividing cells. Lentiviral vectors transduce quiescent cells and are engineered as self-inactivating (SIN) transfer vectors to minimize the risk of emergence of replication-competent virions. The tropism of lentiviral vectors is modified through the use of the promiscuous vesicular stomatitis virus (VSV) glycoprotein as a heterologous envelope. Although lentiviral vectors have been employed for gene transfer to cancer cells in mouse models, such as hepatocellular carcinoma and metastatic prostate cancer, their main use in cancer therapy is in the development of antitumor vaccines.

2.3 *Oncolytic Viruses*

It has long been known that viremia, following infection or vaccination, can result in tumor responses in patients with lymphoma or leukemia. In order to address some of the limitations of non-replicating viral therapeutics, over a dozen distinct families of viruses have been engineered to optimize their "oncolytic" potential — i.e. their ability to replicate in and kill tumor cells without harming normal tissue. Viruses recognized as oncolytic agents can be divided into three categories: (i) naturally occurring viruses (e.g. New Castle Disease virus, VSV, autonomous parvoviruses, some measles virus strains, and reovirus) that selectively replicate in tumor cells, in some instances owing to their dependence on an activated Ras pathway (reovirus) or to their relative resistance to interferon action; (ii) virus mutants in which some genes essential for replication in normal cells but evitable in cancer cells have been deleted (e.g. adenovirus ONYX-015 that replicates only in cells with mutant p53); and (iii) virus mutants modified by the introduction of tissue-specific transcriptional elements that drive viral genes (e.g. adenovirus CV706 that expresses E1A and E1B under the control of a prostate-specific antigen (PSA) promoter and adenovirus adMyc-TK that expresses a thymidine kinase gene under the control of a Myc-Max response element).

2.3.1 *Herpesvirus*

HSV-1 is an enveloped, double-stranded linear DNA virus that can infect replicating and quiescent cells, and infections can be either lytic or latent. The advantages of HSV-1 vectors over other vectors used in cancer gene therapy include (i) potential for incorporation of large amounts of foreign DNA ($\sim$30 kb); (ii) neurotropism; (iii) sensitivity to antiherpetic agents; and (iv) the fact that HSV-1 does not integrate. Recombinant HSV that are used as anti-cancer agents are viruses with mutations in two or more different genes to reduce the potential risk of restoring a wild-type phenotype via recombination with latent host virus. Results from clinical trials using HSV (1716) locally in glioblastoma patients showed viral replication in the tumor site and subtle changes in tumor morphology but no clinical improvement.[5] Injection of the same HSV (1716) in melanoma patients exhibited microscopic tumor necrosis but no systemic efficacy. Intratumor (i.t.) injection of a GM-CSF-armed HSV (Oncovex[GM−CSF], Biovex, Cambridge MA) into cutaneous metastases also did not show systemic efficacy although necrosis in multiple tumor nodules was observed. Intrahepatic artery delivery of a replication-competent, attenuated, genetically engineered HSV-1 vector into patients with hepatic colorectal metastases was well tolerated, but demonstrated only minor effects in 2 out of 12 patients and either stable disease or disease progression in the remainder.[6]

2.3.2 *Adenovirus*

Adenoviruses are double-stranded DNA viruses that have been extensively used in gene therapy studies. Since adenoviruses typically infect cells expressing coxsackie and adenovirus receptors (CARs) and CARs are only expressed by some human tumor cells, such as prostate cancer,[7] several fiber mutants have been generated to overcome the limited tropism. The adenoviral genome is maintained as an extra-chromosomal element that is rapidly lost in dividing cells. Oncolytic adenoviruses are highly immunogenic, which can be advantageous if an antitumor immune response is achieved, but disadvantageous if the immune response blocks viral propagation or leads to toxicity. One modification that confers tumor selectivity is deletion of the E1B region, found in the vector Onyx-015, which restricts viral replication to cells lacking a normal p53 protein (~40% of human tumors), since E1B's ability to sequester p53 is essential to viral replication. The uses of Onyx-015 include i.t. injections into head and neck cancer lesions, i.t. injections into pancreatic cancers and recurrent glioblastomas, intraperitoneal (i.p.) injections into ovarian carcinoma patients, injections into the hepatic artery of colorectal cancer patients and i.v. injections into lung cancer patients, all of which demonstrated local oncolytic viral replication, but no clinical responses.[5]

2.3.3 Poxvirus

Poxviruses are double-stranded enveloped DNA viruses. Vaccinia virus has had a critical role in the eradication of small pox and its highly immunogenic nature has led to its continued use in immunotherapy. Vaccinia virus can accommodate up to 25 kb of foreign DNA, causes timely lysis of infected cells (48–72 hours after infection), and has a broad tumor tissue tropism due to its ability to infect through several membrane fusion pathways. Vaccinia virus DNA does not integrate into the host chromosome but replicates in mini-nuclear structures in the cytoplasm. Poxviruses are highly efficient in spreading to distant tumors within a host and direct infection of cancer cells results in efficient lysis. Furthermore, immune-mediated cell death through release of cellular and viral danger signals and release of virus- and tumor-associated antigens at the site of infection as well as vascular collapse, due to intravascular thrombosis following excessive neutrophil attraction, and direct infection and lysis of tumor-associated endothelial cells contribute to its antitumor efficiency. In addition, the potent inflammatory response initiated by release of virus- and tumor-associated antigens has been augmented by the expression of trans-genes that encode relevant cytokines.[8,9] The tumor selectivity of myxoma virus, for example, relies on disrupted type I interferon (IFN) induction in murine tumor cells,[10] and increased Akt levels in human tumor cells.[11] Non-engineered live vaccinia vaccine strains were used in early trials in the 1970s and 1990s, primarily by superficial tumor injection in melanoma patients and tumor responses at the injection site, but no distant responses were observed.[12] Clinical trial results on a targeted and armed oncolytic poxvirus (JX-594) expressing GM-CSF demonstrated local efficacy in injected melanomas, including some cases of complete responses, and occasional responses in distant skin but not visceral metastases. A second trial in patients with advanced hepatocellular carcinoma demonstrated replication-dependent dissemination in the blood and infection of non-injected tumor sites, significant increases in blood neutrophil concentration due to GM-CSF expression, and tumor cell and tumor vasculature destruction.[13,14]

2.3.4 *Measles virus*

Measles virus (MV) is a negative sense single-stranded RNA paramyxovirus that has oncolytic potential in several types of cancer.[15] MV uses CD46 and signaling-activation molecule (SLAM), which is predominantly found on activated B and T lymphocytes, as receptors for entry into human cells. The tumor selectivity of the attenuated measles vaccine strain used in oncolytic approaches is based on its preferential use of the CD46 receptor, which is overexpressed in tumor cells. In animal studies, an attenuated MV strain engineered to express the human carcinoembryonic antigen (CEA) has demonstrated significant antitumor activity against glioma.[16] A clinical protocol has been developed where MV-CEA is installed into the resection cavity of glioma multiforme and CEA is used to monitor viral gene expression.[17] MVs were also engineered to express the human sodium-iodide symporter, which enhances iodide uptake and allows monitoring of radioiodine concentration in MV-infected cells. In addition, combination with therapeutic doses of radioiodine proved effective in inducing a complete regression in a measles-resistant myeloma tumor model in mice.[18]

2.3.5 *Vesicular stomatitis virus*

Vesicular stomatitis virus (VSV) is a negative sense single-stranded RNA virus that has oncolytic activity when administered systemically.[19] VSV is highly sensitive to the antiviral effects of the interferon system and its propagation is hindered in cells that contain a functional interferon signaling system. Mutant VSVs have been engineered that carry a deletion in the matrix protein therefore blocking the nuclear export of interferon RNA. These variants have been demonstrated to have superior antitumor effects.[20] VSV engineered to express the human sodium-iodide symporter demonstrated increased oncolytic activity in a mouse myeloma model following i.t. or i.v. administration in combination with radioiodine therapy.[21]

3. Targeting the Immune System

A 'tumor surveillance hypothesis' was originally formulated by Burnet in 1970, stating that tumor cells are recognized as foreign by the immune system and subsequently specifically eliminated without damaging their healthy counterparts, much in the same way as for virus infected cells, thus avoiding autoimmunity.[22] Dendritic cells (DC) are considered the most potent antigen-presenting cells (APC) of the immune system. In animal models, vaccination with DCs pulsed with tumor peptides, lysates or RNA or loaded with apoptotic/necrotic tumor cells could induce significant antitumor cytotoxic T lymphocyte (CTL) responses and antitumor immunity. However, results from early clinical trials pointed to a need for additional improvement of DC-based vaccines. In subsequent studies, DCs that expressed transgenes encoding tumor antigens proved to be more potent primers of antitumor immunity both *in vitro* and *in vivo*. Although cancer cells can potentially be distinguished from normal cells through "tumor antigens", they are still considered to be self-antigens by the immune system. Thus, several tolerogenic mechanisms and different modes of active inhibition impede a productive antitumor immune response as exemplified by the presence of high numbers of CTL in a tumor and the periphery without measurable tumor regression.[23] A productive T cell immune response requires specific recognition of an MHC-peptide complex by the T cell receptor (signal 1) along with adequate signaling through co-stimulatory molecules (signal 2). Recently, more and more data indicate that to establish strong antitumor responses, an inflammatory environment (signal 3) is required alongside functional recognition of T cells by APC. This can be achieved by strong activation of the innate arm of the immune system, in particular through toll-like receptors (TLR).[24] In recent years, CD4$^+$CD25$^+$ regulatory T cells have been discovered and found to be potent suppressors of effector T cells. Importantly, depletion of regulatory T cells greatly enhances antitumor responses.[25,26]

3.1 Cancer Vaccines

Plasmid DNA vaccines and recombinant viral vectors are particularly suited to induce CD8[+] T cell responses because they express antigens intracellularly, introducing them directly into the MHC class I antigen processing and presentation pathway. In 2005, a clinical trial with plasmid DNA encoding IL-12 injected into one melanoma lesion per patient over a 10-month period demonstrated local responses in 8 out of 9, and substantial tumor reductions in 5 out of 9 patients.[27] These results were superior to i.t. injection of virally transduced autologous cells expressing IL-12,[28] and i.v. injected recombinant IL-12 protein.[29] Clinical results with plasmid-DNA vaccines expressing CEA and hepatitis B virus surface antigen as well as the melanoma antigens Melan-A/MART and gp 100, however, demonstrated poor immunogenicity in cancer patients,[30] even in the presence of adjuvant. Some improvement has been made using more immunogenic approaches such as electroporation and particle-mediated delivery ('gene gun) for epidermal transduction of Langerhans cells.

3.1.1 Vaccinia virus

The non-replicating, modified vaccinia virus Ankara (MVA) has demonstrated favorable results upon co-expression of MUC-1 and IL-2 in lung and prostate cancer, HPV type 16 E6 and E7 antigens and IL2 in cervical cancer, and the oncofetal antigen 5T4 in colon, breast, ovarian and renal cancers, respectively. Weaker immune responses were observed with a non-replicating avipoxvirus containing MAGE-1 and MAGE-3 melanoma antigen peptides.[31]

3.1.2 Lentivirus

Since the first successful transduction of human monocyte-derived DC with lentivirus vectors in 1999, several groups reported *in vitro* priming of naïve T cells against tumor-associated antigens using lentivirus-transduced human DC. Strong protective immunity against subsequent tumor challenges as well as regression of pre-established tumors was documented *in vivo*. Compared to vaccination with *ex vivo* lentivirus-transduced DC, superior immune responses were generated by direct *in vivo* administration of the virus.[32]

3.1.3 *Adenovirus*

Although recombinant adenovirus has been widely tested as a delivery vector for gene therapy, only a few cancer vaccine studies have been published. Adenoviral vectors expressing the melanoma antigens MART-1 or gp100 or the lung cancer antigen L523S have demonstrated limited immunogenicity, which might be related to high-levels of pre-existing neutralizing antibodies against the adenovirus seroptype used.

3.1.4 *Parvoviruses*

Parvoviruses, such as rat H-1 virus and minute virus of mice (MVM), are lytic viruses that exert their cytotoxic activity preferentially in transformed cells, which is mediated largely by the non-structural protein NS-1. Recombinant MVMp vectors harboring cytokines such as IL2 and IP-10 have been demonstrated to exert moderate antitumor activity in tumor-bearing mice. Recently, novel AAV serotypes have been evaluated for their efficiency in delivering cytokine genes to intracerebral tumors. In a glioblastoma mouse model, AAV.rh8-hIFN-β vectors demonstrated prolonged survival but no cure. Transduction of DC with the human CEA antigen using an AAV-6 vector was shown to result in CEA-specific antibodies and a Th1 immune response in mice.[33]

3.2 *Mesenchymal Stem Cells (MSC) as Delivery Vehicles*

Adenoviruses as well as AAV vectors have been used to transduce MSC with cytokine genes such as IFN-β and IFN-α. Injection into tumor bearing mice demonstrated efficient localization of the MSC to tumors and metastases and efficient secretion of these cytokines. Moderate to strong reductions in tumor cell growth rates *in vitro* and *in vivo* and a moderate increase in animal survival were reported.[34]

3.3 *Adoptive T Cell Transfer*

Studies in murine models demonstrated that CD8$^+$ T cells can mediate human tumor rejection. Expansion of autologous antigen-specific CD8$^+$ T cells *in vitro* and subsequent transfer into patients can elicit tumor regression.[35] Adoptive immunotherapy with T cells expressing a tumor-specific chimeric T cell receptor (TCR) has been explored for the treatment of non-Hodgkin and mantle cell lymphoma. Autologous T cells were transduced *ex vivo* using electroporation with a plasmid encoding a CD20-specific chimeric TCR and reinfused followed by s.c. injections of IL-2. However, clinical responses were modest which was attributed to inadequate chimeric TCR surface expression and lack of co-stimulatory signaling from the chimeric TCR construct.[36]

4. Targeting the Tumor Microenvironment

The promising results of antiangiogenesis therapy using drugs that are able to control solid tumor and metastasis growth are dampened by the side effects of constant drug administration and limited half-life of antiangiogenic peptides. However, it is encouraging that i.m. injected viral vectors that encoded secretable forms of human angiostatin and endostatin demonstrated strong antiproliferative effects on human angiogenesis-dependent tumor xenografts in mice.[37]

5. Challenges and Risks of Cancer Gene Therapy

Many safety issues, such as the dissemination of vectors to remote, non-targeted tissues (even upon local administration), the potential induction of an exaggerated immune response by improperly disabled viruses, and the insertion of viral genomes into sites within the human genome leading to oncogene activation or tumor suppressor gene inactivation, need to be addressed. One of the major problems with viral gene therapy of cancers is that it is not possible at this time to introduce sufficient quantities of the virus into every cell of every tumor a patient might carry. Thus, more efficient delivery systems are needed. In highly invasive cancers, such as glioblastoma, it was estimated that the replicating viral infection travels at a speed about 5-10 times slower than the tumor wave front invasive velocity.[38] Replication of oncolytic viruses might be restricted by immunity, whether innate or acquired. Depending on the route of administration, the immune system can either antagonize the efficacy of intravascular injected oncolytic viruses by limiting viral delivery to the tumor, or augment tumor reduction by redirecting the CTL response from viral antigens to tumor antigens, once the virus reaches the tumor cells. The anti-tumor effects of oncolytic viruses armed with therapeutic transgenes have had limited effects in clinical trials so far. This might be due to the local immunosuppressive nature of the tumor microenvironment or immune editing, which can lead to the escape of tumor cell subpopulations that do not express the target antigens.

The risk of dissemination of a viral vector into the environment via excreta from the treated patient, a phenomenon called shedding, is also a safety concern. The collected data (1619 patient from 100 publications) from trial using retroviral, adenoviral, AAV and pox viral vectors document that shedding of viral vectors occurs in practice, mainly determined by the type of vector and the route of administration.

6. Novel Strategies

6.1 *Prime/Boost Regimens*

After a priming immunization, the antigen-specific T cell population expands to a modest level and then contracts. Over time, a proportion of these cells transform into antigen-specific memory T cells, which have the ability to expand rapidly upon subsequent encounter with the same antigen. In order to circumvent pre-existing immunity to the vector, heterologous prime-boost immunizations were tested using the same antigen delivered in sequence by different vectors. Thus, T cells that specifically target the viral vectors are not boosted and do not activate cell number control mechanisms, therefore allowing for greater expansion of the disease antigen-specific T cell population.[39] The results obtained from studies using plasmid DNA and recombinant MVA, or replicating vaccinia virus followed by a fowlpox virus vector, indicate that enhanced antigen-specific T cell responses are observed in the majority of patients.

6.2 *Immune Cells as Carriers for Viruses*

Neutralizing antiviral antibodies can hinder systemic virotherapy. One exciting development in the field has been the combination of utilizing cytokine induced killer (CIK) cells as carrier vehicles to deliver oncolytic viruses, such as vaccinia virus, to the tumor. This approach potentially eliminates the major limitations of each single-agent approach: inefficient tumor delivery of vaccinia virus and limited tumor cell killing by CIK cells. CIK cells have been demonstrated to display impressive tumor-trafficking potential following systemic delivery. However, large numbers (effector to target ratios of 10 or 20 to 1) are typically required to observe tumor clearance. Other cell types have been explored for their use as carriers for oncolytic viruses such as mesenchymal progenitors, monocytic cell lines, T lymphocytes, and even irradiated tumor cells.[40]

7. Conclusions

So far, gene therapy has failed in cancer patients owing to inefficient delivery of the gene to sufficient numbers of cancer cells locally and systemically, and vaccines have been limited by the immune evasion of tumors and the exclusive reliance on host factors for vaccine efficacy in patients with advanced cancers. However, combinations of viro- and chemo/radiation therapy and novel strategies, such as cell carrier-mediated oncolytic virus delivery, bear the promise of gene therapy becoming an effective tool in the fight against cancer.

References

1. Daud AI, *et al.* (2008). Phase I trial of interleukin-12 plasmid electroporation in patients with metastatic melanoma. *J Clin Oncol* **26**: 5896–5903.
2. Hacein-Bey-Abina S, *et al.* (2003). A serious adverse event after successful gene therapy for X-linked severe combined immunodeficiency. *N Engl J Med* **348**: 255–256.
3. Harsh GR, *et al.* (2000). Thymidine kinase activation of ganciclovir in recurrent malignant gliomas: a gene-marking and neuropathological study. *J Neurosurg* **92**: 804–811.
4. Galanis E, *et al.* (2008). Phase I trial of a pathotropic retroviral vector expressing a cytocidal cyclin G1 construct (Rexin-G) in patients with advanced pancreatic cancer. *Mol Ther* **16**: 979–984.
5. Aghi M, Martuza RL (2005). Oncolytic viral therapies - the clinical experience. *Oncogene* **24**: 7802–7816.
6. Kemeny N, *et al.* (2006). Phase I, open-label, dose-escalating study of a genetically engineered herpes simplex virus, NV1020, in subjects with metastatic colorectal carcinoma to the liver. *Hum Gene Ther* **17**: 1214–1224.
7. Rauen KA, *et al.* (2002). Expression of the coxsackie adenovirus receptor in normal prostate and in primary and metastatic prostate carcinoma: potential relevance to gene therapy. *Cancer Res* **62**: 3812–3818.

8. Kirn DH, Wang Y, Le Boeuf F, Bell J, Thorne, SH (2007). Targeting of interferon-beta to produce a specific, multi-mechanistic oncolytic vaccinia virus. *PLoS Med* **4**: e353.

9. Kim JH, *et al.* (2006). Systemic armed oncolytic and immunologic therapy for cancer with JX-594, a targeted poxvirus expressing GM-CSF. *Mol Ther* **14**: 361–370.

10. Wang F, *et al.* (2004). Disruption of Erk-dependent type I interferon induction breaks the myxoma virus species barrier. *Nat Immunol* **5**: 1266–1274.

11. Wang G, *et al.* (2006). Infection of human cancer cells with myxoma virus requires Akt activation via interaction with a viral ankyrin-repeat host range factor. *Proc Natl Acad Sci USA* **103**: 4640–4645.

12. Mastrangelo MJ, Maguire HC, Lattime EC (2000). Intralesional vaccinia/GM-CSF recombinant virus in the treatment of metastatic melanoma. *Adv Exp Med Biol* **465**: 391–400.

13. Liu TC, Hwang T, Park BH, Bell J, Kirn DH (2008). The targeted oncolytic poxvirus JX-594 demonstrates antitumoral, antivascular, and anti-HBV activities in patients with hepatocellular carcinoma. *Mol Ther* **16**: 1637–1642.

14. Park BH, *et al.* (2008). Use of a targeted oncolytic poxvirus, JX-594, in patients with refractory primary or metastatic liver cancer: a phase I trial. *Lancet Oncol* **9**: 533–542.

15. Nakamura T, Russell, SJ (2004). Oncolytic measles viruses for cancer therapy. *Expert Opin Biol Ther* **4**: 1685–1692.

16. Phuong LK, *et al.* (2003). Use of a vaccine strain of measles virus genetically engineered to produce carcinoembryonic antigen as a novel therapeutic agent against glioblastoma multiforme. *Cancer Res* **63**: 2462–2469.

17. Myers R, *et al.* (2008). Toxicology study of repeat intracerebral administration of a measles virus derivative producing carcinoembryonic antigen in rhesus macaques in support of a phase I/II clinical trial for patients with recurrent gliomas. *Hum Gene Ther* **19**: 690–698.

18. Dingli D, *et al.* (2004). Image-guided radiovirotherapy for multiple myeloma using a recombinant measles virus expressing the thyroidal sodium iodide symporter. *Blood* **103**: 1641–1646.

19. Stojdl DF, *et al.* (2003). VSV strains with defects in their ability to shutdown innate immunity are potent systemic anti-cancer agents. *Cancer Cell* **4**: 263–275.

20. Stojdl DF, *et al.* (2000). Exploiting tumor-specific defects in the interferon pathway with a previously unknown oncolytic virus. *Nat Med* **6**: 821–825.

21. Goel A, *et al.* (2007). Radioiodide imaging and radiovirotherapy of multiple myeloma using VSV(Delta51)-NIS, an attenuated vesicular stomatitis virus encoding the sodium iodide symporter gene. *Blood* **110**: 2342–2350.

22. Burnet FM (1970). The concept of immunological surveillance. *Prog Exp Tumor Res* **13**: 1–27.

23. Lurquin C, *et al.* (2005). Contrasting frequencies of antitumor and anti-vaccine T cells in metastases of a melanoma patient vaccinated with a MAGE tumor antigen. *J Exp Med* **201**: 249–257.

24. Yang Y, Huang CT, Huang X, Pardoll DM (2004). Persistent Toll-like receptor signals are required for reversal of regulatory T cell-mediated CD8 tolerance. *Nat Immunol* **5**: 508–515.

25. Fehervari Z, Sakaguchi S (2004). CD4+ Tregs and immune control. *J Clin Invest* **114**: 1209–1217.

26. Van Meirvenne S, Dullaers M, Heirman C, Straetman L, Michiels A, Thielemans K (2005). *In vivo* depletion of CD4+CD25+ regulatory T cells enhances the antigen-specific primary and memory CTL response elicited by mature mRNA-electroporated dendritic cells. *Mol Ther* **12**: 922–932.

27. Heinzerling L, *et al.* (2005). Intratumoral injection of DNA encoding human interleukin 12 into patients with metastatic melanoma: clinical efficacy. *Hum Gene Ther* **16**: 35–48.

28. Kang WK, *et al.* (2001). Interleukin 12 gene therapy of cancer by peritumoral injection of transduced autologous fibroblasts: outcome of a phase I study. *Hum Gene Ther* **12**: 671–684.

29. Atkins MB, *et al.* (1997). Phase I evaluation of intravenous recombinant human interleukin 12 in patients with advanced malignancies. *Clin Cancer Res* **3**: 409–417.

30. Triozzi PL, Aldrich W, Allen KO, Carlisle RR, LoBuglio AF, Conry RM (2005). Phase I study of a plasmid DNA vaccine encoding MART-1 in patients with resected melanoma at risk for relapse. *J Immunother* **28**: 382–388.

31. van Baren N, *et al.* (2005). Tumoral and immunologic response after vaccination of melanoma patients with an ALVAC virus encoding MAGE antigens recognized by T cells. *J Clin Oncol* **23**: 9008–9021.

32. Breckpot K, Aerts JL, Thielemans K (2007). Lentiviral vectors for cancer immunotherapy: transforming infectious particles into therapeutics. *Gene Ther* **14**: 847–862.

33. Aldrich WA, *et al.* (2006). Enhanced transduction of mouse bone marrow-derived dendritic cells by repetitive infection with self-complementary adeno-associated virus 6 combined with immunostimulatory ligands. *Gene Ther* **13**: 29–39.

34. Ren C, Kumar S, Chanda D, Chen J, Mountz JD, Ponnazhagan S (2008). Therapeutic potential of mesenchymal stem cells producing interferon-alpha in a mouse melanoma lung metastasis model. *Stem Cells* **26**: 2332–2338.

35. Powell DJ Jr., Dudley ME, Hogan KA, Wunderlich JR, Rosenberg SA (2006). Adoptive transfer of vaccine-induced peripheral blood mononuclear cells to patients with metastatic melanoma following lymphodepletion. *J Immunol* **177**: 6527–6539.

36. Till BG, *et al.* (2008). Adoptive immunotherapy for indolent non-Hodgkin lymphoma and mantle cell lymphoma using genetically modified autologous CD20-specific T cells. *Blood* **112**: 2261–2271.

37. Ponnazhagan S, *et al.* (2004). Adeno-associated virus 2-mediated antiangiogenic cancer gene therapy: long-term efficacy of a vector encoding angiostatin and endostatin over vectors encoding a single factor. *Cancer Res* **64**: 1781–1787.

38. Wu JT, Kirn DH, Wein LM (2004). Analysis of a three-way race between tumor growth, a replication-competent virus and an immune response. *Bull Math Biol* **66**: 605–625.

39. Schneider J, *et al.* (1998). Enhanced immunogenicity for CD8+ T cell induction and complete protective efficacy of malaria DNA vaccination by boosting with modified vaccinia virus Ankara. *Nat Med* **4**: 397–402.

40. Ong HT, Hasegawa K, Dietz AB, Russell SJ, Peng KW (2007). Evaluation of T cells as carriers for systemic measles virotherapy in the presence of antiviral antibodies. *Gene Ther* **14**: 324–333.

Chapter 18
Gene Therapy for Autoimmune Disorders

Daniel F. Gaddy, Melanie A. Ruffner
and Paul D. Robbins*

Gene therapy approaches for treating autoimmune disorders such as rheumatoid arthritis and type I diabetes have shown efficacy in animal models. The effective approaches include *ex vivo* methods involving genetically modified dendritic cells as well as direct, *in vivo* gene transfer of immunomodulatory cytokines. Several clinical trials in rheumatoid arthritis have been completed with a recent small, clinical study demonstrating efficacy. The significant progress made towards developing viable gene therapy approaches for treating autoimmune disorders using rheumatoid arthritis and type 1 diabetes as model autoimmune diseases will be discussed.

1. Introduction

Initially gene therapy was developed to treat monogenetic diseases such as muscular dystrophies, SCID and cystic fibrosis. However, local gene transfer of immunomodulatory agents also has shown promise in treating acquired diseases such as cancer and certain autoimmune diseases. For these diseases, gene transfer can modulate the immune response, either to stimulate an immune response to tumor antigens, or dampen the immune response to self-antigens. Moreover, gene transfer can protect cells against apoptosis, preventing loss of ß cells or chondrocytes. This chapter will focus

*Correspondence: University of Pittsburgh School of Medicine, Pittsburgh, PA, USA.
E-mail: probb@pitt.edu

on the significant progress that has been made towards developing clinically useful approaches to treat two types of autoimmune diseases: rheumatoid arthritis and type I diabetes.

2. Rheumatoid Arthritis

2.1 *Background*

Arthritis is the leading cause of disability among adults in the United States, affecting approximately 21% (more than 46 million) adults. Rheumatoid arthritis (RA) affects approximately 1.3 million adults in the United States, and more than 60 million people worldwide.[1] RA is characterized by inflammation of the synovial lining and destruction of extra-articular bone and cartilage.

It is believed that arthritogenic peptides, either from foreign or self-proteins, are presented to T cells preferentially by MHC molecules on antigen-presenting cells. Activated T cells produce a variety of proinflammatory cytokines, including tumor necrosis factor (TNF)-α, interleukin (IL)-1 and IL-6. The inflammatory response induced by these cytokines is directly responsible for the overt RA symptoms, including joint pain, swelling, effusion and stiffness.

2.2 *Existing Therapies*

Despite the prevalence and rising economic burden of RA, effective therapies for this disease remain limited, and there is no cure. Current therapies consist of early, aggressive and continuous treatment with non-steroidal anti-inflammatory drugs (NSAIDs) and disease-modifying antirheumatic drugs (DMARDs). Among the DMARDs, methotrexate has long been the drug of choice for RA, but newer biologic targeted therapies have emerged in recent years. These targeted therapies include abatacept (Orencia®), a fusion protein composed of the extra-cellular domain of cytotoxic T lymphocyte antigen-4 fused to an immunoglobulin (CTLA4-Ig); anakinra (Kineret®), an interleukin-1 receptor antagonist (IL-1Ra); infliximab (Remicade®), a chimeric anti–TNFα antibody; adalimumab (Humira®), a fully human anti–TNFα antibody; and etanercept (Enbrel®), a soluble TNF-α receptor.[2]

2.3 Target Tissues and Routes of Delivery

2.3.1 Local RA Gene Therapy

Synovium, the soft tissue that lines non-cartilaginous surfaces of joints, is the primary target for intra-articular gene therapy because synovium has a large surface area and direct access to the joint space. This is achieved by directly injecting vectors or cells expressing the therapeutic gene of interest into individual diseased joints, which limits the amount of vector needed compared to systemic injections. The drawbacks of this approach are that multiple joints may require injections in patients with polyarticular disease, and that extra-articular manifestations of the disease may not be affected.

One method to overcome this drawback is the use of genetically modified dendritic cells[3] or T cells.[4] When transduced with anti-arthritic genes using viral vectors, these cells demonstrate the ability to home in on multiple joints or extra-articular tissues affected by RA, such as lymph nodes. Similarly, vesicles, known as exosomes, derived from genetically modified dendritic cells, or exosomes from immature dendritic cells that are subsequently treated with immunomodulatory factors, reduce inflammation and severity of disease in arthritis mouse models.[5]

2.3.2 Systemic RA Gene Therapy

Systemic delivery, in which vectors are administered extra-articularly, such as via intramuscular or intravenous injections, has been attempted in multiple studies (Fig. 18.1). Systemic injection will result in relatively high circulating levels of therapeutic proteins and lower concentrations of therapeutic proteins in the joints. It is possible that this method will improve the ability of gene therapy to treat polyarticular or extra-articular manifestations of RA, and do so in a manner that improves upon conventional biologic therapies that require frequent injections or infusions. While this strategy has demonstrated promising results in animal models, it has multiple drawbacks, including the requirement for large doses of vectors to facilitate systemic infection, and the observation that the majority of viral vectors, when administered systemically, will home in primarily on the liver, as opposed to joints or other sites of disease.[6]

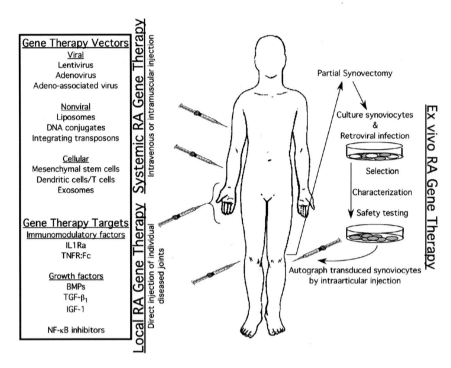

Fig. 18.1 A variety of vectors and approaches for RA gene therapy have been attempted. Viral, nonviral and cellular vehicles for delivery of antiarthritic genes have shown great progress in recent years. In addition, two primary approaches, *in vivo* and *ex vivo* gene therapy, have been attempted. *In vivo* gene therapy can be segregated into either systemic or local gene transfer, and both involve direct administration of vectors to the body. Alternatively, *ex vivo* gene transfer involves isolation of cells from the body and modification, typically by retroviral transduction, in culture. This allows extensive testing of gene expression and safety of the modified cells before return to the patient, but at significantly greater cost.

2.4 *Immunomodulation*

As discussed in Section 2.1, RA is characterized by chronic inflammation of the tissue surrounding joints, and the subsequent deterioration of joint cartilage and bone. This chronic inflammation is mediated by proinflammatory cytokines, particularly TNF-α and IL-1β. As discussed in Section 2.2, most of the biological therapeutics currently available to RA patients target the actions of these cytokines. The inhibition of IL-1β reduces cartilage and bone destruction, and slows RA progression in animal models.[2] Based on these observations, recombinant IL-1 receptor antagonist (IL-1Ra; Kineret®) was approved by the U.S. Food and Drug Administration (FDA) for RA therapy in 2001. Similarly, inhibition of TNF-α activity ameliorates joint disease in animal models.[2] Three drugs, Humira®, Remicade®, and Enbrel®, have been licensed by the FDA for use as TNFα inhibitors for the treatment of RA and other inflammatory diseases.

Other strategies to treat RA via immunomodulation include blocking the activity of IL-6 and NF-κB. As of this writing, several clinical trials, including 5 phase III trials, are underway and indicate that an antibody against the IL-6 receptor shows some therapeutic benefit in patients with RA and systemic-onset juvenile idiopathic arthritis.[7] In addition, a growing body of evidence indicates that NF-κB may be a master regulator of RA. Numerous inhibitors of the IKK complex, the enzymatically active component of the NF-κB complex, have been developed, several of which have entered clinical trials for the treatment of RA.[8]

2.5 *Overview of Preclinical Gene Therapy Studies*

Given the success of biologic targeted therapies, various strategies have been employed to utilize these immunomodulatory agents in RA gene therapy. Multiple preclinical RA gene therapy studies have demonstrated that gene therapy vectors, including retrovirus and adenovirus vectors, expressing IL-1Ra produce high levels of the transgene in target tissues and inhibit inflammation and cartilage loss in animal models. Similarly, numerous studies have demonstrated the effectiveness of blocking TNF-α activity via gene therapy, including the suppression of inflammatory cell infiltration, pannus formation, cartilage and bone destruction, and expression of joint proinflammatory cytokines in animal models.[2]

Furthermore, adeno-associated virus (AAV) expressing a TNFR:Fc fusion gene under the control of an inflammation-inducible NF-κB promoter delayed disease onset and decreased the incidence and severity of joint damage in mouse and rat arthritis models.[2] This type of gene therapy, utilizing a disease-inducible promoter, is of particular interest in autoimmune diseases like RA, which are characterized by flare-ups followed by periods of disease regression. By utilizing the inflammation-inducible NF-κB promoter, high levels of transgene expression are obtained only during disease flares, preventing unnecessary exposure of the patient to immunosuppressive agents during periods of disease regression.

In addition, both adenovirus and AAV have been utilized to deliver NF-κB inhibitors to synovium, successfully preventing expression of proinflammatory cytokines.[9,10] One potential drawback of using non-secreted factors, such as IKK complex inhibitors, is that only directly transduced cells will express the therapeutic gene, whereas secreted factors, such as IL-1Ra and soluble TNFR, will mediate protection to the entire inflamed environment.

A variety of growth factors has also been studied as therapeutics for RA. Bone morphogenetic proteins (BMPs), particularly BMP-2 and BMP-7, have been shown to induce chondrogenesis and osteogenesis when delivered by adenovirus or AAV vectors.[2] Additional growth factors that have been studied in relation to RA gene therapy include transforming growth factor (TGF)-β_1 and insulin-like growth factor (IGF)-1. Adenovirus or AAV vectors expressing TGF-β_1 have been used to transduce mesenchymal stem cells (MSCs) and drive *ex vivo* differentiation of MSCs into chondrocytes, facilitating cartilage repair in animal models.[11] Similarly, *ex vivo* chondrocytes transduced with an adenovirus expressing IGF-1 enhanced matrix synthesis and cartilage repair in horses.[12] These studies suggest important roles for growth factors in RA gene therapy, and indicate that a variety of growth factors may be able to join the list of effective RA therapies.

Non-viral gene transfer systems, including naked DNA, cationic liposomes, histones and polymers, have experienced limited use for the treatment of RA because RA will necessarily require long-term, high-level expression of therapeutic transgenes, which has traditionally not been an option with non-viral systems.[2] However, recent advances may overcome this issue. For example, transient expression of the phage phiC31 integrase allows for efficient, targeted integration of plasmid DNA. Also, plasmids containing vectors based on transposons such as *Sleeping Beauty* or *Himar1* also can confer efficient integration when delivered in conjunction with the appropriate transposase. While these vectors have not been extensively tested for RA gene therapy, phiC31 integrase and *Himar1* transposase have been shown to confer stable transgene expression in cultured human synoviocytes and in the rabbit knee following intra-articular delivery.[2] Thus it is possible that these non-viral, integrating vector systems, when delivered intra-articularly, could be useful for chronic RA gene therapy.

2.6 Overview of Clinical Gene Therapy Studies

As illustrated in Fig. 18.1, early arthritis gene transfer trials used retrovirus vectors in *ex vivo* protocols to deliver IL-1Ra to the metacarpophalangeal joints of RA patients.[13] These safety and feasibility studies illustrated that gene transfer to RA joints could be safe and effective, but the low numbers of patients enrolled prevented any conclusions with regard to efficacy. A similar phase I trial has been initiated by TissueGene, Inc utilizing retroviral vectors expressing TGF-β_1 to transduce human chondrocytes, which are then mixed with normal human chondrocytes and injected into the knee of patients with degenerative joint disease. As of this writing, 16 patients have been treated in the United States and South Korea, with approximately 50% of the treated patients demonstrating symptomatic improvement.[2,14]

Another RA gene therapy trial, sponsored by Targeted Genetics, Inc evaluated the safety and efficacy of a single-stranded rAAV2 virus expressing the complete coding sequence of a TNFR:Fc fusion protein, which is identical to etanercept. This trial received a great deal of publicity in 2007 because of the death of an enrolled patient. Subsequent investigation determined that the death of the subject was not likely due to the viral vector.[2] Despite the death of this subject, the Targeted Genetics trial has shown promising results. The AAV vector appears safe, in that there is no evidence of circulating TNFR:Fc or extra-articular over-expression of TNFR-Fc, which would have indicated vector dissemination and amplification in extra-articular tissues. Clinical response was assessed using patient reported outcomes, revealing moderate improvement in target joint pain and swelling.[15]

3. Type I Diabetes Mellitus

3.1 Background

Type 1 diabetes (T1D), also known as insulin-dependent diabetes mellitus, is recognized as a rapidly growing health threat worldwide. The CDC estimates that 15,000 young people in the United States per year are diagnosed with T1D, with 19 new cases per 100,000 youth each year.[16] Worldwide, the incidence of diabetes varies greatly by geographical location, ranging from 0.1 per 100,000/year in China and Venezuela to 40.9 per 100,000/year in Finland, and has been increasing since the 1950s.[17] From 1990–1999, the incidence of T1D worldwide increased 2.8%, confirming the need for new methods to address this growing public health concern.[17]

T1D is a polygenic autoimmune disease characterized by the T cell-mediated destruction of insulin-producing β cells in the pancreatic islets of Langerhans, leading to insulin deficiency. T1D is a chronic disease that is known to exist for years in a preclinical phase before the classic manifestations are observed. This preclinical phase, known as insulitis, is characterized by the infiltration of leukocytes into the islets. The actual onset of diabetes occurs after most β cells have been killed, resulting in insulin deficiency and hyperglycemia. β cell auto-antigens, such as glutamic acid decarboxylase (GAD) and insulin, are processed by antigen presenting cells (APC), including macrophages, dendritic cells, and B cells, in the pancreatic islets. APC then present processed peptides to autoreactive CD4$^+$ T cells in the peripheral lymphoid tissues. Activated autoreactive CD4$^+$ T cells secrete cytokines and activate β cell-specific cytotoxic CD8$^+$ T cells. Activated CD8$^+$ T cells (CTL) then migrate to the islets and produce cytokines, which activate more CTL, as well as macrophages, further contributing to the destruction of β cells.[18] In general, T$_H$1 cytokines, such as IL-2, interferon (IFN)-γ, and TNF-α, promote T1D development, while T$_H$2 or T$_H$3 cytokines, such as IL-4, IL-10, and TGF-β, prevent T1D onset. β cell death is the direct result of these cytotoxic cytokines and other agents, including oxygen free radicals, granzyme and perforin, released from CTL and macrophages. Additionally, Fas and TNF receptor-mediated apoptosis play roles in β cell death.[18]

3.2 *Existing Therapies*

T1D patients require lifelong insulin replacement therapy, and are at risk of developing significant complications associated with hyperglycemia, such as retinopathy, neuropathy, nephropathy, and accelerated peripheral vascular and coronary artery disease. The goal of existing therapy is to provide tight glycemic control in all diabetic patients in order to minimize the complications associated with hyperglycemia.

Furthermore, while allogeneic islet transplantation has shown some evidence of success, the allo- and autoimmune response against the islets often leads to their destruction.[18,19] Therefore patients receiving grafted islets require lifelong immunosuppressive therapy, which carries significant side effects. Due to the need for lifelong immunosuppression, whole-pancreas transplant is rarely performed alone, and more frequently performed in T1D patients undergoing kidney transplant for diabetic nephropathy. The widespread applicability of both of these techniques is limited by the supply of islets from cadaveric donors, and in general both procedures have been offered only to T1D patients who have hypoglycemic unawareness, severe metabolic or secondary complications, or are unable to follow an insulin regimen.[20]

3.3 *Target Tissues and Routes of Delivery*

As shown in Fig. 18.2, T1D interventions have taken many forms, and as a result, many different target tissues have been tried. Chief among the target tissues investigated has been the pancreatic islets. The vast majority of the studies utilizing gene transfer strategies to enhance islet function or survival have used an *ex vivo* transduction protocol, which follows logically from the islet transplant protocols currently in clinical use. *In vivo* transduction protocols of β cells are more rare, but have been accomplished using AAV vectors.[21]

In some cases, systemic expression of protein has been desired for therapy of T1D. Examples of this may include expression of insulin to regulate the blood glucose or systemic expression of an immunomodulatory agent in order to prevent the onset of diabetes. In these cases, investigators have utilized the relative ease of transduction of other tissues, such as liver or muscle to achieve high transduction efficiencies and better systemic expression of protein.

3.4 *Immunomodulation*

Over the past 30 years, significant efforts have been made to test the efficacy of immunomodulating therapies in T1D. An ideal immunomodulating agent for diabetes would specifically halt β cell destruction without causing systemic immunosuppression or inhibiting the process of β cell regeneration. It has been demonstrated in both humans and non-obese diabetic (NOD) mice that at the onset of clinical diabetes there is still significant residual

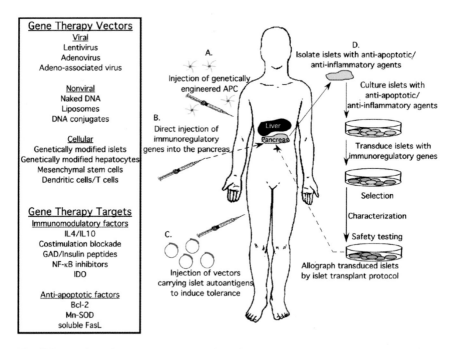

Fig. 18.2 A variety of vectors and approaches for T1D gene therapy have been attempted. Viral, non-viral and cellular vehicles for delivery of genes have been utilized. Both *in vivo* and *ex vivo* gene therapy has been attempted. *Ex vivo* gene transfer involves isolation of cells from the body and modification in culture as shown for antigen presenting cells in A, and pancreatic islet transplant in D. *In vivo* gene therapy involves direct administration of vectors to the body, either directly to the pancreas as shown in B, or to another target such as muscle (as in C) or liver for systemic expression of transgene.

β cell mass, and early intervention with an effective immunotherapy during this period has the potential to restore tolerance and allow endogenous cells to regenerate the islets.[22] However, some immunosuppressive drugs used in common islet transplant protocols have been shown to inhibit spontaneous regeneration of β cell mass *in vivo*, and therefore drugs intended for this application must be carefully tested for their effects on β cell growth *in vivo*.[23] Anti-CD3 has shown some promise in clinical trials, but recent data concerning potentially serious systemic side effects has dampened enthusiasm for this molecule.[24] Trials are also underway investigating the therapeutic potential of rituximab, an anti-CD20 antibody that down regulates the B cell signaling of T cells.

In terms of diabetes antigen-specific studies, systemic or oral administration of insulin given to high-risk relatives had no efficacy in preventing diabetes onset when compared to a matched control group.[26] Other trials examining the benefits of intranasally administered insulin, as well as an intramuscular alum-Gad65 preparation, have shown some initial promise and work is ongoing to investigate the ultimate benefit of these methods.[25,26]

Collectively, these studies illustrate the challenges facing investigators in finding appropriate immunotherapies for T1D. At this point, it remains unclear, which agents are the most beneficial at halting the progression of autoimmunity as many underlying aspects of the pathogenesis of T1D have yet to be elucidated. The clinical heterogeneity of human disease when compared to the homogeneity of inbred animal models, such as the NOD mouse, remains another challenge, as many interventions that have prevented diabetes in preclinical models do not have promising results when translated into human therapy.

3.5 *Overview of Preclinical Gene Therapy Studies*

Until recently, gene therapy studies for T1D had focused primarily on *ex vivo* approaches to modify islets for transplantation. Although many different approaches have been examined, the goals of these modifications are similar: to transfer a gene encoding a protein which would confer some type of islet-protective effect to the grafted islets in order to protect them from allo- and autoimmune attack when transplanted into the patient. When successful, this prolongs graft survival and may potentially reduce the need for systematic immunosuppressive therapy to prevent loss of the graft. Some lingering concerns with the use of transduced islets are safety and efficacy. Particular concerns include whether expression of virally encoded proteins from the islets will have some effect on the cellular function of the transplanted cells or immunological function of the patient receiving the graft.

Another major focus of *ex vivo* gene therapy strategies for T1D has been in the modification of immunological cells to promote tolerance upon adoptive transfer *in vivo*. Dendritic cells have been of particular interest for applications in T1D because of their unique role in regulating T cell responses. Both adenoviral and lentiviral gene transfer of IL-4 to DC have been shown to have protective effects in the NOD model of T1D.[27,28] Others have shown that modifying DC with an adenoviral vector encoding for galectin-1, a lectin with regulatory effects on T cells, could delete diabetogenic T cells *in vivo*.[29] Lastly, approaches using oligonucleotides have been particularly successful. NF-κB decoy oligoneucleotide treatment of DC was shown to inhibit DC activation and maturation, inhibit the expression of co-stimulatory molecules, and prevent the onset of diabetes in the NOD mouse model. These studies have progressed into clinical trials.[30] They are described in further detail in Section 3.6. In addition, retroviral transduction of genes encoding Gad65-IgG or (pro)Insulin-IgG fusion proteins into B cells resulted in a suppressive B cell phenotype which was capable of reducing diabetes incidence in a regulatory T cell-dependent manner on adoptive transfer into NOD mice.[31] Approaches using the patient's autologous antigen-presenting cells are particularly attractive because of the known importance of these cells in regulating autoimmunity, as well as the ease and relative safety of *ex vivo* transduction protocols.

In vivo gene therapy efforts for T1D have primarily focused on addressing the underlying autoimmune destruction and loss of β cell mass rather than on insulin replacement strategies.[32] Many *in vivo* efforts have been directed towards modifying nonpancreatic tissues, such as liver and muscle, to systemically express insulin or immuno-regulatory compounds in order to modulate autoimmunity in models of T1D.[33]

For many years, the potential of gene therapy as a means to treat T1D was hindered by the lack of an effective vehicle to target genes specifically to the pancreas *in vivo*. However, recent studies have demonstrated that a double-stranded AAV vector utilizing the insulin promoter provided stable, long-term expression that is specific to β cells. Use of this system to deliver IL-4 was protective against insulitis and hyperglycemia in the NOD model, and was able to maintain levels of regulatory T cells as the mice aged.[34] *In vivo* transduction of islets could be of major benefit in preserving or enhancing the amount of β cell mass remaining in T1D patients at diagnosis.

3.6 *Overview of Clinical Gene Therapy Studies*

To date, few proposed gene therapy strategies for T1D have progressed to the point of clinical trials. A phase I clinical trial to examine safety is currently ongoing using DC genetically modified using antisense oligonucleotides (AS-ODN) to the co-stimulatory molecules CD40, CD80, and CD86. In pre-clinical studies using the NOD mouse model of T1D, investigators demonstrated that the AS-ODN treated bone-marrow derived DC were able to delay the incidence of diabetes after a single injection, and observed that the

AS-ODN treated DC resulted in an expansion of a $CD4^+CD25^+CD62L^+$ regulatory T cell population.[35] It is believed that the lack of co-stimulatory molecules on the DC result in an anergizing signal to diabetogenic T cells, as well as expansion of regulatory T cells. In the ongoing clinical study, autologous DC are generated from T1D patient's leukocytes *in vitro* after harvest by leukapheresis. The DC are cultured and engineered using the AS-ODN to CD40, CD80, and CD86 and then injected intradermally at a site distal to the pancreas to allow for DC migration into the lymph nodes. At the time of writing, the phase I trial is currently ongoing and is focused on the safety of this approach in a patient population that is over 18 and has had insulin-dependent T1D for greater than five years.[36] If successful, the next step planned by investigators will be a phase II trial focused on efficacy of treatment in newly-diagnosed T1D patients, in fitting with the paradigm that early intervention with immunotherapies is most effective at saving any residual β cell mass.

Some additional phase I/II trials have occurred for gene therapy products, which are not specifically targeted towards T1D, but rather are aimed at treating hyperglycemic complications that can occur with Type 1 or 2 diabetes. For example, Tissue Repair Company has recruited patients for trials of EXCELLERATE, which is a collagen-matrix gel containing adenoviral vectors expressing platelet-derived growth factor for the treatment of chronic lower extremity ulcers in diabetic patients. Another clinical trial is also ongoing looking at the gene transfer of vascular-endothelium derived growth factor for the treatment of diabetic neuropathy.[37] As of this writing, results from these trials are eagerly anticipated, but have not yet been published.

4. Conclusions and Outlook

As outlined in this chapter, considerable progress has been made towards developing gene therapy approaches for autoimmune disease such as RA and T1D, leading to several phase I clinical trials, and one phase II trial in RA. There is also evidence of clinical responses in certain subjects, suggesting that additional trials to establish efficacy are merited. Although not discussed in this chapter, proof of principle also has been established in animal models of Sjogren's syndrome and lupus. Because these diseases, unlike

RA and T1D, do not respond well to present biologics, the development of alternative approaches, such as gene therapy, seems highly appropriate. Furthermore, although osteoarthritis (OA) is not an autoimmune disease, it shares certain pathology with RA. Two clinical trials exploring the use of gene therapy in OA are underway and another is in the pipeline. Given there currently is no effective treatment for OA, gene therapy offers the potential to improve the quality of life for the 10% of the population that will develop this disease. Overall the technology of gene transfer, along with efficacy of gene therapy in animal models of autoimmune diseases, has developed to the point where it is no longer the rate-limiting step for many purposes. Instead, the focus of the field of gene therapy for autoimmune diseases is now on bringing these approaches into the clinic.

References

1. Lundkvist J, Kastäng F, Kobelt G (2008). The burden of rheumatoid arthritis and access to treatment: Health burden and costs. *The European Journal of Health Economics* **8**: S49–S60.
2. Gaddy DF, Robbins PD (2008). Current status of gene therapy for rheumatoid arthritis. *Current Rheumatology Reports* **10**: 398–404.
3. Kim SH, Kim S, Oligino TJ, Robbins PD (2002). Effective treatment of established mouse collagen-induced arthritis by systemic administration of dendritic cells genetically modified to express FasL. *Mol Ther* **6**: 584–590.
4. Nakajima A, *et al.* (2001). Antigen-specific T cell-mediated gene therapy in collagen-induced arthritis. *J Clin Invest* **107**: 1293–1301.
5. Bianco NR, Kim SH, Morelli AE, Robbins PD (2007). Modulation of the immune response using dendritic cell-derived exosomes. *Methods Mol Biol* **380**: 443–455.
6. Evans C, Ghivizzani S, Robbins P (2006). Gene therapy for arthritis: What next? *Arthritis and rheumatism* **54**: 1714–1729.
7. Finckh A, Gabay C (2008). At the horizon of innovative therapy in rheumatology: New biologic agents. *Current opinion in rheumatology* **20**: 269–275.
8. Calzado MA, Bacher S, Schmitz ML (2007). NF-kappaB inhibitors for the treatment of inflammatory diseases and cancer. *Current medicinal chemistry* **14**: 367–376.
9. Amos N, Lauder S, Evans A, Feldmann M, Bondeson J (2006). Adenoviral gene transfer into osteoarthritis synovial cells using the endogenous inhibitor IkappaBalpha reveals that most, but not all, inflammatory and destructive mediators are NFkappaB dependent. *Rheumatology (Oxford, England)* **45**: 1201–1209.
10. Tas SW, *et al.* (2006). Amelioration of arthritis by intra-articular dominant negative Ikk beta gene therapy using adeno-associated virus type 5. *Human gene therapy* **17**: 821–832.

11. Pagnotto M, Wang Z, Karpie J, Ferretti M, Xiao X, Chu C (2007). Adeno-associated viral gene transfer of transforming growth factor-beta1 to human mesenchymal stem cells improves cartilage repair. *Gene therapy* **14**: 804–813.

12. Goodrich L, Hidaka C, Robbins P, Evans C, Nixon A (2007). Genetic modification of chondrocytes with insulin-like growth factor-1 enhances cartilage healing in an equine model. *J Bone Joint Surg Br* **89**: 672–685.

13. Evans C, *et al.* (1996). Clinical trial to assess the safety, feasibility, and efficacy of transferring a potentially anti-arthritic cytokine gene to human joints with rheumatoid arthritis. *Human gene therapy* **7**: 1261–1280.

14. Lee K (2008). Preclincal and early clinical analysis of allogeneic chondrocytes transfected retrovirally with TGF-beta1 gene for degenerative arthritis patients. In *5th International Meeting of Gene Therapy of Arthritis and Related Disorders*, Seattle, WA, USA.

15. Mease P, *et al.* (2008). Safety, Local Tolerability and Clinical Response After Intra-articular Administration of a Recombinand Adeno-associated Vector Containing a TNF Antagonist in Inflammatory Arthritis. In *EULAR Congress 2008*, Paris, Fance.

16. (2008). Centers for Disease Control and Prevention. National Diabetes Fact Sheet: United States 2007. General Information and National Estimates on Diabetes. *Available at: http://wwwcdcgov/diabetes/pubs/pdf/ndfs_2007pdf.*

17. (2006). Incidence and trends of childhood Type 1 diabetes worldwide 1990–1999. *Diabet Med* **23**: 857–866.

18. Marzorati S, Pileggi A, Ricordi C (2007). Allogeneic islet transplantation. *Expert Opin Biol Ther* **7**: 1627–1645.

19. van der Windt DJ, Bottino R, Casu A, Campanile N, Cooper DK (2007). Rapid loss of intraportally transplanted islets: An overview of pathophysiology and preventive strategies. *Xenotransplantation* **14**: 288–297.

20. Meloche RM (2007). Transplantation for the treatment of type 1 diabetes. *World J Gastroenterol* **13**: 6347–6355.

21. Wang Z, *et al.* (2006). Widespread and stable pancreatic gene transfer by adeno-associated virus vectors via different routes. *Diabetes* **55**: 875–884.

22. Pasquali L, Giannoukakis N, Trucco M (2008). Induction of immune tolerance to facilitate beta cell regeneration in type 1 diabetes. *Adv Drug Deliv Rev* **60**: 106–113.

23. Nir T, Melton DA, Dor Y (2007). Recovery from diabetes in mice by beta cell regeneration. *J Clin Invest* **117**: 2553–2561.

24. Gandhi GY, *et al.* (2008). Immunotherapeutic agents in type 1 diabetes: A systematic review and meta-analysis of randomized trials. *Clin Endocrinol (Oxf)* **69**: 244–252.

25. Harrison LC, *et al.* (2004). Pancreatic beta-cell function and immune responses to insulin after administration of intranasal insulin to humans at risk for type 1 diabetes. *Diabetes Care* **27**: 2348–2355.

26. Ludvigsson J, *et al.* (2008). GAD treatment and insulin secretion in recent-onset type 1 diabetes. *N Engl J Med* **359**: 1909–1920.

27. Creusot RJ, *et al.* (2008). Tissue-targeted therapy of autoimmune diabetes using dendritic cells transduced to express IL-4 in NOD mice. *Clin Immunol* **127**: 176–187.

28. Feili-Hariri M, *et al.* (2003). Dendritic cells transduced to express interleukin-4 prevent diabetes in nonobese diabetic mice with advanced insulitis. *Hum Gene Ther* **14**: 13–23.

29. Perone MJ, *et al.* (2006). Dendritic cells expressing transgenic galectin-1 delay onset of autoimmune diabetes in mice. *J Immunol* **177**: 5278–5289.
30. Ma L, *et al.* (2003). Prevention of diabetes in NOD mice by administration of dendritic cells deficient in nuclear transcription factor-kappaB activity. *Diabetes* **52**: 1976–1985.
31. Soukhareva N, Jiang Y, Scott DW (2006). Treatment of diabetes in NOD mice by gene transfer of Ig-fusion proteins into B cells: Role of T regulatory cells. *Cell Immunol* **240**: 41–46.
32. Samson SL, Chan L (2006). Gene therapy for diabetes: Reinventing the islet. *Trends Endocrinol Metab* **17**: 92–100.
33. Siatskas C, *et al.* (2006). Gene therapy strategies towards immune tolerance to treat the autoimmune diseases. *Curr Gene Ther* **6**: 45–58.
34. Rehman KK, Trucco M, Wang Z, Xiao X, Robbins PD (2008). AAV8-mediated gene transfer of interleukin-4 to endogenous beta-cells prevents the onset of diabetes in NOD mice. *Mol Ther* **16**: 1409–1416.
35. Machen J, Harnaha J, Lakomy R, Styche A, Trucco M, Giannoukakis N (2004). Antisense oligonucleotides down-regulating costimulation confer diabetes-preventive properties to nonobese diabetic mouse dendritic cells. *J Immunol* **173**: 4331–4341.
36. Giannoukakis N, Phillips B, Trucco M (2008). Toward a cure for type 1 diabetes mellitus: Diabetes-suppressive dendritic cells and beyond. *Pediatr Diabetes* **9**: 4–13.
37. Prieto J, Fernandez-Ruiz V, Kawa MP, Sarobe P, Qian C (2008). Cells as vehicles for therapeutic genes to treat liver diseases. *Gene Ther* **15**: 765–771.

Chapter 19

Gene Therapy for Inherited Metabolic Storage Diseases

Cathryn Mah*

In this chapter, we will discuss the principles surrounding the study and development of gene therapy for inherited metabolic storage diseases. These rare diseases are caused primarily by a defect in a single gene and result in the excess storage of substrates within affected cells, ultimately rendering the cells dysfunctional. There exist numerous classifications of inherited metabolic storage diseases which include two broad categories: lysosomal storage diseases and glycogen storage diseases. No cures exist for any of the diseases and current therapies are palliative. Gene therapy provides a possible alternative therapy for inherited metabolic storage diseases and here we discuss current gene therapy strategies and considerations for strategy design for representative diseases including MPS I, Gaucher disease, Pompe disease, and glycogen storage disease type Ia.

1. Introduction

The majority of gene therapy studies for inherited metabolic storage diseases have employed viral-based gene therapy vectors, in particular retroviral, lentiviral, adenoviral, and adeno-associated viral vectors. Other viral and non-viral vectors have also been explored to a lesser extent. Treatment strategies have varied substantially, in that different vectors, differing *cis*-elements within each vector construct, methods of vector delivery, and the target tissue(s) have differed, even when approaching the same disease.

*Correspondence: University of Florida College of Medicine, Dept. Pediatrics, Div. Cellular Molecular Therapy, and Powell Gene Therapy Center, 1376 Mowry Road, Rm 493, Gainesville, FL 32610. E-mail: cmah@ufl.edu

However, the primary goals of sufficient transduction of the affected tissue(s), sustained expression of the delivered gene at therapeutic levels, and avoidance of untoward or adverse effects such as immune response to vector or transgene, are common for each.

While at the time of writing of this chapter, a gene therapy strategy to treat an inherited metabolic storage disorder has not yet yielded a complete cure for disease, nor has a gene therapy treatment been approved as a standard of care, data suggest that significantly improved functional outcomes can be achieved using gene therapy as compared to current therapies. Each new study has provided information towards design of improved strategies and further research can only move the field of gene therapy closer toward the goal of a cure.

2. Lysosomal Storage Diseases

Lysosomal storage diseases (LSDs) are a class of inherited metabolic storage diseases encompassing greater than 40 distinct diseases. Lysosomes are intracellular organelles that contain enzymes that play important roles in the break down of complex substrates and macromolecules. Most LSDs are caused by a deficiency in a functional lysosomal enzyme, however deficiencies in nonenzymatic activator proteins, proteins required for proper intracellular trafficking of lysosomal enzymes, or mutations in integral lysosomal membrane proteins can also result in disease. LSDs are characterized by excess storage of undigested or partially digested materials within the lysosomes of affected cells. Accumulation of substrates within lysosomes eventually leads to disruption of normal cellular architecture and ultimately cellular/tissue function. In some cases, the substrates themselves can be cytotoxic. Affected tissues range from skeletal and cardiac muscle to tissues comprising the central nervous system.[1] While gene therapy strategies are being developed for numerous LSDs, two LSDs, mucopolysaccharidosis Type I and Gaucher disease, will be discussed in this chapter as examples.

Mucopolysaccharidoses (MPS disorders) are a group of LSDs characterized by the inability to degrade glycosaminoglycans used to build bone, cartilage, tendons, joint fluid, and other tissues. Mucopolysaccharidosis Type I (MPS I) is caused by a lack of alpha-L-iduronidase (IDUA), resulting in the build-up of dermatan and heparan sulfate. MPS I can range from a severe form (also known as Hurler syndrome) to mild (Scheie syndrome) form. Clinical features can vary greatly with the severity of disease, however hallmark complications can include joint stiffness and skeletal dysplasia, cardiovascular disease, liver and/or spleen enlargement, clouding of the cornea, developmental delay, and characteristic facial and physical dysmorphisms.[1,2]

Gaucher disease is the most common LSD and results in the storage of the lipid glucocerebroside primarily in macrophages and monocytes due to a defect in beta-glucosidase enzyme. Clinical symptoms of Gaucher disease include hepatosplenomegaly, anemia, thrombocytopenia, leukopenia, and severe bone disease. Like with most other LSDs, there exist subtypes of disease that are characterized by severity and differing clinical symptoms. Type 1 Gaucher disease is the most prevalent form of Gaucher disease and is non-neuropathologic. Gaucher disease Types 2 and 3 are characterized by having acute or chronic neurologic symptoms, respectively.[1,2]

Current therapies for both MPS I and Gaucher disease include bone marrow/hematopoietic stem cell transplantation and/or recombinant enzyme replacement therapy (ERT) or substrate reduction therapy (SRT).[3-8] While these therapies have been shown to improve survival and quality of life, in most cases, disease is not cured, and complications can arise. Bone marrow/stem cell transplantation has been employed as blood cells have been shown to express and secrete the deficient enzymes in MPS I and Gaucher disease. The primary acute complication with this form of therapy is rejection of the graft, and while disease progression can be halted or delayed, bone and neurological components of disease cannot be reversed. ERT requires repeated intravenous dosing for the lifetime of the patient and can correct non-neuronopathlogies however, due to the blood-brain barrier, ERT is unable to correct neurological manifestations. SRT utilizes oral dosing of small molecule compounds to inhibit the formation of glycosphingolipids. The potential advantage of SRT compounds is the ability to cross the blood-brain barrier, which theoretically can improve CNS pathology, however safety concerns still exist for current SRT drugs and are used only in patients unable to receive ERT. Gene therapy may provide a more effective alterative to conventional therapies for lysosomal storage diseases as a single administration could lead to expression of the therapeutic enzyme for the lifetime of the individual. Furthermore it is feasible that gene therapy can provide widespread correction in all affected tissues including cells of the CNS.[9-17]

3. Glycogen Storage Diseases

Glycogen storage diseases (GSDs) are a group of nine different inherited metabolic storage disorders. As the name implies, GSDs are characterized by aberrations in normal glycogen metabolism and results in the accumulation of glycogen in affected cells (with the exception of GSD Type 0 which is considered a form of GSD but results in an inability to store glycogen). Much like the lysosomal storage disorders, most GSDs are a result of a lack of a functional enzyme in glycogen metabolism, but some types or subtypes of GSD types can be caused by improper intracellular trafficking of enzymes from proper compartments. Furthermore, GSDs can manifest very different clinical symptoms and pathologies.[2,18] Two example types of GSD will be discussed in this chapter, GSD Types I and II.

Glycogen storage disease Type II (GSDII; Pompe disease) is also a lysosomal storage disorder and results from a lack of lysosomal acid alpha-glucosidase (GAA). GAA is responsible for cleavage of alpha-1,4- and alpha-1,6-glycosidic bonds of glycogen and in Pompe disease, glycogen will build up within the lysosomes to the point of destroying normal cell structure and function.[19,2] The clinical phenotype of Pompe disease is that of a muscular dystrophy with severe progressive weakening of cardiac and skeletal muscles as the most outwardly affected tissues, however evidence suggests a neurological component exists as well. In 2006, ERT was approved for the treatment of Pompe disease. Like with ERT for other lysosomal storage diseases, therapy significantly improves survival and quality of life, however, there exist shortcomings and therapy is not curative.[20] Gene therapy provides the potential for sustained endogenous expression of the therapeutic enzyme, thus eliminating the need for serial treatments. Furthermore endogenous enzyme expression may lead to more efficient processing, post-translational modification, and intracellular trafficking of the enzyme, thereby creating a more potent therapeutic agent.

Glycogen storage disease Type I (GSDI, von Gierke disease) is an inherited metabolic storage disorder resulting from the inability to break down glycogen to glucose for energy. As such, affected individuals suffer from severe hypoglycemia, which in turn results in other profound metabolic disturbances such as lactic acidemia and hyperlipidemia. There exist subtypes of GSDI (types Ia and Ib) in which the molecular mechanism of disease differs. GSDIa is caused by a lack of functional glucose-6-phosphatase (G6Pase) enzyme whereas type Ib is caused by improper transport of the enzyme within the cell. Therapy for GSDI is solely palliative and consists of strict dietary supplementation with uncooked cornstarch as a slow-release source of glucose. Gene therapy is currently the only option to correct the basic metabolic defect by providing the necessary G6Pase to the affected cells.[21–23]

4. Animal Models

While most inherited metabolic storage diseases are caused primarily by a single protein defect and may result in biochemical abnormalities in isolated tissues, clinical symptoms tend to manifest in more global impaired functioning of affected individuals. The metabolic processes of the body are strongly tied to one another in that changes in one small aspect can cause a cascade of downstream events far removed from the original founding problem. Preclinical studies in animal models of disease provide unique complex systems in which therapeutic success and outcomes of gene therapy can be assessed as a whole.

While animal models do not exist for every form of disease, there are numerous naturally-occurring as well as genetically engineered animal models of inherited metabolic storage disease.[14,17] Mouse models of disease, when available, provide an invaluable resource as a system in which to assess proof-of-principle studies. Most mouse models are inbred with identical genetically-defined backgrounds, thereby reducing variability in response to therapy and outcome measures. In addition, mice are relatively easy to breed in that they have multiple pups per litter and gestation ranges from approximately 18–22 days, depending on the strain, thus large numbers of animals can be evaluated in a relatively shorter span of time than larger animal models. Mouse models of disease, however, do not always necessarily mimic all aspects of human disease and due to their homogenous genetic backgrounds, may not accurately represent therapeutic outcomes that will be seen in the general patient population where the genetic makeup of each individual can be vastly different. While the biochemical and even molecular basis for most of the inherited metabolic storage diseases remains the same between individuals with a particular disease, other underlying factors (such as the existence of other congenital abnormalities or other modifying factors that are not directly related to disease) may impact the course of disease pathology.

Larger animal models of disease tend to be naturally-occurring and are usually originally identified by their clinical symptoms as they relate to human disease.[14–17] These higher vertebrates tend to be more complex than their rodent counterparts with more heterogeneous genetic makeup and in some cases, more accurately represent physiological and pathological phenotypes seen in the patient population. In addition, more clinically relevant vector delivery methods or functional testing measures can be performed in the larger animal models of disease.

Murine, feline and canine models exist for MPS I.[24–25] The MPS I knockout mouse model was generated by the targeted disruption of the murine *Idua* gene and exhibits most hallmark clinical features of disease such as altered facial and paw morphology, and abnormal lysosomal storage in most analyzed tissues including liver, spleen, bone, muscle, and brain. Animals exhibit skeletal dysplasia, abnormal cardiac function, and reduced vision and hearing capabilities. While primary complications of MPS I are noted in these mouse model, they do not precisely mimic human disease either. For example, despite having impaired vision, corneal clouding is not noted, animals also do not have shortened lifespans, and while MPS I mice have impaired cardiac function, the specific regions of cardiac abnormalities differ between mice and humans. The feline and canine models of MPS I are caused by naturally-occurring mutations in the *Idua* gene. Both exhibit facial deformities, lysosomal storage of dermatan and heparan sulfates, cardiac murmurs, corneal clouding, joint disease, and skeletal dysplasia. MPS I dogs have more pronounced skeletal disease than cats with MPS I and the course of pathology most closely resembles the moderate Hurler-Scheie form of disease.

Gene therapy studies have been performed in all three animal models of MPS I and comparison of a single strategy performed in each revealed different outcomes, thus demonstrating that strategies may not translate in a linear fashion as we move from preclinical to clinical studies. Intravenous delivery of a retroviral vector encoding the canine *IDUA* gene under the control of liver-specific promoter resulted in very high levels of serum cIDUA that was sustained throughout the study (8 months) with significant storage reduction and complete correction and normalization of cardiac function, vision, and bone mass density in the mouse model.[26] In the MPS I dog, the same gene therapy strategy still resulted in 25% of the serum cIDUA level seen in mice and the dogs had only moderate clinical improvement.[27] In the MPS I cat, gene expression was similar to that in the dogs but expression lasted only 1–3 months as a result of a robust cytotoxic T lymphocyte response against cIDUA-expressing cells, in the absence of a humoral anti-cIDUA response.[14,28] Several factors including the difference of vector transduction (infection and subsequent gene expression) efficiency as it relates to each different species of animal, the maturity and reactivity of each species immune system to the vector and the expressed enzyme, and the subtle differences in disease pathology may have played roles in the difference responses to the gene therapy.[29–31]

Several independent mouse models of Gaucher disease have been developed either by point mutation or postconditional knockout of *gba*.[14,32,33] Both models exhibit pathology, however the point mutation model does not develop clinical disease whereas the conditional knockout exhibits more clinical symptoms of disease, including infiltration of Gaucher cells (reticuloendothelial cells with excess storage of glucocerebroside) in bone marrow, liver, and spleen and anemia, one year post-induction of the knockout condition. Up until recently, mouse models of neuronopathic Gaucher disease (Types 2 and 3) resulted in animals with limited viability and usually died within hours after birth. In 2007, a model in which acid beta-glucosidase enzyme deficiency was restricted to neural and glial cells was generated and resulted in mice with significant neurological abnormalities similar to Types 2 and 3 Gaucher. These animals have only slightly longer lifespans of 3-4 weeks of age however; this timeframe does provide opportunities to test some therapies.[34]

Animal models of Pompe disease include knockout mouse models and a naturally-occurring Japanese quail model.[14,16,17] Although independent mouse models of Pompe disease all lack expression of acid alpha-glucosidase (GAA) and present with glycogen storage in skeletal and cardiac muscle, only the exon 6-disrupted model ($6^{neo}/6^{neo}$) manifests the clinical phenotype of progressive skeletal muscle weakening. The difference is, in part, attributed to the difference of background mouse strains (FVB versus C57BL/6 for $6^{neo}/6^{neo}$ mice) used to develop the models. There still exist however, differences between the mouse and human forms of disease, the most striking being that although the mouse model completely lacks GAA enzyme activity, mice survive to adulthood and exhibit progressive disease similar to juvenile and adult-onset patients, whereas patients who lack functional enzyme normally do not survive past infancy.[16]

There exist a knockout mouse and naturally-occurring canine model of glycogen storage disease type Ia. Both models are unable to survive past weaning age without intense dietary intervention to prevent severe hypoglycemia and are characterized by massive hepatomegaly, hyperlipidemia, and hyperuricemia, however the canine model also develops lactic acidemia which is a primary complication of disease in the patient population.[22,23]

Animal models of disease are invaluable resources in which to assess novel therapeutics and when available, preclinical proof-of-concept and toxicological studies are required prior to moving toward clinical trials. When designing preclinical gene therapy studies, an understanding of the interplay from the molecular level to whole animal physiology is important in developing effective treatment strategies. Other factors such as species-related differences and genetic background should also be taken into consideration

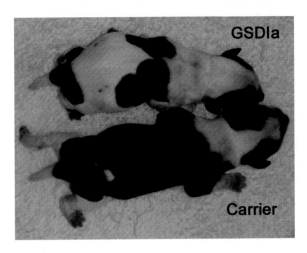

Fig. 19.1 Newborn affected and carrier GSDIa dogs. Note the gross hepatomegaly that is evident in the affected puppy even at birth.

when evaluating therapeutic strategies and outcomes. For study end-point measurements, it is important to characterize the immediate biochemical outcomes (is the transgene expressed and how much is being expressed) and the effects on the target cells (is substrate storage reduced) and tissues (tissue histology/morphology), however improvements in physiologic and/or behavioral function are the true indicators of therapeutic success as these are the markers that would ultimately translate to improved survival and quality of life in the patient.

5. Cross-Correction Strategies

For most inherited metabolic storage diseases, levels of expression of the therapeutic protein, as small as 1% of normal levels, can result in significant benefits and improvement of the clinical phenotype.[14,16,17] A unique characteristic of lysosomal storage diseases is that most arise from a deficiency of a soluble hydrolase. While the majority of synthesized lysosomal enzymes are targeted to the endosomal system via binding of the enzyme to the mannose-6-phosphate receptor followed by intracellular trafficking of the complex, a small percentage of expressed enzyme does not traffic to the endosome/lysosome and instead is secreted from the cell, which can be taken up by distal cells and trafficked to the lysosome to perform its enzymatic function.[16,17] Current ERT and bone marrow transplantation therapies take advantage of this secretion/uptake process. A challenge for cross-correction-based strategies is the efficient penetration of circulating enzyme into the affected tissue. This challenge becomes more prominent in diseases that have a neural component, due to the blood-brain barrier. However, gene therapy has the potential to both directly correct affected cells and tissues and also provide an endogenous system for cross-correction, thus providing correction that is more similar to the native function and trafficking of the enzyme.

Due to ease of vector delivery by intramuscular (IM) injection, muscle has been one target as an endogenous source of enzyme, however strategies have been ineffective so far and is thought to be partially attributed to the inefficiency of those secretion processes in muscle.[14,16,17] In addition, data from multiple gene therapy studies suggests that the site of vector delivery plays an important role in the individual to mount an immune response

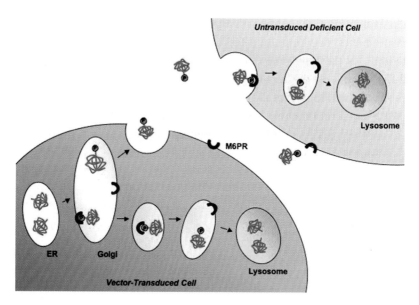

Fig. 19.2 Schematic of cross-correction. The expressed lysosomal enzyme (green) are glycosylated in the rough endoplasmic reticulum followed by phosphorylation of the mannose-6-phosphate in the Golgi apparatus (orange circle) where they can bind the mannose-6-phosphate receptor (blue, M6PR). The majority of enzyme is then trafficked within the cell to the mature lysosome. A small percentage of enzyme is secreted from the cell where it can bind M6PR on distal cells and be endocytosed followed by trafficking to the lysosome.

(either to the vector itself, the therapeutic protein, or the transduced cells) and that muscle tissue may be more prone to the development of humoral and/or cytotoxic immune response.[35]

In contrast to muscle cells, hepatocytes synthesize and secrete myriad proteins and may provide a more appropriate milieu as a center for therapeutic enzyme production. Administration of liver-targeting vectors to neonatal animals has been quite successful for several diseases including MPS I and MPS VII, however treatment of adult animals has been more challenging due to the mature immune system being able to mount an immune response. In the mouse model of Pompe disease, studies showed that when antibodies were elicited in response to gene therapy, cross-correction was completely abrogated.[36,23,35] However, in animals that did not elicit an immune response, cross-correction of skeletal and cardiac muscle could be achieved.

Numerous *in vivo* studies in varying models of disease ranging from muscular dystrophy to hemophilia have shown that restriction of gene expression to only target cells can eliminate unwanted immunological reactions resulting from expression of the therapeutic protein in cells that do not express the gene in wild-type status or from expression directly in antigen presenting cells. Use of tissue-specific promoters does not completely preclude the possibility of immune reactivity, suggesting that other factors such as mechanisms of immune evasion or thresholds of immune tolerance, as well as the nature of the animal model itself likely play a role in the immune response.

Ex vivo bone marrow or hematopoietic stem cell gene therapy has been explored for the treatment of lysosomal storage disease. These strategies provide an advantage in which autologous bone marrow or hematopoietic stem cells are transduced with a gene therapy vector to express the therapeutic protein and are then introduced to the affected individual. The potential benefit of this is a reduction in graft rejection and other complications associated with the use of allogeneic or unrelated cells. Murine studies have resulted in varying levels of efficacy and as studies moved to the larger animal models, transplantation of *ex vivo*–transduced bone marrow has not been successful, suggesting greater complexities. Both preclinical and clinical studies have revealed that a degree of immune conditioning (in which bone marrow cells are ablated or partially eradicated) prior to transplantation is required for the successful engraftment of *ex vivo* transduced cells.[4,10,14,16] Other factors such as the nature of disease itself, animal model, degree of disease progression, and the levels of gene expression also play a role with the trend being that less disease pathology and the greater the levels of transduction lead to the most clinical improvement.

6. Direct Correction of Target Tissues

Direct delivery of gene therapy vectors to affected tissues is the most straightforward gene therapy strategy however most inherited metabolic storage diseases involve pathologies in several tissue or organ systems. Due to the inherent differences between each affected cell type, it may be necessary to treat specific tissue types separately to achieve global correction. Significant correction of the prominently affected tissues could result in substantial amelioration of clinical symptoms and in some cases, such as when both CNS and peripheral organs are affected, separate correction for each aspect is probably required. Multiple considerations exist in developing gene therapy strategies for direct correction of target tissues and include vector type, promoter and other *cis*-elements in the vector construct, and method of vector delivery. Selection of each component will vary for each tissue type and disease as well as the ultimate goals of the study.

The majority of gene therapy studies to-date involving direct correction of CNS for lysosomal storage diseases have been performed using AAV-based vectors, however other vectors have also shown promise and therapeutic strategies will evolve with each subsequent study. Intrathecal and intracranial administration of rAAV vector resulted in widespread biochemical correction in the brain with significant histological correction for several lysosomal storage diseases including MPS I, MPS IIIB, MPS VII, Sandhoff diseases, Niemann-Pick A, globoid cell leukodystrophy, and metachromatic leukodystrophy and in some cases, improvement in behavioral symptoms were noted.[10,16,37] Differences in success between each study suggest that interplay between the specific disease (whether it is the altered cellular structure or physiology, or basic cellular mechanics or biochemistry) and specific vector type also affects therapeutic outcomes. While a single or few injections in the mouse can result in correction throughout the brain, it is likely that as studies progress towards humans, modifications either to the vector itself or in the basic delivery method may need to be done to ensure widespread distribution, however the fact that vectors can be transported along neuronal connections to distal sites in the CNS and that secreted enzymes can be transported antero- and retrograde to cross-correct cells away from the site of injection provide advantages to CNS-targeted gene therapy.

The challenge in gene therapy strategies to directly correct muscle is the ability to correct all muscle fibers within the body. As body-wide direct injection is not practical, studies have focused more on specific muscles groups that may provide the most clinical benefit. For example, Pompe disease strategies have targeted the diaphragm for respiratory function or hindlimbs for ambulation.[23,38] The development of novel vectors with distinct kinetics of expression and *in vivo* biodistribution studies have resulted in vectors that can mediate wide-spread transduction of skeletal and cardiac muscle after simple intravenous injection. For example, intravenous injection of a therapeutic AAV serotype 1 vector to neonate Pompe mice resulted in long-term biochemical and histological correction of skeletal muscle, diaphragm, and cardiac tissues with corresponding functional improvement in each.[39]

Systemic intravenous delivery of vectors has been explored to treat many forms of inherited metabolic storage diseases. Ideally, gene therapy strategies should be optimized so that a single peripheral injection of vector would be sufficient to cure disease and that the inherent nature of the vector itself would selectively infect and completely correct all negatively affected cells in the body. While that level of correction has not yet been achieved, understanding of the basic biology of vectors provides avenues for manipulation to design a better vector. The biodistribution of vector transduction is primarily centered around the mechanism of transduction; (1) whether it is taken up by a receptor-mediated process or endocytosed, (2) the intracellular fate of the vector genome as it is released in the cell, (3) the availability of the vector genome to transcription machinery as well as the presence of transcription factors specific to each construct, and (4) efficient translation (and when necessary post-translational modification) to a functional product. These factors can vary due to species, age, gender, and/or other genetic modifiers of the subject it is also possible that the aberrant functioning of diseased cells could affect biodistribution and expression. Furthermore, other factors such as immune status with respect to the vector or the transgene product also affect transduction.

Direct correction of liver is the primary strategy for the treatment of GSDIa as the therapeutic enzyme is a transmembrane protein and therefore not amenable to cross-correction strategies. To-date all gene therapy studies for GSDIa have involved neonatal delivery of vector due to the very short average lifespans (1–4 weeks) for both mouse and dog models of disease.[22,23,40] Intense dietary therapy has been shown to prolong life in the GSDIa dog and future studies examining the effects of gene therapy in older animals would be of interest, as much of the current patient population approaches adulthood. AAV vectors have mediated expression of hepatic G6Pase, prolonged survival, and reduction in accumulated glycogen and most importantly, the ability to sustain short periods of fasting without suffering severe hypoglycemia. Like with the lysosomal storage diseases, only small amounts of enzyme activity (as little as 10–11% wild-type activity in both the GSDIa mouse and dog) resulted in substantial physiologic correction.[22,23,40] Interestingly, despite the use of constitutive promoters in some of the aforementioned studies, no untoward effects have been noted however this may be due to subphysiologic levels of expression. As such metabolic processes are tightly controlled, regulatable promoters will likely become important as newer vectors give rise to more efficient transduction levels.

7. Conclusions and Outlook

There exists minimal clinical experience with gene therapy for inherited metabolic storage diseases with only two phase I/II studies that investigated the potential of *ex vivo* retrovirally-transduced autologous cells for the treatment of MPS II or Gaucher disease (Clinicaltrials.gov NCT00004294, NCT00004454).[16] Unfortunately low expression and no improvement in disease phenotype were noted. Since the initiation of those studies, a surge of improvements in vector design as well as the identification of novel vectors occurred and preclinical gene therapy studies have shown very promising results with significant correction of the clinical phenotype of numerous inherited metabolic storage diseases. It is possible that a combination of different gene therapy strategies together or with current therapeutic modalities will further improve clinical outcomes. Nonetheless, while no one therapy has been shown to be absolutely perfect so far, significant improvements in survival and quality of life may be achievable with the current technology. Further investigation and development of vectors as well as a more refined understanding of the host-vector interactions will only lead to better gene therapy approaches.

References

1. Neufeld EF (1991). Lysosomal storage diseases. *Annu Rev Biochem* **60**: 257–280.
2. Scriver CR, *et al.* (2001). *The Metabolic and Molecular Basis of Inherited Disease.*
3. Beck M, (2007). New therapeutic options for lysosomal storage disorders: enzyme replacement, small molecules and gene therapy. *Hum Genet* **121**: 1–22.
4. Boelens JJ (2006). Trends in haematopoietic cell transplantation for inborn errors of metabolism. *J Inherit Metab Dis* **29**: 413–420.
5. Brady RO (2006). Enzyme replacement for lysosomal diseases. *Annu Rev Med* **57**: 283–296.

6. Grabowski GA (2008). Treatment perspectives for the lysosomal storage diseases. *Expert Opin Emerg Drugs* **13**: 197–211.

7. Pastores GM, Barnett NL (2005). Current and emerging therapies for the lysosomal storage disorders. *Expert Opin Emerg Drugs* **10**: 891–902.

8. Winchester B, Vellodi A, Young E, (2000). The molecular basis of lysosomal storage diseases and their treatment. *Biochem Soc Trans* **28**: 150–154.

9. Barranger JM, Novelli EA (2001). Gene therapy for lysosomal storage disorders. *Expert Opin Biol Ther* **1**: 857–867.

10. Biffi A, Naldini L, (2005). Gene therapy of storage disorders by retroviral and lentiviral vectors. *Hum Gene Ther* **16**: 1133–1142.

11. Cabrera-Salazar MA, Novelli E, Barranger JA (2002). Gene therapy for the lysosomal storage disorders. *Curr Opin Mol Ther* **4**: 349–358.

12. Caillaud C, Poenaru L, (2000). Gene therapy in lysosomal diseases. *Biomed Pharmacother* **54**: 505–512.

13. Cheng SH, Smith AE (2003). Gene therapy progress and prospects: gene therapy of lysosomal storage disorders. *Gene Ther* **10**: 1275–1281.

14. Ellinwood NM, Vite CH, Haskins ME (2004). Gene therapy for lysosomal storage diseases: the lessons and promise of animal models. *J Gene Med* **6**: 481–506.

15. Grabowski GA (2003). Perspectives on gene therapy for lysosomal storage diseases that affect hematopoiesis. *Curr Hematol Rep* **2**: 356–362.

16. Hodges BL, Cheng SH (2006). Cell and gene-based therapies for the lysosomal storage diseases. *Curr Gene Ther* **6**: 227–241.

17. Sands MS, Davidson BL (2006). Gene therapy for lysosomal storage diseases. *Mol Ther* **13**: 839–849.

18. Wolfsdorf JI, Weinstein DA (2003). Glycogen storage diseases. *Rev Endocr Metab Disord* **4**: 95–102.

19. Raben N, Plotz P, Byrne BJ (2002). Acid alpha-glucosidase deficiency (glycogenosis type II, Pompe disease). *Curr Mol Med* **2**: 145–166.

20. Kishnani PS, *et al.* (2007). Recombinant human acid [alpha]-glucosidase: major clinical benefits in infantile-onset Pompe disease. *Neurology* **68**: 99–109.

21. Chou JY, Matern D, Mansfield BC, Chen YT (2002). Type I glycogen storage diseases: disorders of the glucose-6-phosphatase complex. *Curr Mol Med* **2**: 121–143.

22. Chou JY, Mansfield BC (2007). Gene therapy for type I glycogen storage diseases. *Curr Gene Ther* **7**: 79–88.

23. Koeberl DD, Kishnani PS, Chen YT (2007). Glycogen storage disease types I and II: treatment updates. *J Inherit Metab Dis* **30**: 159–164.

24. Clarke LA, *et al.* (1997). Murine mucopolysaccharidosis type I: targeted disruption of the murine alpha-L-iduronidase gene. *Hum Mol Genet* **6**: 503–511.

25. Russell C, *et al.* (1998). Murine MPS I: insights into the pathogenesis of Hurler syndrome. *Clin Genet* **53**: 349–361.

26. Liu Y, *et al.* (2005). Liver-directed neonatal gene therapy prevents cardiac, bone, ear, and eye disease in mucopolysaccharidosis I mice. *Mol Ther* **11**: 35–47.

27. Traas AM, *et al.* (2007). Correction of clinical manifestations of canine mucopolysaccharidosis I with neonatal retroviral vector gene therapy. *Mol Ther* **15**: 1423–1431.

28. Ponder KP, *et al.* (2006). Mucopolysaccharidosis I cats mount a cytotoxic T lymphocyte response after neonatal gene therapy that can be blocked with CTLA4-Ig. *Mol Ther* **14**: 5–13.

29. Chung S, Ma X, Liu Y, Lee D, Tittiger M, Ponder KP (2007). Effect of neonatal administration of a retroviral vector expressing alpha-L-iduronidase upon lysosomal storage in brain and other organs in mucopolysaccharidosis I mice. *Mol Genet Metab* **90**: 181–192.

30. Kobayashi H, *et al.* (2005). Neonatal gene therapy of MPS I mice by intravenous injection of a lentiviral vector. *Mol Ther* **11**: 776–789.

31. Ma X, *et al.* (2007). Improvements in mucopolysaccharidosis I mice after adult retroviral vector-mediated gene therapy with immunomodulation. *Mol Ther* **15**: 889–902.

32. Enquist IB, *et al.* (2007). Murine models of acute neuronopathic Gaucher disease. *Proc Natl Acad Sci USA* **104**: 17483–17488.

33. Berglin EI, Nilsson E, Mansson JE, Ehinger M, Richter, J, Karlsson S, (2009). Successful low-risk hematopoietic cell therapy in a mouse model of type 1 Gaucher disease. *Stem Cells*, in press

34. Sinclair GB, *et al.* Jevon G, Colobong KE, Randall DR, Choy FY, Clarke LA (2007). Generation of a conditional knockout of murine glucocerebrosidase: utility for the study of Gaucher disease. *Mol Genet Metab* **90**: 148–156.

35. Warrington KH Jr, Herzog RW (2006). Treatment of human disease by adeno-associated viral gene transfer. *Hum Genet* **119**: 571–603.

36. Cresawn KO, *et al.* (2005). Impact of humoral immune response on distribution and efficacy of recombinant adeno-associated virus-derived acid alpha-glucosidase in a model of glycogen storage disease type II. *Hum Gene Ther* **16**: 68–80.

37. Sands MS, Haskins ME (2008). CNS-directed gene therapy for lysosomal storage diseases. *Acta Paediatr* Suppl **97**: 22–27.

38. Mah C, Fraites TJ Jr., Cresawn KO, Zolotukhin I, Lewis MA, Byrne BJ (2004). A new method for recombinant adeno-associated virus vector delivery to murine diaphragm. *Mol Ther* **9**: 458–463.

39. Mah C, *et al.* (2007). Physiological correction of Pompe disease by systemic delivery of adeno-associated virus serotype 1 vectors. *Mol Ther* **15**: 501–507.

40. Koeberl DD, *et al.* (2008). AAV vector-mediated reversal of hypoglycemia in canine and murine glycogen storage disease type Ia. *Mol Ther* **16**: 665–672.

Chapter 20
Retinal Diseases

Shannon E. Boye*, Sanford L. Boye
and William W. Hauswirth

A variety of retinal diseases can potentially be treated with gene therapy. Because of its safety, long term expression and ability to transduce differentiated cells, recombinant Adeno-associated virus (AAV) has emerged as the most optimal gene delivery vehicle to treat retinal disease. Recombinant AAV can be engineered to have either broad or selective tropism based on the choice of promoter sequence or serotype used. The ability to target specific cells affords researchers the opportunity to treat diseases affecting specific cells of the retina including rod and cone photoreceptors, retinal ganglion cells and the retinal pigment epithelium. Proof of concept experiments have demonstrated the efficacy of AAV-mediated therapy in a variety of animal models of retinal disease. In the case of one form of inherited retinal dystrophy, Leber's congenital amaurosis-2 (LCA2), these studies laid the groundwork for the first AAV-mediated ocular gene therapy in patients. A summary of events which bridged the gap from bench (pre-clinical, proof-of-principle and safety studies) to bedside (phase I clinical trials for LCA2) are also discussed.

1. Introduction

The use of recombinant viral vectors to treat disease has increased in recent years. Although of limited impact presently, they have perhaps the most potential to treat inherited and acquired ocular disease. Several factors make the eye a most amenable organ for gene therapy, including its immune privileged status, physical compartmentalization and ease of

*Correspondence: University of Florida, Gainesville, Fl.
E-mail: shaire@ufl.edu

surgical accessibility. Certain organs such as the eye have developed an evo-lutionary adaptation which protects them from the damaging effects of most inflammatory immune responses. This allows intraocular administration of certain gene therapy vectors without eliciting a cell-mediated response. The small size and compartmentalization of this organ permits delivery of small volumes of therapeutic vector without systemic exposure, thereby prevent-ing unwanted, systemic side effects. In addition, the eye is easily accessible for a surgeon to both administer vector and make phenotypic determinants of therapy *in-vivo*, through fundus imaging or electroretinography (ERG). Additionally, for example if one eye is treated, the contralateral eye may serve as a control. Lastly, many well characterized animal models of reti-nal degenerations exist which allow development and testing of potential therapeutic interventions.

This chapter will focus on attempts to target therapy to two layers of the eye whose healthy interaction is crucial in maintaining vision, the neu-ral retina and the RPE. The neural retina can be subdivided into several

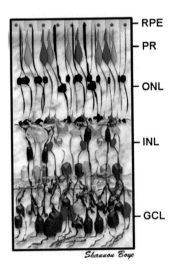

Fig. 20.1 Simple Anatomy of the Retina. Retinal Pigment Epithelium (RPE) is the most posterior layer, laying just behind the photoreceptor (PR) outer segments. The outer nuclear layer (ONL) is comprised of PR cell bodies (black = rods, the 3 subtypes of cones are colored according to the approximate wavelength of light they recognize). The inner nuclear layer (INL) is comprised of the cell bodies of horizontal, bipolar and amacrine cells. The innermost portion of the retina contains retinal ganglion cells (RGC). Mueller cells span the entire retinal thickness.

classes of neurons including rod and cone photoreceptors, horizontal, bipolar, amacrine and ganglion cells.

Rod and cone photoreceptors support the initial event in the phototransduction cascade, the conversion of electromagnetic, light energy into electrochemical signals. Proper function of both rod and cone photoreceptor cells is dependent on the support of the retinal pigment epithelium. Among other things, the RPE is involved in the phagocytosis of shed photoreceptor outer segment tips and the replenishment of chromophore (Vitamin A) to rods and cones. The neural output originating from the visual signal in rods and cones is sent through secondary order neurons (horizontal and bipolar cells) to the retinal ganglion cells (RGCs). The process of vision culminates in the retina with transmission of these signals from the RGCs to the visual cortex. It is therefore not surprising that the health of photoreceptors, RPE and RGCs are often compromised in many types of inherited retinal degenerations, and, for this reason these three retinal layers are frequently targeted by gene therapies.

There are a number of viral vectors capable of delivering therapeutic transgenes to mammalian rods, cones, RPE and/or RGCs, including adenovirus, lentivirus and adeno-associated virus (AAV). Each vector is unique in its tropism, efficiency, persistence, carrying capacity and immunogenicity. The occurrence of two high-profile adverse events in humans involving adenoviral and lentiviral vectors has impacted their utility in clinical studies.[1,2] Some changes have been made in these vectors to increase safety. For example, non-integrating HIV lentiviral vectors have been developed which are capable of long-term transgene expression with reduced risk of insertional mutagenesis. While these lentiviral vectors have a greater packaging capacity than AAV-based vectors, they are still only capable of efficiently transducing the RPE. AAV is the only vector capable of efficiently transducing rods, cones and the RPE and is still considered a safer viral vector. For this reason, it is the most commonly used vector for retinal gene therapy and will be the focus of this chapter.

To date, ocular gene therapy has focused primarily on 'loss of function' mutations in genes encoding proteins that mediate critical functions of photoreceptors, retinal ganglion cells and RPE. The potential for treatment in these cell types has been aided by the cloning and mapping of more than 120 retinal disease genes. The focus of this chapter is AAV-mediated ocular gene therapies in animal models which have shown the most potential in bridging the gap from bench to bedside. These therapies will be divided by the retinal cell type which is targeted for treatment. For each example, AAV-mediated somatic gene transfer resulted in significant functional improvement, as assessed by ERG or behavior, and/or regeneration or stabilization of retinal structure. One study will receive special attention because it laid the groundwork for ongoing human clinical trials. Analysis of the research, safety studies and regulatory review process leading up to the clinical trial will also be discussed.

2. Rod and Cone Photoreceptors

One of the most prevalent types of retinal degeneration is X-linked retinoschisis (XLRS). Affecting between 1 in 15,000 to 1 in 30,000 people worldwide, XLRS is the leading cause of juvenile macular degeneration in males. Female carriers are usually unaffected. This retinal dystrophy is caused by mutations in the retinoschisis (*RS1*) gene on Xp22.2 which encodes the 224 amino acid, secretable retinal protein retinoschisin (RS1).[3] This protein consists of a 23 amino acid N-terminal leader sequence, a 39 amino acid Rs1 domain, a 157 amino acid discoidin domain and a five amino acid C-terminal segment. Following cleavage of the leader sequence in the ER lumen, the protein is assembled into a homo-octameric complex linked by disulfide bonds. Subsequently, this complex is secreted from cells, specifically for rod and cone photoreceptors.[4] Upon secretion, RS1 is localized to the surface of photoreceptors and bipolar cells. Over 130 mutations in *RS1* (intragenic deletions, nonsense and missense mutations, frame shift insertions and deletions and splice site mutations) have been associated with disease. Molecular and cellular studies highlight three mechanisms through which these mutations may lead to XLRS; mutations in the discoidin resulting in misfolding and retention in the endoplasmic reticulum, cysteine mutations in the Rs1 domain or C-terminal segment which cause defective octamer assembly, and mutations in the leader sequence which prevent insertion of the polypeptide into the ER membrane.[4] It has been determined that it is the lack of functional RS1 protein and not the presence of a mutated protein that is responsible for XLRS.[4]

There is great variation in disease severity and rate of progression among affected patients, even in those carrying the same causative mutation or among those in the same family. While some patients present at school age with poor vision, the disease can also be detected in infants as young as 3 months. Visual impairment is also variable, with best corrected visual acuity from 20/20 to 20/600. Typical features of XLRS are a loss in central vision in the first decade of life and the presence of retinal schisis (splitting) which appears as 'spoke wheel-like' cystic cavities radiating from the central retina. Peripheral schisis is observed in approximately 50% of patients and is often accompanied by peripheral vision loss. This schisis, or retinal splitting, is caused by cystic cavities or gaps, most often in, but not limited to the inner retina. Blood vessels which traverse these gaps are fragile and at risk of hemorrhage. Visual function in XLRS patients is often stable until the fourth decade of life, but severe complications such as vitreal hemorrhage (up to a third of patients) and retinal detachment (up to 20% or patients) may lead to severe visual impairment.

Diagnosis of XLRS is typically made when specific problems are found in the patient's electroretinogram (ERG). Affected patients exhibit a characteristic reduction in b-wave amplitudes which is best detected after dark adaptation. These reduced b-wave amplitudes are indicative of a problem with the inner retina. Up to one third of XLRS patients also exhibit reductions in a-wave amplitudes over time, indicative of a progressive photoreceptor abnormality. Because these characteristic ERG patterns are not unique to XLRS, further diagnostic tests are necessary. Optical coherence tomography (OCT) is a non-invasive procedure which produces a two dimensional, cross-sectional image of structures in the eye. OCT scans of the central and peripheral retina can detect splitting and the eventual central retinal thinning due to photoreceptor cell loss. Scanning laser ophthalmoscopy in conjunction with OCT confirms that splitting may occur in any layer of the retina. Currently, there is no treatment for XLRS.

While the exact function of RS1 in the eye remains to be elucidated, it is generally thought to play an important role in the structural integrity and maintenance of retinal cell adhesion and architecture. Recent studies suggest that this protein interacts with extracellular $\beta2$ laminin, known for its role in the development and stabilization of neuronal synapses. In fact, deletion of the laminin $\beta2$ chain leads to reductions in ERG signal amplitudes like those seen in XLRS.[5]

Three separate mouse models have been created in which the mouse homolog to human *RS1*, *Rs1h*, has been knocked out, and all display morphological and functional phenotypes similar to human XLRS.[6,7,8] *Rs1h* knockout mice have highly disorganized retinas, cystic cavities and gaps between bipolar cells and disrupted outer nuclear layers that lead to eventual loss of photoreceptors. In addition, they exhibit the characteristic negative ERG response, indicating a disruption in the integrity of photoreceptor-bipolar cell synapses. Because the *Rs1h*-deficient mouse shares these important diagnostic features with XLRS patients, it has been a valuable model for evaluating potential therapeutic interventions.

Thus far, two groups have attempted to restore retinal structure and function in the *Rs1h* KO mouse.[7,9] The most recent study will be highlighted here as it reported the most complete therapeutic outcome. Several choices concerning vector design were made in that study with the goal of targeting expression of human RS1 cDNA to photoreceptors. Of the two well characterized serotypes known to efficiently target photoreceptors (AAV2 and AAV5), AAV5 has demonstrated higher efficiency and a more rapid onset of transgene expression when delivered into the subretinal space. Secondly, a photoreceptor-specific opsin promoter was chosen based on its ability to target transgene expression to rods and cones. At postnatal day 15 (P15), the AAV5-opsin promoter-*Rs1h* vector was delivered subretinally to *Rs1h* KO mice. Histological analysis showed that RS1 expression was restored to both photoreceptors and the inner retina, a pattern identical to that seen in wild type mice. Importantly, vector-mediated gene expression led to a progressive and significant improvement in retinal function (ERG).[9] Significant improvements were found in both rod- and cone-mediated aspects of the ERG signal. This functional recovery was evident for at least 5 months, the longest time point evaluated. In vivo imaging and histological analysis revealed that therapy also conferred structural improvements, with treated retinas exhibiting almost a complete lack of retinal schisis. These results are more complete than an earlier study which showed only functional improvements.[7] This difference is most likely attributable to the choice of vector serotype (AAV2) and promoter (CMV) used in the earlier study which likely resulted in lower levels of therapeutic

RS1 expression in treated eyes. These results of successful gene replacement therapy in the *Rs1h* KO mice demonstrate that this approach is capable of preserving retinal structure and visual function in an animal model of XLRS and provides support for the development of clinical trials for this disease.

3. Cone Photoreceptors

The human retina contains approximately 6 million cones. Cone photoreceptors are responsible for central, high-resolution, color vision. Unlike rod photoreceptors which support vision in dim, night-time conditions, we rely on cones to perform daily tasks in ambient to bright lighting. Diseases which compromise the integrity of this cell type are debilitating because they can induce daytime blindness, preventing the patient from performing 'normal' tasks such as driving. One such disease is rod monochromatism, better known as complete achromatopsia. Affecting approximately 1 in 30,000 individuals worldwide, this recessively inherited disease is characterized by permanent central vision loss, a lack of cone-mediated ERG signal and complete color blindness. Patients afflicted by achromatopsia typically have visual acuity of approximately 20/200 or less and are extremely sensitive to light. There is currently no cure for achromatopsia. All that is available for the management of symptoms are tinted contact lenses or sunglasses which limit exposure of the retina to bright light.

The three genes found to be associated with human achromatopsia thus far are *CNGB3*, *CNGA3* and *GNAT2*.[10,11,12,13] *CNGB3* and *CNGA3* encode the β and α subunits of the cone cyclic nucleotide-gated cation channel, respectively. This cone-specific cGMP-gated cation channel plays a pivotal role in cone phototransduction by opening and closing in response to light. These positions, open or closed, allow for or prevent the flow of ions into the cell, thereby establishing the electrical state of the cone photoreceptor, either depolarized or hyperpolarized, respectively. *GNAT2* encodes the cone-specific α subunit of transducin, another essential phototransduction protein. Transducin bridges the gap between photoisomerized opsin proteins and the breakdown of cGMP via phosphodiesterase. This ultimately leads to the closure of cGMP-gated channels.

To date, two mouse models and one canine model of achromatopsia have been characterized. They each have a different gene mutated, *GNAT2* and *CGNA3* individually in mice and *CNGB3* in dogs, with all models imitating the human form of the disease very well. This chapter will focus on gene therapy in the mouse model of the disease, the $Gnat2^{cpfl3}$ mouse. This mouse contains a point mutation in cone α-transducin rendering it inactive. Consequently, this animal has no recordable cone ERG but a normal rod ERG.[14] Light and electron microscopy reveal that the $Gnat2^{cpfl3}$ mouse retina appears to have relatively normal structure.

As with other retinal gene therapies, specific choices were made about the nature of the vector, this time with the goal to target expression to cone photoreceptors in the $Gnat2^{cpfl3}$ mouse. Again, a serotype 5 vector was chosen based on its superior ability to target transgene expression in the retina. To specifically target cone photoreceptors, a cone-preferential promoter was chosen. This type of specific targeting is done in order to avoid the potentially toxic side effects that expression in the wrong cell type might elicit. A 2.1 kb fragment of the human red/green cone opsin promoter (PR2.1) was chosen based on previous evidence that mice transgenic for sequences upstream of the red/green opsin gene containing a core promoter and locus control regions directed reporter gene expression to both classes of cones. Additionally, this promoter, in conjunction with a serotype 5 vector was shown to target reporter gene, green fluorescent protein (GFP), expression preferentially to cones in squirrel monkeys, mice, rats, ferrets, guinea pigs, dogs and monkeys. A serotype 5 vector containing the PR2.1 promoter and the wild type, human *Gnat2* cDNA (AAV5-PR2.1-*hGnat2*) was administered to the subretinal space of the $Gnat2^{cpfl3}$ mice, a delivery route intended to optimize infection of photoreceptors. Following treatment, cone function in treated and untreated eyes were assessed by ERG assays specific for cone function. Additionally, visually evoked behavior was assessed by determining visual acuity through optomotor (head movement) responses to a rotating sine-wave grating. The results of both tests showed that AAV5-PR2.1-*hGnat2* treated eyes responded significantly better than untreated eyes in the $Gnat2^{cpfl3}$ mice. Responses in treated eyes were equivalent to those seen in normal sighted, wild type mice.[14] This study was the first successful attempt demonstrating that cone-targeted AAV therapy is capable of restoring cone-mediated function and visual acuity in an animal model of achromatopsia. Currently, an analogous approach is being taken to correct vision in another mouse model with the *CGNA3* form as well as a canine model with the *CNGB3* form of this disease.

More recently, investigators have reported success in another cone-based disorder, red-green color blindness.[15] This disease is the result of the absence of either the long- (L) or middle- (M) wavelength-sensitive visual photopigments and is the most common single locus human genetic disorder. AAV containing a human L-opsin gene under the control of the L/M-opsin enhancer and promoter was delivered to the photoreceptors of adult squirrel monkeys in three separate subretinal injections (100 μl each). Behavior-based color vision tests were used to compare subjects before and after treatment. In spite of doubts that the absence of neural connections normally established during development would prevent appropriate processing post-treatment, investigators showed that addition of a third opsin to these adult red-green colour deficient primates was sufficient to restore trichromatic color vision behavior.[15] This result suggests that gene-based therapy in the retina has the potential to cure adult cone vision disorders.

The success of these studies supports the development of other cone-targeted therapies in which cures for debilitating human conditions such as late stage retinitis pigmentosa, age-related macular degeneration and diabetic retinopathy are needed. While more efficacy and safety data is necessary, cone-targeting clinical trials for human achromatopsia and red-green color blindness may be reasonably considered.

4. Retinal Ganglion Cells

Retinal ganglion cells (RGC) are located in the innermost portion of the retina and are, on average, larger than most other retinal neurons. They are responsible for receiving electrical signals originally generated by photoreceptors and second order neurons and passing that information through the optic nerve to the brain. Diseases which affect the health of RGC can have deleterious effects on vision. One such disease is glaucoma, the second most common cause of blindness in the world.[16] A major component of this disease is apoptotic death of RGCs. It was once thought that this neuronal loss in glaucoma patients was attributed solely to the long term elevation of intraocular pressure (IOP). However, clinical trials revealed that when the

IOP was normalized, some patients still continued to lose RGCs. Currently, two main hypotheses are commonly accepted as explanations for glaucomatous RGC death; long term elevation in intraocular pressure (IOP) or the presence of genetic factors which predispose a patient to RGC loss in the absence of elevated IOP. Regardless of the cause of initial damage, it results in a cascade of biochemical events which can further affect RGCs. Such secondary events include uncontrollable increases in levels of biochemical compounds such as glutamate and nitric oxide that are released by dying neurons. These compounds can eventually reach concentrations that are neurotoxic. Additionally, the initial insult may cause an interruption in normal axonal transport which results in deprivation of target-derived trophic factors and consequent triggering of intracellular changes that may lead to apoptotic death of affected RGCs.

One neurotrophic factor that is especially important for RGC health, both during development and in adult life, is brain-derived neurotrophic factor (BDNF). It is produced in the brain and is retrogradely transported from RGC axons to cell bodies in microsomal vesicles. It was found that when IOP is elevated in rats and monkeys that retrograde transport of BDNF is obstructed at the optic nerve head.[17] This finding led to the hypothesis that BDNF deprivation plays a role in glaucomatous RGC death. This theory was further supported by evidence that intravitreal administration of BDNF in a rat model of glaucoma temporarily slowed RGC loss. However, the need for repeated administration of BDNF protein limits its clinical utility for persistent diseases like glaucoma.

Multiple viral vector systems have been employed in attempts to achieve long term expression of BDNF in RGCs. Initial studies using adenoviral vectors to deliver the BDNF cDNA to the intravitreal space of rats with axotomized RGCs revealed that expression of the therapeutic transgene was capable of temporarily promoting RGC survival.[18] However, due to limitations of adenoviral gene therapy, including the inability to elicit long term transgene expression and the tendency to promote inflammation, a more optimal vector system was necessary. More recently an AAV vector was used to deliver the BDNF cDNA to the intravitreal space of an experimental rat model of glaucoma.[19] Two weeks later, glaucoma was induced in this model through an elevation in IOP via laser treatment of the trabecular meshwork (the mesh-like structure near the cornea and iris which facilitates removal of aqueous humor from the eye into the blood stream). At four weeks post treatment, AAV-mediated BDNF therapy promoted survival of a statistically significant fraction of RGCs (as estimated by RGC axon counts) in treated eyes relative to untreated controls. Transgene expression was found predominantly in the desired target cell, the RGC. One limitation of this therapy however was that RGC survival was not quantified beyond 4 weeks. In addition, BDNF is known to have high affinity for its receptor TrkB, an antigen expressed on multiple retinal cell types, not just RGCs. This in theory permits the AAV-delivered transgene to exert its effects in a broad fashion which could have implications for other aspects of retinal structure and/or function. In fact, it is known that application of neurotrophic factors in the retina can cause upregulation of nitric oxide synthase activity and suppression of heat shock proteins, both of which could have detrimental effects. In addition, it has been shown that delivery of a related molecule, ciliary neurotrophic factor (CNTF), has detrimental effects on retinal function.[20] With this in mind, it was thought that intervening in pathways downstream of neurotrophin administration might have 'cleaner' therapeutic effects. It is known that BDNF binding to its receptor, TrkB, stimulates multiple signaling pathways including the extracellular signal-related kinase 1/2 (ERK1/2) which has been implicated in RGC survival. Using an RGC preferential serotype 2 AAV vector to deliver MEK1 (the upstream activator of ERK1/2) to the intravitreal space of an experimental rat model of glaucoma, researchers were able to stimulate the ERK1/2 pathway in cells transduced with vector.[21] At five weeks post treatment, it was found that MEK1 gene transfer markedly increased RGC survival in this model. Through the use of a serotype 2 AAV vector and delivery to the vitreous cavity, this therapy specifically targeted RGCs. This study is further proof that gene therapy is capable of promoting RGC survival in animal models of glaucoma.

5. Retinal Pigment Epithelium

The retinal pigment epithelium (RPE) is the pigmented cell monolayer positioned between the neurosensory retina and the choroidal blood supply. The RPE sustains the metabolic needs of the underlying neural retina by controlling the transfer of small molecules between the blood stream and the

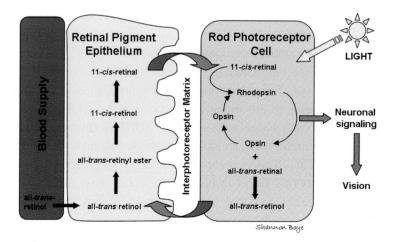

Fig. 20.2 The Vitamin A Cycle: all-*trans*-retinol (vitamin A) is transported from the blood stream to the retinal pigment epithelium (RPE). There it is esterified into all-*trans*-retinyl esters. On demand, these esters can be hydrolyzed and isomerized (by RPE65) to form 11-*cis*-retinol which is oxidized to form 11-*cis*-retinal. 11-*cis*-retinal is transported to photoreceptor outer segments where it binds with opsin to form the visual pigment, rhodopsin in rods and cone opsin in cones. Absorption of a photon of light catalyzes the isomerization of 11-*cis*-retinal to all-*trans*-retinal which releases it from the opsin molecule. This photo-isomerization triggers a biochemical cascade in photoreceptors which eventually leads to the generation of a neural signal in RGCs that is interpreted in the brain as vision.

retina. It is also responsible for the phagocytosis of shed photoreceptor outer segment tips and plays a crucial role in the Vitamin A cycle by isomerizing all trans-retinol to 11-cis retinal (Fig. 20.2).

RPE65 (retinal pigment epithelium-specific 65 kDa protein), expressed predominantly in RPE cells, has been identified as the all-trans to 11-cis isomerase.[22] Improper functioning of RPE65 results in a lack of 11-cis retinal production and an inability to form visual pigment (rhodopsin and cone opsin). Without these light-sensitive opsins, photoreceptor cells of the retina lack the ability to usefully absorb photons and initiate the conversion of light into a visual signal. Mutations in *Rpe65* are associated with Leber's congenital amaurosis type 2 (LCA2), one of the earliest and most severe forms of inherited retinal dystrophies.[23] This disease, which is most often inherited in an autosomal recessive fashion, is usually diagnosed at birth or within the first few months of life when patients present with severely impaired vision, pendular nystagmus and abnormal ERG and pupillary light reflexes. Patients progress to near total blindness by adolescence. Despite these severe, early symptoms there are often not any obvious ophthalmo-scopic abnormalities present.

In order to circumvent the underlying defect in the retinoid cycle (the lack of 11-cis retinal and failure to regenerate visual pigment), three therapies have been evaluated; RPE cell transplantation, retinoid delivery and gene based intervention. RPE cell transplantation is mechanically difficult and invasive. Oral administration of retinoid is not optimal because it requires repeated administration of drug. Of the three, gene-based intervention has had the most success. AAV was used to deliver functional copies of RPE65 to the naturally occurring *Rpe65* mutant Briard dogs, a naturally occurring RPE65 mutant (*rd12*) mouse and the transgenic knockout *Rpe65*$^{-/-}$ mouse. Taken together, these studies showed that AAV-mediated RPE65 expression confers partial restoration of ERG, retinal structure, and visually evoked behaviors in these animals.[24,25,26] These studies laid the groundwork for Phase I Clinical trials for LCA2.

6. LCA2 Gene Therapy, a Perspective on Translational Research

The University of Florida — University of Pennsylvania LCA Consortium, established through a NEI Cooperative Clinical Research Grant, had the goal of establishing a safe and efficacious AAV gene therapy vector and employing it in a clinical trial for LCA2 within 5 years. The Consortium's first step was to expand on the proof of concept work that began with AAV-RPE65 treatments of Briard dogs.[24] The first of these dogs to be treated and subsequently shown to have restored vision, 'Lancelot', was instrumental in raising public awareness of vision research. Eventually AAV-RPE65 vectors were evaluated in two other animal models of LCA2, the transgenic knock-out *Rpe65*$^{-/-}$ mouse and *rd12* mutant mouse.[26,25] Restoration of visual responses was achieved by subretinal injection of AAV-RPE65 vector in both mouse models. In addition, in the Briard dogs vision-evoked cortical responses had been restored similar to behavioral characteristics correlating with the functional vision that were documented in treated *rd12* mice. These experiments provide further evidence that therapy could be successful in restoring 'functional' vision.[27,25] Vision restoration has persisted unchanged for more than 8 years after a single treatment in a Briard dog (Lancelot).

A dose response study in affected Briard dogs revealed the therapeutically effective dose range of the vector. Safety within this range was further validated in Cynomolgous monkeys. Both short term and long term biodistribution and pathology studies were performed in vector-treated normal vision rats and monkeys. In order to ensure that the monkey data was robust and conclusions impartial, it was done by a contract research organization in accordance with Good Laboratory Practices (GLP).[28] Ophthalmological examinations carried out in dogs and monkeys post vector administration indicated that there were some minor to moderate post surgical complications, most of which resolved within a week post surgery and, aside from anterior chamber flare, were not related to vector treatment (i.e. also occurred in vehicle treated animals). Signs of retinal thinning localized to the injection site of treated dogs and monkeys was likely a consequence of the surgical procedure, and have been previously documented. Other measures of ocular health such as visual function as assessed by ERG did not indicate a loss of visual function in monkeys when comparing pre and post vector treatment time points. In affected Briard dogs, vector treatment was correlated with improved ERG responses, as had already been established in prior proof of concept experiments.

Pathological examinations of all non-ocular tissues were unremarkable in all of the studies. Immunology experiments were carried out in the monkey studies. Circulating antibodies to AAV2 capsid protein were evaluated by ELISA. For the 7 day monkey study, antibody titers were low before treatment and remained so. However, given the short duration of this experiment and the time necessary to mount a humoral response, these results are inconclusive. In the 3 month monkey study, despite some variation in pre-treatment antibody titers, there were no consistent changes in antibody levels over the 3 month period following vector administration.

Biodistribution studies in dogs, rats and monkeys looked for the presence of vector genomes in specific tissues. Tissues of particular interest included the injected eye (where vector DNA is expected), tissues of the visual pathway and brain (to determine whether retrograde vector transport along the optic nerve occurred), and gonad (necessary to rule out the possibility of germ-line vector transmission). All studies indicated vector was rarely and only sporadically found outside of the injection site. Importantly, vector was generally not found in the optic nerve or other tissues of the optic pathway leading to the brain, and never in the brain itself. Gonads, both male and female, were found to be negative indicating that the potential for germ-line transfer of vector is minimal.

Taking the conclusions from the safety studies and translating them into a protocol for a LCA2 clinical trial highlighted several key points: 1) Details of the surgical technique are of critical importance. Because limited injection site damage was noted proximal to the retinotomy, vector should be delivered adjacent to target areas of best retained retina, thereby allowing vector to spread into them by flow into the subretinal space. 2) The vector dose range would be defined by the results in dogs which provided efficacy but no established toxicity. Based on the efficacy and safety studies summarized above, approval to proceed with a clinical trial was given by the US FDA, the US Recombinant DNA Advisory Committee, an NIH/NEI constituted Data Safety and Monitoring Committee, the Institutional Biosafety Committees of the University of Florida and the University of Pennsylvania and the Institutional Review Boards of the University of Florida (through the Western IRB) and the University of Pennsylvania.

Presently, 3–12 month results from three separate clinical studies (including the aforementioned University of Florida and University of Pennsylvania, NEI funded trial) have been reported involving a total of nine young adults with LCA2.[29,30,31,32,33] Although the primary purpose of all three studies was to determine safety, each trial reported some measure of therapy in at least one of their patients. Neither humoral nor T-cell mediated immune responses to vector were detected in any patient. The fact that LCA2 patients received 2 to 4 orders of magnitude lower AAV2 vector doses than had been reported for patients in other non-ocular clinical trials together with the relative immune privileged status of the eye makes this lack of immune response understandable. One patient from the Bainbridge *et al.* (2008) study and all three patients from the Hauswirth *et al.* (2008) and Cideciyan *et al.* (2008) study experienced significant increases in light sensitivity within the retinal area treated, with the best response being a 63,000-fold improvement over pre-treatment levels.[32] A key conclusion of all three trials was that the subretinal vector delivery procedure and any transient post-surgical complications did not promote vision loss. A striking observation was noted in one patient one year post treatment.[33] For the first time in her life, the patient reported an ability to read the illuminated numerical clock display on the dashboard of the family vehicle. This new perception was accompanied by a shift in fixation into the treated supero-temporal retina. In other words, the patient had developed the ability to fixate on an image with a treatment-created pseudo-fovea, an area which now had better sensitivity than the untreated foveal region. The slow development of this pseudo-fovea suggests an underlying experience-dependent plasticity in the adult visual system. Put simply, this patient's brain had 'learned' how to use the treated pseudo-foveal retinal region for fixation over the 12 months post treatment. This result raises the possibility that other retinal gene-based therapies have the potential to improve vision in previously unexpected ways. In summary, early results suggesting efficacy and safety in three AAV2-mediated gene therapy trials for RPE65 LCA portend a hopeful future for retinal gene therapy with AAV vectors.

References

1. Raper SE, Chirmule N, Lee FS, Wivel NA, Bagg A, Gao GP, *et al.* (2003). Fatal systemic inflammatory response syndrome in a ornithinetranscarbamylase deficient patient following adenoviral gene transfer. *Mol Genet Metab* **80**: 148–158.
2. Williams DA (2006). Vector insertion, mutagenesis and transgene toxicity. *Mol Ther* **14**: 457.
3. Sauer CG, Gehrig A, Warneke-Wittstock R, Marquardt A, Ewing CC, Gibson A, *et al.* (1997). Positional cloning of the gene associated with X-linked juvenile retinoschisis. *Nat Genet* **17**: 164–170.
4. Wu WW, Molday RS (2003). Defective discoidin domain structure, subunit assembly, and endoplasmic reticulum processing of retinoschisin are primary mechanisms responsible for X-linked retinoschisis. *J Biol Chem* **278**: 28139–28146.
5. Libby RT, Lavallee CR, Balkema GW, Brunken WJ, Hunter DD (1999). Disruption of laminin beta2 chain production causes alterations in morphology and function in the CNS. *J Neurosci* **19**: 9399–9411.
6. Weber BH, Schrewe H, Molday LL, Gehrig A, White KL, Seeliger MW, *et al.* (2002). Inactivation of the murine X-linked juvenile retinoschisis gene, Rs1h, suggests a role of retinoschisin in retinal cell layer organization and synaptic structure. *Proc Natl Acad Sci USA* **99**: 6222–6227.
7. Zeng Y, Takada Y, Kjellstrom S, Hiriyanna K, Tanikawa A, Wawrousek E, *et al.* (2004). RS-1 Gene Delivery to an Adult Rs1h Knockout Mouse Model Restores ERG b-Wave with Reversal of the Electronegative Waveform of X-Linked Retinoschisis. *Invest Ophthalmol Vis Sci* **45**: 3279–3285.
8. Jablonski MM, Dalke C, Wang X, Lu L, Manly KF, Pretsch W, *et al.* (2005). An ENU-induced mutation in Rs1h causes disruption of retinal structure and function. *Mol Vis* **11**: 569–581.

9. Min SH, Molday LL, Seeliger MW, Dinculescu A, Timmers AM, Janssen A, *et al.* (2005). Prolonged recovery of retinal structure/function after gene therapy in an Rs1h-deficient mouse model of x-linked juvenile retinoschisis. *Mol Ther* **12**: 644–651.

10. Kohl S, Baumann B, Rosenberg T, Kellner U, Lorenz B, Vadala M, *et al.* (2002). Mutations in the cone photoreceptor G-protein alpha-subunit gene GNAT2 in patients with achromatopsia. *Am J Hum Genet* **71**: 422–425.

11. Kohl S, Varsanyi B, Antunes GA, Baumann B, Hoyng CB, Jagle H, *et al.* (2005). CNGB3 mutations account for 50% of all cases with autosomal recessive achromatopsia. *Eur J Hum Genet* **13**: 302–308.

12. Kohl S, Marx T, Giddings I, Jagle H, Jacobson SG, Apfelstedt-Sylla E, *et al.* (1998). Total colourblindness is caused by mutations in the gene encoding the alpha-subunit of the cone photoreceptor cGMP-gated cation channel. *Nat Genet* **19**: 257–259.

13. Sundin OH, Yang JM, Li Y, Zhu D, Hurd JN, Mitchell TN, *et al.* (2000). Genetic basis of total colourblindness amoung Pingelapese islanders. *Nat Genet* **25**: 289–293.

14. Alexander JJ, Umino Y, Everhart D, Chang B, Min SH, Li Q, *et al.* (2007). Restoration of cone vision in a mouse model of achromatopsia. *Nat Med* **13**: 685–687.

15. Mancuso K, Hauswirth WW, Li Q, Connor TB, Kuchenbecker JA, Mauck MC, *et al.* (2009). Gene therapy for red-green colour blindness in adult primates. *Nature*, Sep 16 (Epub ahead of print).

16. Quigley HA (1996). Number of people with glaucoma worldwide. *Br J Ophthalmol.* **80**: 389–393.

17. Pease ME, McKinnon SJ, Quigley HA, Kerrigan-Baumrind LA, Zack DJ (2000). Obstructed axonal transport of BDNF and its receptor TrkB in experimental glaucoma. *Invest Ophthalmol Vis Sci* **41**: 764–774.

18. Di Polo A, Aigner LJ, Dunn RJ, Bray GM, Aguayo AJ (1998). Prolonged delivery of brain-derived neurotrophic factor by adenovirus-infected Muller cells temporarily rescues injured retinal ganglion cells. *Proc Natl Acad Sci USA* **95**: 3978–3983.

19. Martin KR, Quigley HA, Zack DJ, Levkovitch-Verbin H, Kielczewski J, Valenta D, *et al.* (2003). Gene therapy with brain-derived neurotrophic factor as a protection: retinal ganglion cells in a rat glaucoma model. *Invest Ophthalmol Vis Sci* **44**: 4357–4365.

20. McGill TJ, Prusky, GT, Douglas, RM, Yasumura, D, Matthes, MT, Nune, G, *et al.* (2007). Intraocular CNTF Reduces Vision in Normal Rats. *Invest Ophthalmol Vis Sci* **48**: 5756–5766.

21. Zhou Y, Pernet V, Hauswirth WW, Di Polo A (2005). Activation of the extracellular signal-regulated kinase 1/2 pathway by AAV gene transfer protects retinal ganglion cells in glaucoma. *Mol Ther* **12**: 402–412.

22. Moiseyev G, Chen Y, Takahashi Y, Wu BX, Ma JX (2005). RPE65 is the iosmerohydrolase in the retinoid visual cycle. *Proc Natl Acad Sci USA.* **102**: 12413–12418.

23. Marlhens F, Bareil C, Griffoin JM, Zrenner E, Amalric P, Eliaou C, *et al.* (1997). Mutations in RPE65 cause Leber's congenital amaurosis. *Nat Genet.* **17**: 139–141.

24. Acland GM, Aguirre GD, Ray J, Zhang Q, Aleman TS, Cideciyan AV, *et al.* (2001). Gene therapy restores vision in a canine model of childhood blindness. *Nat Genet* **28**: 92–95.

25. Pang JJ, Chang B, Kumar A, Nusinowitz S, Noorwez SM, Li J, *et al.* (2006). Gene therapy restores vision-dependent behavior as well as retinal structure and function in a mouse model of RPE65 Leber congential amaurosis. *Mol Ther* **13**: 565–572.

26. Dejneka NS, Surace EM, Aleman TS, Cideciyan AV, Lyubarsky A, Savchenko A, *et al.* (2004). In utero gene therapy rescues vision in a murine model of congenital blindness. *Mol Ther* **9**: 182–188.

27. Aguirre GK, Komáromy AM, Cideciyan AV, Brainard DH, Aleman TS, Roman AJ, *et al.* (2007). Canine and human visual cortex intact and responsive despite early retinal blindness from RPE65 mutation. *PLoS Med.* **4**: e230.

28. Jacobson SG, Boye SL, Aleman TS, Conlon TJ, Zeiss CJ, Roman AJ, *et al.* (2006). Safety in nonhuman primates of ocular AAV2-RPE65, a candidate treatment for blindness in Leber congenital amaurosis. *Hum Gene Ther* **17**: 845–858.

29. Bainbridge JW, Smith AJ, Barker SS, Robbie S, Henderson R, Balaggan K, *et al.* (2008). Effect of gene therapy on visual function in Leber's congenital amaurosis. *N Engl J Med* **358**: 2231–2239.

30. Maguire AM, Simonelli F, Pierce EA, Pugh EN Jr, Mingozzi F, Bennicelli J, *et al.* (2008). Safety and efficacy of gene transfer for Leber's congenital amaurosis. *N Engl J Med* **358**: 2240–2248.

31. Hauswirth W, Aleman TS, Kaushal S, Cideciyan AV, Schwartz SB, Wang L, *et al.* (2008). Phase I Trial of Leber Congenital Amaurosis due to RPE65 Mutations by Ocular Subretinal Injection of Adeno-Associated Virus Gene Vector: Short-Term Results. *Hum Gene Ther*, 2008 Sep 7 [Epub ahead of print].

32. Cideciyan AV, Aleman TS, Boye SL, Schwartz SB, Kaushal S, Roman AJ, *et al.* (2008). Human gene therapy for RPE65 isomerase deficiency activates the retinoid cycle of vision but with slow rod kinetics. *Proc Natl Acad Sci USA* **105**: 15112–15117.

33. Cideciyan AV, Hauswirth WW, Aleman TS, Kaushal S, Schwartz SB, Boye SL, *et al.* (2009). Vision 1 year after gene therapy for Leber's congenital amaurosis. *N Engl J Med* **361**: 725–727.

Chapter 21
A Brief Guide to Gene Therapy Treatments for Pulmonary Diseases

Ashley T. Martino,* Christian Mueller
and Terence R. Flotte

Pioneering studies in the field of gene therapy were focused on the common monogenic lung disease, cystic fibrosis. These efforts were largely unsuccessful. Subsequent to these early studies gene therapy for pulmonary disorders has expanded well beyond the initial targets into diseases such as lung cancer, infectious diseases, and asthma. The emergence of new vectors and concepts of preparative treatment have allowed gene correction to advance despite initial setbacks. Moreover hybrid technologies such as vector-mediated RNAi therapy and cell-based genetic manipulation have added to the therapeutic armamentarium. Overall, the various advances increased the number of potential drug candidates that might ultimately have a lasting impact on outcomes of patients with lung diseases.

1. Introduction

Initially pulmonary gene therapy studies involved cystic fibrosis (CF) and brought gene therapy into the spotlight. There have been 25 phase I/II clinical trials for CF using viral and non-viral gene therapy methods.[1] Unfortunately, the goal of effective, long-term cystic fibrosis transmembrane conductance regulator (CFTR) gene correction was not achieved with

*Correspondence: University of Florida, Dept. Pediatrics, Gainesville, FL.
E-mail: amartino@peds.ufl.edu

the initial trials. Human trials have also been pursued for the genetic disease: Alpha-1 anti-trypsin (A1AT), with no significant therapeutic benefit. The challenges encountered in these first clinical trials were critical in designing strategies to overcome the early setbacks. Data from these studies prompted improvements in the technology that yielded new generations of viral and non-viral delivery systems and provided insight into how to improve efficiency, many of which are currently either moving towards or are being used in clinical studies.

The transformation of gene therapy for genetic lung diseases bolstered the research potential of genetic manipulation for lung disorders. The full spectrum of gene therapy based trials has expanded and now includes multiple genetic diseases and non-hereditary conditions like lung cancer and asthma. In addition, multiple strategies, such as cell-based therapy and RNAi mediated gene silencing, have emerged.

Although many critical advances have been made, there are still issues to solve and new avenues to pursue before gene therapy becomes an accepted treatment for lung diseases. This chapter will review common diseases, evolution of vectors and strategies for their effective delivery, hurdles to overcome, successes and the foreseeable future.

2. Common Disorders

2.1 *Cystic Fibrosis*

Cystic fibrosis (CF) is the most common genetic disease among Caucasians. Interest in CF gene therapy treatments materialized after the 1989 publication by Riordan *et al.* revealed that CF was linked to a homozygous gene defect in the Cystic Fibrosis Transmembrane Conductance Regulator (CFTR) gene.[2] The CFTR gene encodes a membrane-bound chloride channel protein and regulates the depth and composition of the airway surface liquid (ASL), in which the cilia of the respiratory epithelium are immersed.[3]

It is widely accepted that a CFTR deficiency induces an alteration in the ASL.[4] This alteration leads to a build up of mucus, which impairs ciliary function and compromises pathogen clearance from the lung. Defective bacterial clearance results in the colonization of *Pseudomonas aeruginosa*, as well as other pathogens. This chronic state of persistent lung infection

Table 21.1.   Pathological consequences related to cystic fibrosis.

Organs/System Affected	Pathological Condition
Lung/respiratory system	Chronic endobronchial infection with *Staphylococcus aureus, Pseudomonas aeruginosa, Burkholderia cepacia*, and other specific organisms, small and large airways obstruction, bronchiectasis, allergic bronchopulmonary aspergillosis, hyper innate immune response, pneumonia, hemoptysis, fibrosis, pulmonary hypertension, heart failure, respiratory failure; chronic rhinosinusitis, nasal polyposis
Sweat gland	Increased chloride and sodium loss due to impairment of ductal reabsorption
Liver	Ductal obstruction, biliary cirrhosis, multi-lobular cirrhosis, metabolic complications, fibrosis
Pancreas	Pancreatic insufficiency, ductal obstruction, loss of insulin producing beta-cells, cystic fibrosis related diabetes (CFRD)
Digestive System	Gut obstruction, malabsorption, malnutrition
Reproductive system	Infertility, absence of vas deferens (99% of men), thickened cervical mucus (minor in women)

promotes a constant influx of cellular infiltrates from the immune system, which ultimately leads to lung tissue destruction that may progress to respiratory failure and premature death. This severe pathology is complicated by a collection of consequences in other organs and biological systems (Table 1).

Why did CF become a popular candidate for gene therapy? Forty years ago cystic fibrosis patients died in early childhood. There have been many advances in treatments unrelated to gene therapy that have dramatically increased life expectancy, but premature death still occurs. Correcting the CFTR gene deficiency is of great interest due to the potential to completely reverse the lung pathology.

The bulk of the mortality comes from the lung condition and it has been generally accepted that a minor CFTR correction in the deficient lung, 6–10% of normal levels, will be effective in reversing the pathophysiology.[5] There is also the potential that correction of a single organ, the lung, will lead to long-term benefit.[6] Since newborns with CF usually have healthy lungs, there may be an opportunity to prevent symptoms before they begin. Furthermore, the lungs are relatively accessible as compared with other organs, since the airway may serve as a route of vector delivery. Most vectors can be readily aerosolized for inhaled delivery to the lung, an option not available to other target organs, such as the brain, pancreas, liver, etc.

2.2 *Alpha-1 Antitrypsin (A1AT)*

A1AT is caused by a homozygous gene defect in the SERPINA1 gene. The encoded protein, alpha-1 antitrypsin (A1AT), is a protease inhibitor predominately secreted by the liver. This protein is critical in regulating the levels of pro-inflammatory elastases (i.e. neutrophil elastase) that are present at sites of inflammation by degranulation of polymorphonuclear cells such as neutrophil granulocytes.[7]

The common lung conditions that develop from the decrease in systemic A1AT levels are chronic obstructive pulmonary disease (COPD) and emphysema, both of which lead to respiratory insufficiency. The progression to these conditions is linked to the secretion of elastases in the lung environment without proper regulation offered by A1AT. Unchecked, the elastases linger and eventually lead to cellular damage. The lung is notably vulnerable to the deficiency in circulating A1AT levels and is augmented by smoking.

3. Development of Viral Vectors for Lung Disease

The evolution of viral vectors for genetic manipulation in the lung has progressed via the early clinical trials related to CFTR gene correction for cystic fibrosis treatment. These trials involved adenovirus (Ad) and a non-pathogenic, helper-dependent adeno-associated virus (AAV2) administered by nasal installation, lung installation and aerosolization.

3.1 *Adenoviral Vectors*

Adenovirus was chosen because infections occur naturally in the lung and intestines. Unfortunately, these infections lead to a host of conditions including pneumonia, diarrhea and vomiting. Steps were taken to ensure safety when using Ad vectors. With early indication showing that the E1 genetic region was responsible for the cytotoxicity of Ad, the first clinical trial involved an E1 deficient Ad vector expressing full length CFTR.[8] This initial adenoviral trial revealed a transient pulmonary condition categorized as pneumonia. Later efforts to address these safety issues by deleting the E2 and E4 regions along with the E1 region did not prevent this pulmonary condition.[9]

3.2 *Adeno-Associated Viral Vectors*

The AAV2 trials utilized the full-length CFTR coding region flanked by the AAV2 viral integrated terminal repeats (ITRs). Since the CFTR coding region is 4.4 kb, the small genome restraints (4.7 kb maximum capacity) of the AAV2 viral vector prevented the use of expression enhancing elements. Therefore transcription was relegated to the minor innate ability of the viral ITRs. Wild-type AAV infections occur without pathology, and with safety being a primary concern, the AAV viral vector delivery system was preferable to the Ad delivery system.

3.3 *Early Conclusions*

Unfortunately, the initial trials of CF gene therapy fell short of expectations. Viral vector transduction was not as efficient in the lungs of CF patients as was predicted from the preliminary cell culture and animal models.[10,11] However, safe administration of rAAV vectors was promising, and there was correlation between presence of vector-derived CFTR cDNA and physiologic correction of the CFTR chloride channel defect in cells recovered from patients after rAAV gene therapy.[12] The most difficult issues remaining were efficiency and duration of gene transfer and the inability to redose vectors after neutralizing antibody responses were elicited.

Therefore, to increase efficiency, alternate forms of these viral vectors were created.

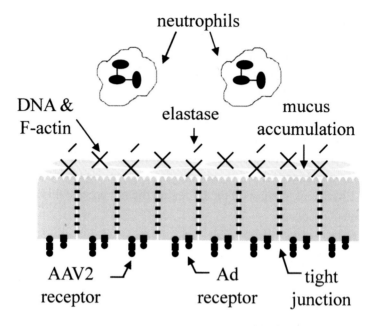

Fig. 21.1 Schematic illustration of limits on vector efficiency in the lung.

4. Enhancing Efficiency

Improving efficiency requires a multi-faceted approach to overcome suboptimal vectors augmented by innate lung barriers associated with the pathogen clearance mechanisms (Fig. 21.1). Target cells for gene transfer are at the differentiated epithelium with a basal and apical surface. The apical surface is exposed to the airway lumen while the basal lateral surface is directed toward the interior of the tissue. Accessing the basal lateral surface from the airway requires the gene therapy vector to cross epithelial tight junctions. Unfortunately, the primary tropic factors for AAV2 and Ad vectors are present only on the basal lateral surface.

4.1 *Alternative AAV Viral Vector Serotypes*

Targeting the apical surface was desirable to optimize gene delivery and an extensive library of AAV serotypes have been cloned and tested for tropic potential in the lung and other organs. These efforts have been crucial in the evolution of AAV viral vectors. AAV5 and AAV1 serotypes have been chosen in the newest generations of AAV viral vectors to target the apical surface of the lung. It was discovered that the receptors for AAV5 serotype are the platelet derived growth factor (PDGF) and 2,3 sialic acid residues. Both these receptors are located on the apical surface of airway epithelial cells.[13,14] Notably, AAV1 is more efficient than AAV2 and AAV5 at transducing human airway cultures and also enters through the apical surface of the cell.[15-18]

4.1.1 *Addition of Expression Enhancing Elements*

Adding promoters to enhance expression of the packaged CFTR transgene was envisioned but due to packaging constraints would be problematic using the AAV vector delivery system. To bypass this problem the idea of packaging a functional CFTR mini-gene along with a continuative promoter was investigated.

In 1995 it was discovered that the removal of a large portion of the first trans-membrane domain of the CFTR protein (TMD1) could potentially yield a functional protein. The truncation had negligible effect on the stability and function of the CFTR protein.[19] A more focused study deleting the entire TMD1 and the deletion of the first 264 amino acids in 2004 produced a viable truncated CFTR protein.[17] This deletion facilitated the incorporation of an efficient universal, hybrid promoter using the constitutive chicken beta-actin promoter (CBA), and the cytomegalovirus (CMV) enhancer downstream of the $\Delta 264$ CFTR mini-gene flanked by the two viral ITRs. The result was a very effective CFTR expression cassette that was packaged into the AAV5 capsid. Further tests in non-human primates and CF disease mouse models confirmed the improved efficiency with no evidence of lung inflammation.[20]

4.2 *Adenoviral Vectors*

Discovering an alternative adenoviral vector that transduces through the apical surface has been unsuccessful. Alternate strategies have been pursued to enhance basal lateral transfer by temporarily compromising the tight junctions using a pretreatment of ethylene glycol tetraacetic acid (EGTA). To complement this strategy, less immunogenic second and third generation Ad viral vectors were developed. The second contained deletions or mutations of the E2a, E4 and E3 regions, while the third, helper-dependent (HD-Ad), had all Ad coding sequences deleted.[21]

Although efficiency was improved, obvious focal pulmonary inflammation occurred, which included high levels of lymphocyte and macrophage infiltrates.[22] Such inflammation is problematic to a CF lung already heavily burdened by elevated cellular infiltrates and bacterial colonization and could potentially increase lung exacerbations. Moreover, pulmonary edema by compromised tight junctions or sepsis by bacteria crossing and entering the vascular system is a legitimate concern.[23] Unfortunately, these concerns have marginalized the role of Ad vectors for clinically relevant CFTR gene replacement.

4.3 *Physiological Hurdles in the Lung Environment*

Innate physiological impediments within the lung environment provide a potent barrier to vector transmission. The mucus of a CF patient, characterized as extremely thick and sticky, is extremely obstructive compared to the sputum of a healthy subject. Additionally, chronic innate immune responses burden the CF airways with neutrophils that release large quantities of elastase, DNA and F-actin. This extensive web of overly viscous mucus laden with DNA, F-actin and mucus glycoproteins causes a physical barrier that limits the transduction of the epithelium. Pretreatment with Dnases, mucolytic enzymes, and the elastase inhibitor, Alpha-1 antitrypsin (A1AT) have shown promising results in breaking down this barrier.[24]

5. Non-Viral Vectors

5.1 *Cationic Liposomes*

Cationic lipid delivery has been aggressively pursued in CFTR gene replacement studies and has led to many clinical trials. Cationic liposomes have proportionately lower efficiency and elicit innate immune responses. Yet, cationic liposomes are unlikely to produce an adaptive immune response, providing an advantage over viral vectors.

The absence of an adaptive response perpetuated repetitive lung delivery of the CFTR cDNA as a viable option to improve efficiency. This strategy is less likely using viral vectors. Unfortunately, the inefficiency of cationic liposome gene therapy could not be overcome by compounding gene correction with repetitive delivery.[25]

Improvements in cationic liposomes to increase efficiency have also been pursued. Variations in the hydrophobic body as well as additions to increase the positive charge of the hydrophilic head and the use of different neutral co-lipids have been tested. Although an improved lipid cation and co-lipid combination has surfaced (DC-Chol ((3b[N-[1](N_,N_-dimethylaminoethane)-carbamoyl]cholesterol)) with DOPE), no significant increase to overall efficiency has been achieved.[26]

5.2 *Compacted DNA Nanoparticles*

Compacted DNA nanoparticle delivery is a new player in CFTR gene correction. Briefly, polylysine or polyethylenimine can compact DNA into nanoparticles. These vectors have the capability of transducing non-dividing cells and have shown encouraging results in delivery to the lung epithelium in animal models.[27]

The results from the only phase I clinical trial, using nasal installation with the CMV enhancer to induce expression, were published in 2004.[28] Although there were no detectable levels of vector-derived CFTR mRNA, the overall prognosis was positive. There was no evidence of clinical toxicity and partial correction of ion transport occurred in some patients. Still, it seems unlikely that this delivery system will overcome the lung barriers that have obstructed the success of other vectors. Improvements to increase expression and transduction efficiency will surely have to been considered.

6. Gene Therapy Development for Alpha-1 Anti-trypsin

The first human trial for A1AT involved nasal delivery of A1AT using cationic liposomes. Although, nasal delivery showed an increase of A1AT, the levels were not significant enough to be of any therapeutic benefit.[29]

Early results redirected studies towards reversing the global deficiency in A1AT levels. The protective threshold of circulating A1AT is 600–800 ug/ml and the primary goal clinical of studies has been to achieve this level.[30] Using the skeletal muscle as a protein factory for recombinant A1AT is a suitable model for elevating levels of secreted protein in serum. The intramuscular delivery approach has minimal invasive properties and avoids the physiological lung barriers that limit local gene delivery to the airway epithelium. This and other strategies have been effective in augmenting systemic levels of the transgene product in a host of animal studies, which include direct liver injections, intramuscular injections and intravenous injections using AAV2, AAV1, AAV8, retrovirus, adenovirus and naked DNA.

Since the first human trial involving lung delivery of A1AT, there have been two clinical trials, one completed and one ongoing, both targeting A1AT delivery to skeletal muscle. The completed phase 1 study utilized an AAV2-A1AT viral vector.[31] Elevated antibody titers against the AAV2 capsid were the only observed immune response and detection of recombinant A1AT protein was complicated by the protein replacement therapy many of the patients were on prior to entering the trial. To bypass this, protein therapy was suspended prior to vector instillation but the washout period for the recombinant protein was longer than anticipated. Consequently, the A1AT baseline levels where elevated by the residue protein. The active phase 1 study is a follow up to the first study employing an AAV1 vector. A host of studies have confirmed that AAV1 transduces skeletal muscle substantially better than AAV2. AAV1 coupled with CBA/CMV hybrid promoter represents the most advanced AAV vector for muscle expression. Despite the development of neutralizing immune responses to the AAV1 viral vector in this trial, there was sustained vector specific A1AT expression, 0.1% of normal, for one year in the highest-dose patient group.[32]

An added complication occurs in A1AT patients with a PiZ genotype. Secretion of the PiZ protein from liver hepatocytes is severely impaired. The PiZ protein fails to be degraded and aggregate lesions develop in the hepatocytes. This pathology progresses to liver disease and cannot be treated by A1AT gene correction. This condition requires an additional strategy to reduce the PiZ proteins levels which can also be achieved by RNAI using AAV to deliver shRNA or miRNAs that specifically target these transcripts.

7. Lung Cancer Gene Therapy Development

The development of gene therapy has added another alternative for the aggressive push to treat lung cancer. Assisted by the dissection of cancer genetics, gene delivery offers the development of gene-based vaccines against tumor cells, oncogene inhibition by RNAi delivery and gene correction of tumor suppressor genes. This has been a welcome advancement due to resistance against current chemotherapies in some types of lung cancer like esophageal squamous cell carcinoma (SCC).

Preferred candidate genes have been elucidated, which include *ras* and *myc* as oncogenes to inhibit and p53 and FUS1 as suppressor genes to enhance. Both p53 and FUS1 gene replacement strategies have been used in several clinical trials using Ad vectors, lentiviral vectors and cationic liposomes with encouraging results.[32] In addition, combined therapies are being pursued and p53/FUS1 dual gene therapy has shown improved success in tumor regression in preclinical animal studies and is moving towards a clinical trial.[33]

The collection of results from the early body of work related to gene correction (including p53 and FUS1) has demonstrated success in tumor regression and increased apoptosis in tumor cells. This area of gene therapy continues to evolve with strong enthusiasm. Efforts towards materializing the potential of RNAi gene therapy targeted at oncogenes (ras and myc) for lung cancer are underway.[34] More time is necessary to build a strong body of work in order to make this field more concrete.

8. Cystic Fibrosis Animal Models

Developing a confident animal model that presents with the entire panel of disease manifestations is a critical step to developing effective treatments. The absence of a reliable, living model can severely retard treatment progress. This hurdle has been a long-standing obstacle for developing cystic fibrosis treatments.

The first CFTR null mouse was developed after the discovery of the CFTR gene.[35] Unfortunately, this mouse and subsequent CFTR deficient mice strains do not present with the characteristic CF lung pathologies including chronic bacterial colonization, which is the center of lung disease in cystic fibrosis. Although progress has been made using these *cftr* mice strains, the development of a more comprehensive model is crucial.

Efforts are ongoing to generate CFTR deficiencies in animals that have anatomies that strongly resemble that of humans. These advances include ferrets and pig models. The CFTR null ferrets are still under development, while data from the initial study to test whether CFTR deficient newborn pigs develop CF pathology has recently been published.[36] Although the CFTR defective newborn pigs present with several comparable manifestations in other organs (pancreas and intestine), no signs of lung inflammation were evident. It remains to be seen whether lung disease occurs in aged CFTR null pigs. This may seem discouraging, but a long-standing, unanswered question in cystic fibrosis lung pathology has been whether there is an increased inflammatory condition in the absence of lung pathogens. Controlled bacterial challenging of these newborn pigs may help unmask this perennial enigma.

9. Cell-Based Therapy for Cystic Fibrosis

Genetic manipulation of pluripotent stem cells is the focus of a new field of study to treat lung disease in cystic fibrosis. Isolated stem cells can be expanded and infected for genetic manipulation by *in vitro* experiments and can be transferred back to the original donor. This strategy requires successful *in vivo* transformation of these manipulated stem cells so lung cell repopulation can occur (Fig. 21.2). The ease of *ex vivo* gene transfer is a welcome departure from the complications involving *in vivo* transduction. Unfortunately, differentiation of plastic cells to fully functional epithelial cells and successful engraftment into the lung is not without unique challenges.

Progressive studies have led to some progress towards cellular based therapy that encompasses a range of stem cells. ES cells have demonstrated the potential to differentiate into a class of airway cells called Clara Cells under normal tissue culture conditions.[37] An early animal study revealed that systemic transplantation of hematopoietic stem cells (HSCs) could engraft into the lung epithelium.[38] Unfortunately, subsequent studies from other groups failed to replicate these results. Mesenchymal stem cells (MSCs) have been successfully directed to repopulate the lung environment, which resulted in improvement of lung function after an induced lung injury.[39] Finally, with specific emphasis on CFTR gene correction, MSCs from CF patients were successfully isolated, expanded and CFTR corrected. These MSCs possess the capability to differentiate into airway epithelial cells and combat CF lung pathology.[40]

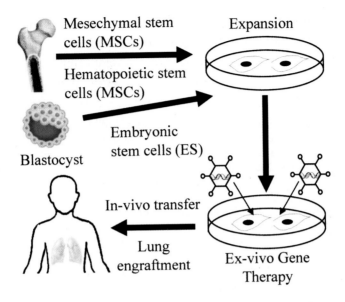

Fig. 21.2 Schematic overview of cell-based therapy.

Initial studies have generated a favorable opinion about the potential of cell-based therapy for lung disease. This promising new field has the potential to restructure the field of genetic manipulation. Unfortunately, ethical issues involving stem cell therapies have slowed progress.

10. Conclusion and Outlooks

Twenty years of clinically relevant gene therapy advancement for lung disease has compiled a diverse collection of studies that expanded from single-gene, hereditary diseases into non-hereditary disorders, multi-gene combination therapy and the merger of stem cell and genetic manipulation studies. During this tenure, gene therapy vectors have evolved along with strategies for genetic modification. Favorable vectors and approaches have emerged specific to the disorder being treated, while some have been precluded. It is definite that unique tailoring of gene therapy approaches is necessary to accommodate the diverse group of lung diseases that are targets for gene correction. The adaptive response of the gene therapy field to these new challenges has been slow but productive. With many of the improved strategies moving towards or being used in clinical studies, this is a dynamic time that may define gene correction for lung diseases.

References

1. Bush AA (2006). *Cystic fibrosis in the 21st century*, Karger: Basel; New York.
2. Riordan JR, Rommens JM, Kerem B, Alon N, Rozmahel R, Grzelczak Z, *et al.* (1989). Identification of the cystic fibrosis gene: Cloning and characterization of complementary DNA. *Science* **245**: 1066–1073.
3. Welsh MJ, Smith AE (1993). Molecular mechanisms of CFTR chloride channel dysfunction in cystic fibrosis. *Cell* **73**: 1251–1254.
4. Boucher RC (2003). Regulation of airway surface liquid volume by human airway epithelia. *Pflugers Arch* **445**: 495–498.
5. Johnson LG, Olsen JC, Sarkadi B, Moore KL, Swanstrom R, Boucher RC (1992). Efficiency of gene transfer for restoration of normal airway epithelial function in cystic fibrosis. *Nat Genet* **2**: 21–25.
6. Wilson JM (1995). Gene therapy for cystic fibrosis: Challenges and future directions. *J Clin Invest* **96**: 2547–2554.
7. Cruz PE, Mueller C, Flotte TR (2007). The promise of gene therapy for the treatment of alpha-1 antitrypsin deficiency. *Pharmacogenomics* **8**: 1191–1198.
8. Boucher RC, Knowles MR, Johnson LG, Olsen JC, Pickles R, Wilson JM, *et al.* (1994). Gene therapy for cystic fibrosis using E1-deleted adenovirus: A phase I trial in the nasal cavity. The University of North Carolina at Chapel Hill. *Hum Gene Ther* **5**: 615–639.
9. Zuckerman JB, Robinson CB, McCoy KS, Shell R, Sferra TJ, Chirmule N, *et al.* (1999). A phase I study of adenovirus-mediated transfer of the human cystic fibrosis transmembrane conductance regulator gene to a lung segment of individuals with cystic fibrosis. *Hum Gene Ther* **10**: 2973–2985.
10. Flotte T, Carter B, Conrad C, Guggino W, Reynolds T, Rosenstein B, *et al.* (1996). A phase I study of an adeno-associated virus-CFTR gene vector in adult CF patients with mild lung disease. *Hum Gene Ther* **7**: 1145–1159.
11. Conrad CK, Allen SS, Afione SA, Reynolds TC, Beck SE, Fee-Maki M, *et al.* (1996). Safety of single-dose administration of an adeno-associated virus (AAV)-CFTR vector in the primate lung. *Gene Ther* **3**: 658–668.
12. Flotte TR, Schwiebert EM, Zeitlin PL, Carter BJ, Guggino WB (2005). Correlation between DNA transfer and cystic fibrosis airway epithelial cell correction after recombinant adeno-associated virus serotype 2 gene therapy. *Hum Gene Ther* **16**: 921–928.
13. Di Pasquale G, Davidson BL, Stein CS, Martins I, Scudiero D, Monks A, *et al.* (2003). Identification of PDGFR as a receptor for AAV-5 transduction. *Nat Med* **9**: 1306–1312.
14. Walters RW, Yi SM, Keshavjee S, Brown KE, Welsh MJ, Chiorini JA, *et al.* (2001). Binding of adeno-associated virus type 5 to 2,3-linked sialic acid is required for gene transfer. *J Biol Chem* **276**: 20610–20616.

15. Virella-Lowell I, Zusman B, Foust K, Loiler S, Conlon T, Song S, *et al.* (2005). Enhancing rAAV vector expression in the lung. *J Gene Med* **7**: 842–850.

16. Liu X, Luo M, Trygg C, Yan Z, Lei-Butters DC, Smith CI, *et al.* (2007). Biological Differences in rAAV Transduction of Airway Epithelia in Humans and in Old World Non-human Primates. *Mol Ther* **15**: 2114–2123.

17. Sirninger J, Muller C, Braag S, Tang Q, Yue H, Detrisac C, *et al.* (2004). Functional characterization of a recombinant adeno-associated virus 5-pseudotyped cystic fibrosis transmembrane conductance regulator vector. *Human Gene Therapy* **15**: 832–841.

18. Muller C, Braag SA, Herlihy JD, Wasserfall CH, Chesrown SE, Nick HS, *et al.* (2006). Enhanced IgE allergic response to Aspergillus fumigatus in CFTR-/- mice. *Lab Invest* **86**: 130–140.

19. Carroll TP, Morales MM, Fulmer SB, Allen SS, Flotte TR, Cutting GR, *et al.* (1995). Alternate translation initiation codons can create functional forms of cystic fibrosis transmembrane conductance regulator. *J Biol Chem* **270**: 11941–11946.

20. Fischer AC, Smith CI, Cebotaru L, Zhang X, Askin FB, Wright J, *et al.* (2007). Expression of a truncated cystic fibrosis transmembrane conductance regulator with an AAV5-pseudotyped vector in primates. *Mol Ther* **15**: 756–763.

21. Palmer DJ, Ng P (2005). Helper-dependent adenoviral vectors for gene therapy. *Hum Gene Ther* **16**: 1–16.

22. Chu Q, St George JA, Lukason M, Cheng SH, Scheule RK, Eastman SJ (2001). EGTA enhancement of adenovirus-mediated gene transfer to mouse tracheal epithelium *in vivo*. *Hum Gene Ther* **12**: 455–467.

23. Kitson C, Angel B, Judd D, Rothery S, Severs NJ, Dewar A, *et al.* (1999). The extra- and intracellular barriers to lipid and adenovirus-mediated pulmonary gene transfer in native sheep airway epithelium. *Gene Ther* **6**: 534–546.

24. Virella-Lowell I, Poirier A, Chesnut KA, Brantly M, Flotte TR (2000). Inhibition of recombinant adeno-associated virus (rAAV) transduction by bronchial secretions from cystic fibrosis patients. *Gene Ther* **7**: 1783–1789.

25. Hyde SC, Southern KW, Gileadi U, Fitzjohn EM, Mofford KA, Waddell BE, *et al.* (2000). Repeat administration of DNA/liposomes to the nasal epithelium of patients with cystic fibrosis. *Gene Ther* **7**: 1156–1165.

26. Caplen NJ, Alton EW, Middleton PG, Dorin JR, Stevenson BJ, Gao X, *et al.* (1995). Liposome-mediated CFTR gene transfer to the nasal epithelium of patients with cystic fibrosis. *Nat Med* **1**: 39–46.

27. Gautam A, Densmore CL, Golunski E, Xu B, Waldrep JC (2001). Transgene expression in mouse airway epithelium by aerosol gene therapy with PEI-DNA complexes. *Mol Ther* **3**: 551–556.

28. Konstan MW, Davis PB, Wagener JS, Hilliard KA, Stern RC, Milgram LJ, *et al.* (2004). Compacted DNA nanoparticles administered to the nasal mucosa of cystic fibrosis subjects are safe and demonstrate partial to complete cystic fibrosis transmembrane regulator reconstitution. *Hum Gene Ther* **15**: 1255–1269.

29. Brigham KL, Lane KB, Meyrick B, Stecenko AA, Strack S, Cannon DR, *et al.* (2000). Transfection of nasal mucosa with a normal alpha1-antitrypsin gene in alpha1-antitrypsin-deficient subjects: comparison with protein therapy. *Hum Gene Ther* **11**: 1023–1032.

30. American Thoracic Society and the European Respiratory Society (2003). American Thoracic Society/European Respiratory Society statement: Standards for the diagnosis and management of individuals with alpha-1 antitrypsin deficiency. *Am J Respir Crit Care Med* **168**: 818–900.

31. Brantly ML, Spencer LT, Humphries M, Conlon TJ, Spencer CT, Poirier A, *et al.* (2006). Phase I trial of intramuscular injection of a recombinant adeno-associated virus serotype 2 alphal-antitrypsin (AAT) vector in AAT-deficient adults. *Hum Gene Ther* **17**: 1177–1186.

32. Brantly ML, Chulay JD, Wang L, Mueller C, Humphries M, Spencer LT, *et al.* (2009). Sustained transgene expression despite T lymphocyte responses in a clinical trial of rAAV1-AAT gene therapy. *Proc Natl Acad Sci USA* **106**: 16363–16368.

33. Toloza EM, Morse MA, Lyerly HK (2006). Gene therapy for lung cancer. *J Cell Biochem* **99**: 1–22.

34. Deng WG, Kawashima H, Wu G, Jayachandran G, Xu K, Minna JD, *et al.* (2007). Synergistic tumor suppression by coexpression of FUS1 and p53 is associated with down-regulation of murine double minute-2 and activation of the apoptotic protease-activating factor 1-dependent apoptotic pathway in human non-small cell lung cancer cells. *Cancer Res* **67**: 709–717.

35. Yin JQ, Gao J, Shao R, Tian WN, Wang J, Wan Y (2003). siRNA agents inhibit oncogene expression and attenuate human tumor cell growth. *J Exp Ther Oncol* **3**: 194–204.

36. Collins FS, Wilson JM (1992). Cystic fibrosis. A welcome animal model. *Nature* **358**: 708–709.

37. Rogers CS, Stoltz DA, Meyerholz DK, Ostedgaard LS, Rokhlina T, Taft PJ, *et al.* (2008). Disruption of the CFTR gene produces a model of cystic fibrosis in newborn pigs. *Science* **321**: 1837–1841.

38. Ali NN, Edgar AJ, Samadikuchaksaraei A, Timson CM, Romanska HM, Polak JM, *et al.* (2002). Derivation of type II alveolar epithelial cells from murine embryonic stem cells. *Tissue Eng* **8**: 541–550.

39. Krause DS, Theise ND, Collector MI, Henegariu O, Hwang S, Gardner R, *et al.* (2001). Multi-organ, multi-lineage engraftment by a single bone marrow-derived stem cell. *Cell* **105**: 369–377.

40. Ortiz LA, Gambelli F, McBride C, Gaupp D, Baddoo M, Kaminski N, *et al.* (2003). Mesenchymal stem cell engraftment in lung is enhanced in response to bleomycin exposure and ameliorates its fibrotic effects. *Proc Natl Acad Sci USA* **100**: 8407–8411.

41. Wang G, Bunnell BA, Painter RG, Quiniones BC, Tom S, Lanson NA Jr., *et al.* (2005). Adult stem cells from bone marrow stroma differentiate into airway epithelial cells: potential therapy for cystic fibrosis. *Proc Natl Acad Sci USA* **102**: 186–191.

Chapter 22
Cardiovascular Disease

Darin J. Falk, Cathryn S. Mah and Barry J. Byrne*

In this chapter we will discuss gene therapy-based applications for the treatment of cardiovascular disease. Although much progress has been made in tackling the disease, it still remains the number one cause of death and disability in the United States and the majority of European countries. It is estimated that 80 million Americans alone have one or more types of cardiovascular disease. Given the worldwide prevalence, the development of an effective strategy for cardiac repair and protection is critical. Numerous approaches have been investigated and to date, there is no single treatment that has proven to be successful. Inevitably, the overall goal is to improve our understanding of the molecular basis of myocardial injury in order to develop treatment paradigms to improve cardiovascular disease related outcomes. Gene therapy based applications provide a unique tool to understand and elucidate the basic underlying causes of disease and presents a promising avenue to circumvent the prevalence of cardiovascular related death and disability.

1. Introduction

Cardiovascular disease (CVD) accounts for ~35% of all deaths and has been the primary cause of death in the United States in each year since 1900 (with the exception of 1918).[1] Much progress has been made during this time with the advent of new technologies and innovative research which has led to rapid diagnosis and treatment for CVD. However, the rank of CVD has remained unchanged and a rapid increase in previously unconventional

*Correspondence: University of Florida College of Medicine, Dept. Pediatrics, Div., Cellular Molecular Therapy, and Powell Gene Therapy Center, 1600 SW Archer Road, ARB-165, Gainesville, FL 32610. E-mail: bbyrne@ufl.edu

diseases (Type II diabetes, obesity, etc.) will invariably lead to increased incidence of CVD disease. The financial impact of CVD constitutes an overwhelming share of burden on the health care system and the estimate for the US alone in 2009 stands at over $475 billion. Despite the significant resources expended on the treatment of this disease, outcomes still remain poor. The five-year survival rate for individuals diagnosed with heart failure is less than 50%, and in end-stage heart failure, the one-year survival rate may be as low as 25% regardless of medical therapy.[1]

A vast amount of preclinical work has laid the foundation for the treatment of CVD disease. Uncovering the cascade of events that occur during myocardial insult has identified the potential targets and timing for intervention. The majority of treatments prescribed for CVD is pharmacologically based and restricts therapeutic options for patients. The recurrence of myocardial injury and lack of regenerative capacity in cardiac tissue has highlighted the limitations of this approach. Patients are required to follow lifelong dosing regimens with repeated visits and monitoring. Therefore, the development of alternative therapies will inevitably enhance outcome of diagnosis and improve the overall quality of life in the affected population.

2. Therapeutic Targets

Presently, there is no effective treatment for these disorders but several novel strategies for replacing or enhancing the defective gene are in development. An overwhelming number of preclinical studies have shown positive outcomes in conjunction with a genetic based approach. At the same time, some investigations are hampered by several limitations but in the end have led to the optimization of applications. Still, we should move forward cautiously to ensure the optimal therapy for affected individuals is prescribed appropriately.

2.1 *Congenital Heart Disease*

The muscular dystrophies consist of a vast group of genetically inheritable diseases characterized by progressive and debilitating muscle atrophy. Over 40 diseases bear similarities to muscular dystrophy and are described by a manifestation of a single or combinatorial defect typically found in structural muscle-membrane associated proteins, ranking them as the most common inheritable childhood muscle disease.[2] One of the hallmarks of the most prevalent form is dilated cardiomyopathy making it one of the two leading causes of death in Duchenne Muscular Dystrophy (DMD) patients. Currently the prognosis is grim for patients and is based upon the type and progression of the disorder. Clinical trials have gained momentum recently although most regimens are pharmaceutical based and only cause a delay in the progression and materialization of pathogenic abnormalities (clinicaltrials.gov).

Accordingly, much interest has been shifted to gene therapy where a wide range of vectors have been employed (plasmid DNA, Ad, lentiviral, antisense oligonucleotides, and AAV) and have shown promise in circumventing the cascade of events defining the dystrophic process.[3] The majority of pre-clinical models show restoration of the dystrophic cardiomyopic abnormalities associated with the disease (Table 1). Recently, stop codon suppression with antisense oligonucleotides has shown substantial success in mdx mice, dystrophic dogs and isolated cells from DMD patients and may move to a clinical trial in the near future.[4-6]

Ongoing clinical trials are investigating AAV-mediated transgene expression for DMD (rAAV2.5-CMV-Mini-Dystrophin) and LGMD-2 (rAAV1.tMCK.hαSG) (clinicaltrials.gov) and while both of these dystrophies affect cardiac phenotype and function, neither treatment is currently addressing cardiac parameters. However, these trials will generate important findings by determining the safety and efficacy of AAV-mediated gene replacement in skeletal muscle for both dystrophies.

While the most commonly known muscular dystrophies are caused by abnormalities in proteins involved in tissue structure and function, inherited metabolic diseases can lead to cardiac pathologies as well. Two examples of this include Fabry disease, an X-linked recessive disorder due to a deficiency of alpha-galactosidase A (alpha-gal A), and Pompe disease, a glycogen storage disease caused by a lack of functional lysosomal acid alpha-glucosidase (GAA).[7,8] Pathology in both diseases results from progressive accumulation of cellular products. The storage of excess glycosphingolipids in Fabry can lead to impaired arterial circulation whereas the storage of lysosomal glycogen in Pompe disease can result in physical damage to heart and muscle cells. In either disease, patients often succumb to cardiac complications.[7,9]

Currently, only enzyme replacement therapy (ERT) is available for both diseases, however there is a short half-life requiring repeated administration which incur substantial health care costs. These qualities have invoked the need for a long-term correction of the enzyme deficiency which has gained the attention of gene therapy.

Work by Ogawa *et al.* investigated the effects of AAV delivery of human alpha-gal A to a neonatal Fabry mouse model. Activities of alpha-gal A 25 weeks after a single injection produced a 45 fold increase above wild-type basal levels and that the high level of cardiac alpha-gal A was attributed to direct transduction of the heart rather than uptake of secreted protein.[10] Interestingly, the authors also noted that enzyme levels varied with respect to age at administration and/or gender of the animals, suggesting that these attributes are also major factors in the degree of success garnered by gene therapy.

Clinical studies of ERT for Pompe disease have shown marked improvement in myocardial pathology and prognosis. However, Pompe disease is a multi-system disorder and improvements in skeletal muscle function and phenotype have been met with less success. Patients initially appear to ward off the rapid progression of the disease but eventually succumb to the cardiac and skeletal pathogenesis.

Recent attempts have focused on improving the delivery of hGAA to affected tissues where the incorporation of enzyme enhancement (chaperone) and protein modification (increased affinity for M6P receptor) have increased the payload to affected tissues.[8] These studies are ongoing and clinical trials are currently enrolling patients to establish the safety and efficacy of these options.

Ideally, treatment for Pompe disease would consist of a single treatment that would reverse and restore normal phenotype and function of organ systems. Gene therapy has consistently shown these characteristics while receiving enhanced development and refinement. Our laboratory and others have repeatedly shown that administration of GAA via AAV or Ad can promote regression of cardiomyopathy.[8] Specifically AAV-GAA has been shown to increase cardiac function (i.e. PR interval, cardiac output, ejection fraction, end diastolic mass), decrease left ventricular mass and reduce myocardial glycogen accumulation. The preclinical evidence for a potential curative option is mounting; however there are still additional hurdles (i.e. immunotolerance) to overcome.

2.2 Coronary Artery Disease and Ischemia/Reperfusion Injury

Myocardial ischemia/reperfusion injury and infarction originally result from inefficient bloodflow to the heart, with myocardial ischemia generally occurring first. Repeated and/or prolonged periods of ischemia can result in injury, and, if allowed to progress, ultimately cause myocardial infarction, or permanent damage, death, and necrosis of cardiomyocytes and scarring of tissue. Atherosclerosis of the coronary arteries is the most common cause of myocardial ischemia. The majority of gene therapy studies for the treatment of CVD have focused on correction of myocardial ischemia by therapeutic angiogenesis in which new vessels are promoted to form, creating a bypass around the restricted or occluded vessels and allowing for continued efficient blood supply to the heart.[11,12] Angiogenesis is the natural response to ischemia, however, in many cases, the native response is inadequate to overcome progressive disease. Clinical studies in which recombinant angiogenic factors were administered to treat coronary or peripheral artery disease have not been successful and are attributed in part to the short half-life of the recombinant proteins. Gene therapy is an attractive alternative mode of therapy, offering the potential for providing a sustained source of therapeutic protein expression. Preclinical gene transfer studies utilizing plasmid DNA vector or adenovirus-based vectors encoding proangiogenic factors including isoforms of vascular endothelial growth factor (VEGF), fibroblast growth factor (FGF), and hepatocyte growth factor (HGF) have resulted in neovascularization and enhanced bloodflow in localized areas of vector delivery.[11] The success of these preclinical studies warranted testing in clinical trials (Table 2), some studies of which are still currently active.

2.3 Oxidative Stress

Myocardial ischemic events lasting beyond 20 minutes result in irreversible damage and death to cardiomyocytes. During such ischemic periods there is a substantial increase in the levels of free radicals, reactive oxygen/nitrogen species and in the rise and duration of intracellular calcium concentration. Depending on the extent and severity of ischemia and reperfusion (IR) events puts the myocardium at risk for oxidative-mediated damage. The concomitant production of these reactive species overwhelms cellular defense mechanisms and the subsequent cascade of events may diminish contractility and cardiac pump function. Fortunately there are innate mechanisms within cardiomyocytes reducing the susceptibility to ischemia-related injury. The presence of cytoprotective proteins involved with calcium handling and antioxidant buffering capacity allow for the maintenance of cellular homeostasis.

2.4 *Antioxidants*

As shown by Sweeney and colleagues, early gene therapy studies manipulating cellular antioxidant defense mechanisms were effective in reducing and minimizing infarct size and cardiac contractile dysfunction during IR injury.[13] Since this time there has been an expansion on initial studies and the outcome for vector delivery of antioxidant species has consistently agreed upon previous findings.[14-16] In a rabbit model of IR injury, extracellular-SOD (Ad5 with and without heparin) reduced the size and area at risk of infarction with a ~5 fold increase in cardiac SOD (total) activity.[15,16]

Modulation of HO-1, and to some extent HO-2, has been under intense study for potential in mediating or preventing the progression of a wide range of CVD (I/R, cardiomyopathy, hypertension, organ transplantation). Upon the degradation of heme, HO-1 exerts antioxidant properties with the release of biliverdin, bilirubin and carbon monoxide. These moieties have shown valuable antioxidant properties in various models of CVD while inhibition of HO-1 promotes lesion formation.[18] Direct augmentation of HO-1 or its breakdown products confers cardioprotection after intramyocardial delivery of HO-1 against IR injury. Ad-mediated transfer (vascular delivery) of HO-1 was shown to maintain arterial vasodilation and exhibited antiproliferative effects in a swine model of vascular injury.[19] Administration of rAAV-HO-1 to rodents prior to I/R injury was associated with significant decreases in lipid peroxidation and apoptotic/inflammatory markers resulting in 70–80% reduction in infarct size.[20] These studies shed light not only on the effectiveness of HO-1 therapy but help delineate the mechanism responsible for IR injury as well.

Preferably, an approach similar to Phillips and colleagues would allow investigators to express the transgene of choice under conditions of cellular stress.[21,22] Using novel responsive promoter and transactivator elements in gene therapy strategies, they demonstrated the feasibility of switching on a transgene only during bouts of IR injury *in vivo* showing preservation of myocardial function and integrity, decreases in oxidative stress, inflammation and apoptotic markers under periods of prolonged hypoxia.

2.5 *Cardiac Contractility*

Calcium dysregulation is considered a primary factor of myocardial pathogenesis. Loss of calcium homeostasis can be attributed to dysfunction of the sarcoplasmic endoplasmic reticulum Ca^{2+}ATPase (SERCA) in cardiomyocytes and directly effects cardiac contractility and relaxation properties. Importantly, phospholamban, the endogenous inhibitor of SERCA pump activity, has also shown to be amenable to gene therapy intervention and improvement of cardiac contractility and calcium handling *in vivo*.[23] Alternatively, modulation of the calcium binding proteins S100A1 and parvalbumin have shown similar results in augmenting cardiac function in senescent and aortic banding rodent models.[24,25] These studies highlight the critical nature of restoring calcium homeostasis *in vivo* where prolonged and sustained damage to the myocardium may result in a rapid upregulation of calcium activated proteases (i.e. calpain), ubiquitin-proteasome machinery and apoptotic signaling. Pharmacological inhibition of these mechanisms results in preservation of cardiac function and reduction in area of infarction. A plausible solution may include siRNA mediated knockdown of key regulators within these signaling cascades or upregulation of chaperone and anti-apoptotic mediators to lower myocardial injury.

3. Animal Models

There exist several animal models of CVD. Progressive CVD manifestations can be observed in some models of congenital disease such as the muscular dystrophies. However, many models require induced physical damage such

Table 22.1.   Selected models of diseases used in preclinical AAV gene therapy studies.

Condition	Animal	Promoter	Vector	Route	Ref.
MD					
Duchenne	Mouse	CMV, CK6	AAV2, 5, 6, 9	IV, IM	36–38
Delta- sarcoglycan	Hamster	CMV	AAV1, 8, 9; dsAAV2, 8	IP, IV	39
Alpha- sarcoglycan	Mouse	MCK	AAV1	IM	40
Fabry	Mouse	CAG	AAV	IV	10
GSD II	Mouse	CMV, CBA, DES	AAV1, 8, 9	IP, IV, IM	8
Angiogenesis/ Ischemia- Reperfusion	Mouse, Rat, Pig	CMV, MHC, MLC2V, CBA	AAV1, 2, 5, 6, 8, 9	IV, IM	11

Abbreviations: MD, Muscular Dystrophy; LSD, Lysosomal Storage Disease; CMV, cytomegalovirus; CK6, creatine kinase; MCK, creatine kinase; CAG, CBA, chicken beta-actin; DES, desmin; MHC, Myosin heavy chain; MLC2V, myosin light chain; AAV, adeno-associated virus; IP, intraperitoneal; IV, intravascular; IM, Intramyocardial.

as aortic banding or ligation of vessels to create regions of ischemia/infarct. While useful in studying correction of severe acute damage, these models lack the pathology related to chronic disease that may impact on therapeutic success. Table 22.1 lists several examples of animal models of diseases that have been treated using AAV vectors. The same models as well as routes of delivery have been tested using other gene therapy vectors as well.

4. Vector Delivery

To-date, the most successful cardiovascular gene therapy applications have utilized non-viral plasmid DNA (pDNA) and viral recombinant (Ad) aden-ovirus and adeno-associated virus (AAV)-based vectors. Both naked pDNA and Ad vectors have progressed to testing in clinical trials for therapeutic angiogenesis applications and AAV vectors are currently being assessed for gene transfer of calcium-handling proteins (Table 22.2). Recently several

Table 22.2.   Gene therapy clinical trials for CVD.

Target	Delivery	Vector	Trial
Angiogenesis	Intracoronary	Ad-VEGF165	KAT
		Ad-FGF-4	AGENT-2,3,4
		Ad-FGF-4	AWARE
		Ad-AC6	Adenylyl cyclase
	Intramyocardial	Ad-VEGF161	REVASC
		pDNA VEGF165	Euroinject one
		pDNA VEGF-2	Genasis
		Ad-VEGF121	Northern
		Ad-VEGF121	NOVA
Heart Failure	Intracoronary	AAV-SERCA2a	SERCA Gene Therapy
		AAV-SERCA2a	CUPID
Angina	Intramyocardial	pDNA VEGF-2	VEGF therapy
In-stent restenosis	Local delivery after stent implantation	Anti-sense oligo against c-myc	Italics

Abbreviations: Ad, Adenovirus; VEGF, Vascular endothelial growth factor; FGF, Fibroblast growth factor; AC, Adenyl Cyclase; pDNA, plasmid DNA; AAV, Adeno-associated virus; SERCA, Sarcoplasmic endoplasmic reticulum calcium ATPase; oligo, oligonucleotide.

lentiviral vectors studies have also shown potential, thus warranting continued investigation of these vectors for cardiovascular applications.[26]

In vivo delivery methods for cardiovascular gene therapy broadly include direct delivery of vector to the target tissue, local intravascular delivery, and systemic delivery of vector. Briefly, direct delivery methods entail administration of vector directly into the heart musculature or onto walls of the vasculature, most commonly with a needle or other device. Coronary delivery is the most common form of local intravascular delivery and requires transient isolation of the circulation by cross-clamping or balloon angioplasty followed by injection of vector into the coronary circulation. Systemic delivery usually entails simple IV injection of vector at a peripheral site.

From a theoretical, but not necessarily technical, standpoint, the simplest mode of delivery is direct injection of vector into the target tissue, however specific techniques can vary. For example, in rodents, blind injections through the chest wall or through the diaphragm from the abdominal side have been used successfully to transduce myocardium in various disease models. In larger animal models, ultrasound or electromechanical mapping technologies to visualize the heart have allowed for more accurate targeted delivery to the affected tissue. Following injection, overall transduction efficiencies then become dependent on the ability of the vector to efficiently enter the target cell and subsequently, cellular expression of the therapeutic transgene. The obvious advantage of this method is the direct exposure of target cells to the vector, thereby maximizing the potential for efficient transduction in those cells. Furthermore, restriction of vector to the sites of injection reduces the potential for spread of vector to non-target tissues and organs thereby reducing the potential of adverse events resulting from inappropriate transgene expression. A limitation of direct injection is the technical difficulty of the procedure, as not all regions of the heart are easily accessible for direct injection. As such, procedure-related adverse events are of concern, especially for highly invasive methods. Expression from direct injection is generally limited to regions of delivery which may not be as effective in diseases that may benefit from more global cardiac transduction; however, applications better suited for localized transduction of the heart do exist.

Direct injection in the myocardium has been shown to be effective in animal models of disease and as such has been utilized in several phase II/III randomized control clinical trials. Direct intramyocardial injection of Ad vectors encoding for $VEGF_{121}$ (REVASC trial) or naked pDNA vectors encoding for VEGF-2 or $VEGF_{165}$ (Genasis and Euroinject One trials, respectively) are currently being assessed for the treatment of coronary artery disease. Despite impressive results in animal models with pDNA vectors, current results in clinical trials thus far have been disappointing. The REVASC trial had more successful results with improvement in time to 1mm ST-segment depression and time to level 2 angina in treated groups, as well as first anatomical evidence that $AdVEGF_{121}$ could mediate neovascularization in humans.[1,11,27–29]

Gene gun, or particle bombardment technology, is a modified form of direct injection in which vector is linked to micron-sized heavy metal particles. The particles are then propelled under high pressure into targeted tissue. This method has been used with pDNA vector for cardiac applications in several preclinical studies but has not been shown to be significantly better than traditional direct injection techniques.[26]

The use of solid support structures to mediate local direct delivery has been investigated, and in particular the potential of vector-eluting stents. Over 1 million stents were placed in patients in the United States in 2006 and restenosis can occur in up to 30% of patients.[27,30] Local gene transfer may provide a sustained supply of therapeutic protein to prevent restenosis and may have the additional advantage of being able to correct the local underlying molecular defect(s). Preclinical studies in which pDNA, Ad, or AAV vectors are tethered to stents have shown promise, giving rise to efficient local transduction.[27,31]

Injection of vector into the pericardial space, rather than directly into the heart musculature has also been employed in preclinical studies utilizing pDNA, Ad, and AAV vectors, leading to moderate transduction in regions adjacent to the pericardial space. Transmural transduction however was inefficient and could thus far be achieved only with concomitant treatment with enzymes such as collagenase or proteases to disrupt the extracellular matrix.[26,29]

For some CVD models, direct delivery of vector to tissue distal from the heart (such as skeletal muscle or liver) has been utilized. This strategy relies upon secretion of the therapeutic protein from the transduced tissue and uptake into the affected myocardium. For example, for Pompe disease, GAA expressed in liver or skeletal muscle can be secreted, then taken up by the heart, resulting in correction of cardiac biochemical abnormalities in a mouse model of disease.[32]

Local intravascular delivery in which vector delivery is isolated to the heart circulation is attractive as the vasculature structure and dense capillary network allows for improved dispersion of vector and access to target cells.[1,26,33] Restriction of delivery to the local vasculature can be achieved by injection into heart *ex vivo* (for transplanted hearts), injection into the vasculature after cardiac bypass with induced cardioplegia and cross-clamping, where blood flow into and/or out of the heart is blocked by clamping or balloon angioplasty, or cross-clamping followed by vector delivery into the left ventricle of the beating heart. These methods have been shown to be very effective in achieving efficient global cardiac transduction in preclinical studies; however the technical difficulty and invasive nature of the procedures may limit clinical use. The most successful application of local intravascular delivery thus far has been the use of intracoronary catheters. In the phase II Kuopio Angiogenesis Trial (KAT), pDNA and Ad vectors encoding $VEGF_{165}$ were delivered locally via an infusion-perfusion catheter during percutaneous coronary intervention procedures. At 6 months post-treatment, there was significant increase in myocardial perfusion in the group treated with Ad vector. No difference was seen with the pDNA vector. The Angiogenic Gene Therapy Trials (AGENT) assessed intracoronary injection of AdFGF-4 vectors and while smaller cohort studies demonstrated some improvement in exercise time and reduction of ischemic defect size, differences were not noted in the larger studies. Post-hoc analysis, however, did reveal that when subject age or gender was a factor, significant improvements in response to therapy were evident and as such, a study assessing a more focused patient population (Angiogenesis in women with angina pectoris who are not candidates for revascularization; AWARE) is currently ongoing.[1,26,28,34] In addition to Ad vectors, recombinant AAV vectors have recently shown great promise as vectors for CVD in preclinical studies and the first AAV trial for cardiac failure entailing delivery of vector encoding SERCA2a via percutaneous intracoronary injection (Calcium Up-Regulation by Percutaneous Administration of Gene Therapy in Cardiac Disease; CUPID) has been initiated and results presented at the 2009 American Society of Gene Therapy meeting so far suggest evidence of biological activity and safety of the vector.[28]

Improvements to current local intravascular delivery methods are being investigated in preclinical animal studies and include modification of physical delivery means such as increasing delivery/injection pressure and vector dwell times, inclusion of adjuvants to vector preparations which increase microvascular permeability, or use of electroporation or sonoporation to increase cell membrane permeability.[35] Vector modifications are also being investigated as potential means to improve uptake into target cells. Such vector modifications also lend themselves well to systemic delivery strategies and will be discussed further below.

Simple intravenous (IV) injection for systemic delivery of vector is the idealized mode of gene therapy in which the vector itself provides all the means to transduce only the affected target cells and mediate correction. The clear advantage to such methods is the ease of vector delivery and avoidance of procedure-related complications. Current disadvantages however include the probable transduction of non-target tissues which could result in severe immune response to the vector itself, the therapeutic transgene product, and the transduced cells themselves. As such, control of each step of vector transduction is likely necessary, beginning from initial attachment to target cells, to tissue/cell-type specific transgene expression, and regulation of said expression.

Systemic delivery of naked pDNA vectors has met with little success for cardiac transduction, however pDNA preparations utilizing lipid or other polymer complexes/coatings and in particular, coatings in which specific peptide ligands or antibodies have been incorporated, have been shown to enhance uptake in target tissues. A physical method of targeted delivery has used microbubble technology in which vectors (pDNA or viral) have been encapsulated in gas microbubbles which can then be delivered systemically. Ultrasound in the heart or vasculature will then result in destruction of the microbubbles in the local area, releasing the vectors in the targeted region. For enveloped viral vectors, pseudotyping envelopes to include targeting moieties have also been employed. For non-enveloped viruses, molecular adapters (cell-specific antibodies, ligands, targeting peptides, etc.) have been either physically conjugated to the viral capsid to allow for re-direction of the vector to target cells, or, in some cases such as with AAV vectors, directly incorporated into the capsid structure itself.

Recently, natural tissue tropism of novel vectors has been exploited in gene therapy strategies. For example, recombinant AAV pseudotype 9 (rAAV2/9) vectors have been shown to be naturally tropic for the heart and globally transduce myocardium after IV delivery and have shown great promise in models of congenital cardiac disease, IR injury, and calcium dysregulation. Several studies have investigated the potential of directed evolution of AAV vectors in which viral capsid proteins are randomly mutated or capsid proteins from different serotypes are mixed together to create novel vectors with different bioactivities and tissue tropisms. These novel vectors can then be screened for tissue tropism for the cell-types of choice and could yield vectors with stringent tropisms.

373

5. Conclusions and Outlook

Since the first preclinical CVD gene therapy studies in the late 1980's, the development of improved vector systems, cell-specific promoters, and routes of delivery have enabled investigators to target a myriad of disease-associated factors affecting the myocardium and the translational nature is now being realized. Initial clinical studies have been promising however it is clear that continued development and understanding of the molecular basis for disease as well as vector biology is necessary to develop effective gene therapies. Questions with regard to dosing, timing of delivery and long-term safety remain. Well-designed clinical trials have the potential to complement and even stimulate basic science and preclinical research. In most cases there likely will not be a single target that may provide an all end cure, and refinement of treatments may likely include a multifactorial approach including gene therapy and additional strategies. Better understanding of the molecular mechanisms will invariably provide platforms on which to develop effective countermeasures for this devastating disease. Gene therapy provides an attractive alternative or complement to current therapies for efficient and effective treatment for CVD. The examples provided throughout this chapter represent some of the many pioneering studies contributing to the characterization, development and implementation of therapy for CVD and the authors regret that only a minor portion could be included.

Acknowledgments

This work was supported in part by NIH PO1 HL059412 and NIH PO1 DK058327 (B.J.B., C.S.M., D.J.F.), American Heart Association, National Center (C.S.M.), NRSA FHL095282A (D.J.F).

References

1. Lloyd-Jones D, *et al.* (2009). Heart disease and stroke statistics–2009 update: A report from the American Heart Association Statistics Committee and Stroke Statistics Sub-committee. *Circulation* **119**: 480–486.
2. Emery AEH (2001). *The Muscular Dystrophies*, Oxford University Press: Oxford.
3. Muir LA, Chamberlain JS (2009). Emerging strategies for cell and gene therapy of the muscular dystrophies. *Expert Rev Mol Med* **11**: e18.

4. Mann CJ, *et al.* (2001). Antisense-induced exon skipping and synthesis of dystrophin in the mdx mouse. *Proc Natl Acad Sci USA* **98**: 42–47.

5. McClorey G, Moulton HM, Iversen PL, Fletcher S, Wilton SD (2006). Antisense oligonucleotide-induced exon skipping restores dystrophin expression *in vitro* in a canine model of DMD. *Gene Ther* **13**: 1373–1381.

6. van Deutekom JC, *et al.* (2001). Antisense-induced exon skipping restores dystrophin expression in DMD patient derived muscle cells. *Hum Mol Genet* **10**: 1547–1554.

7. Clarke JT, Iwanochko RM (2005). Enzyme replacement therapy of Fabry disease. *Mol Neurobiol* **32**: 43–50.

8. Schoser B, Hill V, Raben N (2008). Therapeutic approaches in glycogen storage disease type II/Pompe Disease. *Neurotherapeutics* **5**: 569–578.

9. van der Ploeg AT, Reuser AJ (2008). Pompe's disease. *Lancet* **372**: 1342–1353.

10. Ogawa K, *et al.* (2009). Long-term inhibition of glycosphingolipid accumulation in Fabry model mice by a single systemic injection of AAV1 vector in the neonatal period. *Mol Genet Metab* **96**: 91–96.

11. Rissanen TT, Yla-Herttuala S (2007). Current status of cardiovascular gene therapy. *Mol Ther* **15**: 1233–1247.

12. Yla-Herttuala S, Rissanen TT, Vajanto I, Hartikainen J (2007). Vascular endothelial growth factors: Biology and current status of clinical applications in cardiovascular medicine. *J Am Coll Cardiol* **49**: 1015–1026.

13. Zhu HL, Stewart AS, Taylor MD, Vijayasarathy C, Gardner TJ, Sweeney HL (2000). Blocking free radical production via adenoviral gene transfer decreases cardiac ischemia-reperfusion injury. *Mol Ther* **2**: 470–475.

14. Woo YJ, *et al.* (1998). Recombinant adenovirus-mediated cardiac gene transfer of super-oxide dismutase and catalase attenuates postischemic contractile dysfunction. *Circulation* **98**: II255–260; discussion II260–251.

15. Li Q, Bolli R, Qiu Y, Tang XL, Guo Y, French BA (2001). Gene therapy with extra-cellular superoxide dismutase protects conscious rabbits against myocardial infarction. *Circulation* **103**: 1893–1898.

16. Li Q, Bolli R, Qiu Y, Tang XL, Murphree SS, French BA (1998). Gene therapy with extracellular superoxide dismutase attenuates myocardial stunning in conscious rabbits. *Circulation* **98**: 1438–1448.

17. Peterson SJ, Frishman WH (2009). Targeting heme oxygenase: Therapeutic implications for diseases of the cardiovascular system. *Cardiol Rev* **17**: 99–111.

18. Juan SH, *et al.* (2001). Adenovirus-mediated heme oxygenase-1 gene transfer inhibits the development of atherosclerosis in apolipoprotein E-deficient mice. *Circulation* **104**: 1519–1525.

19. Duckers HJ, *et al.* (2001). Heme oxygenase-1 protects against vascular constriction and proliferation. *Nat Med* **7**: 693–698.

20. Melo LG, *et al.* (2002). Gene therapy strategy for long-term myocardial protection using adeno-associated virus-mediated delivery of heme oxygenase gene. *Circulation* **105**: 602–607.

21. Tang YL, Qian K, Zhang YC, Shen L, Phillips MI (2005). A vigilant, hypoxia-regulated heme oxygenase-1 gene vector in the heart limits cardiac injury after ischemia-reperfusion *in vivo*. *J Cardiovasc Pharmacol Ther* **10**: 251–263.

22. Tang Y, Schmitt-Ott K, Qian K, Kagiyama S, Phillips MI (2002). Vigilant vectors: Adeno-associated virus with a biosensor to switch on amplified therapeutic genes in specific tissues in life-threatening diseases. *Methods* **28**: 259–266.

23. Andino LM, Takeda M, Kasahara H, Jakymiw A, Byrne BJ, Lewin AS (2008). AAV-mediated knockdown of phospholamban leads to improved contractility and calcium handling in cardiomyocytes. *J Gene Med* **10**: 132–142.

24. Schmidt U, Zhu X, Lebeche D, Huq F, Guerrero JL, Hajjar RJ (2005). *In vivo* gene transfer of parvalbumin improves diastolic function in aged rat hearts. *Cardiovasc Res* **66**: 318–323.

25. Sakata S, *et al.* (2007). Restoration of mechanical and energetic function in failing aortic-banded rat hearts by gene transfer of calcium cycling proteins. *J Mol Cell Cardiol* **42**: 852–861.

26. Gaffney MM, Hynes SO, Barry F, O'Brien T (2007). Cardiovascular gene therapy: Current status and therapeutic potential. *Br J Pharmacol* **152**: 175–188.

27. Lyon AR, Sato M, Hajjar RJ, Samulski RJ, Harding SE (2008). Gene therapy: Targeting the myocardium. *Heart* **94**: 89–99.

28. Nordlie MA, Wold LE, Simkhovich BZ, Sesti C, Kloner RA (2006). Molecular aspects of ischemic heart disease: Ischemia/reperfusion-induced genetic changes and potential applications of gene and RNA interference therapy. *J Cardiovasc Pharmacol Ther* **11**: 17–30.

29. Yla-Herttuala S (2006). An update on angiogenic gene therapy: Vascular endothelial growth factor and other directions. *Curr Opin Mol Ther* **8**: 295–300.

30. Davis J, *et al.* (2008). Designing heart performance by gene transfer. *Physiol Rev* **88**: 1567–1651.

31. Fishbein I, Stachelek SJ, Connolly JM, Wilensky RL, Alferiev I, Levy RJ (2005). Site specific gene delivery in the cardiovascular system. *J Control Release* **109**: 37–48.

32. Raben N, Plotz P, Byrne BJ (2002). Acid alpha-glucosidase deficiency (glycogenosis type II, Pompe disease). *Curr Mol Med* **2**: 145–166.

33. Williams ML, Koch WJ (2004). Viral-based myocardial gene therapy approaches to alter cardiac function. *Annu Rev Physiol* **66**: 49–75.

34. Giacca M (2007). Virus-mediated gene transfer to induce therapeutic angiogenesis: Where do we stand? *Int J Nanomedicine* **2**: 527–540.

35. Mayer CR, Bekeredjian R (2008). Ultrasonic gene and drug delivery to the cardiovascular system. *Adv Drug Deliv Rev* **60**: 1177–1192.

36. Townsend D, Blankinship MJ, Allen JM, Gregorevic P, Chamberlain JS, Metzger JM (2007). Systemic administration of micro-dystrophin restores cardiac geometry and prevents dobutamine-induced cardiac pump failure. *Mol Ther* **15**: 1086–1092.

37. Bostick B, Yue Y, Lai Y, Long C, Li D, Duan D (2008). Adeno-associated virus serotype-9 microdystrophin gene therapy ameliorates electrocardiographic abnormalities in mdx mice. *Hum Gene Ther* **19**: 851–856.

38. Athanasopoulos T, Graham IR, Foster H, Dickson G (2004). Recombinant adeno-associated viral (rAAV) vectors as therapeutic tools for Duchenne muscular dystrophy (DMD). *Gene Ther* **11 Suppl 1**: S109–S121.

39. Vitiello C, *et al.* (2009). Disease rescue and increased lifespan in a model of cardiomyopathy and muscular dystrophy by combined AAV treatments. *PLoS One* **4**: e5051.
40. Pacak CA, Conlon T, Mah CS, Byrne BJ (2008). Relative persistence of AAV serotype 1 vector genomes in dystrophic muscle. *Genet Vaccines Ther* **6**: 14.

Index